Mastering Electronics Math

2nd Edition

R. Jesse Phagan

SECOND EDITION
FIRST PRINTING

Library of Congress Cataloging-in-Publication Data

Phagan, R. Jesse.
Mastering electronics math / by R. Jesse Phagan.
p. cm.
Rev. ed. of: Electronics math. © 1986.
Includes index.
ISBN 0-8306-6589-7 (h) ISBN 0-8306-3589-0 (p)
1. Electronics—Mathematics. I. Phagan, R. Jesse. Electronics math. II. Title.
TK7835.P49 1991 91-4705
621.381′0151—dc20 CIP

TAB Books offers software for sale. For information and a catalog, please contact TAB Software Department, Blue Ridge Summit, PA 17294-0850.

Acquisitions Editor: Roland S. Phelps
Technical Editor: B.J. Peterson
Production: Katherine G. Brown
Book Design: Jaclyn J. Boone
Cover Design and Illustration: Greg Schooley, Mars, PA.

Contents

Introduction *ix*

1 **Scientific and engineering notation** *1*
Significant figures *1*
Rounding numbers *2*
Positive and negative numbers *3*
Scientific notation *5*
Engineering notation *8*
Chapter summary *14*

2 **Circuit math: dc** *15*
Ohm's law *15*
Power formulas . . . P = I × E *18*
Direct/indirect relationships in formulas *22*
Series circuits *23*
Parallel circuits *30*
Complex dc circuits *38*
Voltage-divider circuits *47*
Dual-battery circuits *50*
Wheatstone bridge *52*
Maximum transfer of power *53*
Chapter summary *54*

3 **Reading a volt-ohm-milliammeter** *59*
Reading a multimeter *59*
Range *59*

Scale (DCV, ACV, and DCmA) *59*
Ohms *61*
Sample multimeter readings *62*
Practice multimeter readings *66*

4 Reading an oscilloscope *75*
The oscilloscope screen *75*
Voltage measurements *75*
Sample oscilloscope measurements *76*

5 The sine wave *85*
Producing a sine wave *85*
Plotting a sine wave *87*
Converting values in rms, peak, peak-to-peak, and average *90*
Converting values in frequency, period, and wavelength *93*
Using an oscilloscope to measure a sine wave *95*

6 Transformers *105*
Turns ratio, voltage ratio, current ratio *105*
Power relationships in a transformer *108*
Transformer efficiency *108*
Testing a transformer with an ohmmeter *109*
Chapter summary *110*

7 Inductors and capacitors *111*
Properties of inductance *111*
Inductors in series *112*
Inductors in parallel *114*
Properties of capacitance *116*
Capacitors in parallel *118*
Capacitors in series *118*
Capacitive voltage dividers *119*
Chapter summary *121*

8 Time constants *123*
Universal time constant curve *123*
L/R time constant *128*
High voltage produced by opening an *R/L* circuit *131*
RC time constant *132*
High current produced at first instant of discharge *134*
Chapter summary *134*

9 RC waveshaping *139*

Differentiation/integration *139*
Short time constant *140*
Medium time constant *143*
Long time constant *144*
Chapter summary *147*

10 Inductive reactance *149*

Inductive reactance, X_L *149*
Series or parallel inductive reactances *151*
Series X_L and R *152*
Parallel X_L and R *160*
Power in a reactive circuit *166*
Measuring phase angle with an oscilloscope *169*
Chapter summary *174*

11 Capacitive reactance *183*

Capacitive reactance *183*
Series or parallel capacitive reactance *185*
Series X_C and R *185*
Parallel X_C and R *192*
Power in a reactive circuit *198*
Measuring phase angle with an oscilloscope *200*
Chapter summary *205*

12 ac circuits and the *j*-operator *213*

Net reactance *213*
Power in an *RLC* circuit *219*
The *j*-operator *220*
Complex numbers in rectangular form *221*
Complex numbers in polar form *222*
Calculations with complex numbers *223*
Chapter summary *231*

13 Resonance *233*

The resonance effect *233*
Calculating the resonant frequency *234*
Series resonance *236*
Parallel resonance *238*
Q of a resonant circuit *239*
Bandwidth of a resonant circuit *240*
Chapter summary *242*

14 Power supplies *245*
The PN junction diode *245*
Half-wave rectifier *248*
Full-wave rectifier *249*
Bridge rectifier *252*
Calculating the rectifier output *252*
Filtering a rectifier circuit *256*
Zener regulator *261*
Chapter summary *267*

15 Transistors *271*
NPN and PNP transistors *271*
Current relationships in a transistor *272*
Range of transistor operation *274*
dc conditions of a transistor amplifier *276*
Base bias with emitter feedback *279*
Collector feedback *282*
Voltage-divider bias—beta independent *284*
ac conditions of a common-emitter amplifier *288*
Amplifier configurations *291*

16 Binary, octal, and hexadecimal numbering systems *297*
Digits used *297*
Place values *298*
Converting from one system to another *300*
Converting octal *302*
Converting hexadecimal *303*
Combine octal, hexadecimal, binary, and decimal *304*

Appendix Answers to practice problems *305*

Index *325*

Introduction

ELECTRONICS IS A RAPIDLY CHANGING FIELD, AFFECTING MOST EVERY product available. It is exciting to be a student, knowing that many more developments are yet to come. Now is a good time to be a part of the electronics field.

When learning electronics, the ability to understand and perform related math will mean the difference between a weak understanding of electronics theory and being able to grow with the field. For this reason, a math book demonstrating an abundance of sample problems is an important learning tool.

The first edition of *Electronics Math* has been very successful in classroom use, with the author's own class giving suggestions for improvements to this second edition. Where the subject matter is difficult, sections have been rewritten to simplify understanding. Clearer graphs and drawings are also included.

Four new chapters have been added, covering multimeters, oscilloscopes, transistors, and computer numbering systems. Mathematics related to the use of test equipment is too often misunderstood, preventing the proper use of the equipment. These chapters should help you develop better skills in making test measurements. Transistors and numbering systems have been added to extend the useful range of this book.

1 Scientific and engineering notation

THERE ARE TIMES WHEN IT IS NECESSARY TO DEAL WITH VERY LARGE OR very small numbers, but working with numbers like these using ordinary arithmetic is somewhat cumbersome. The term *scientific notation* suggests that scientific and technical fields have a definite need for such a mathematical system. In electronics, it is especially appropriate to use a system that makes using numbers easier. In electronics and other engineering fields, a modification of scientific notation, called *engineering notation,* is used.

Significant figures

Mathematics, with its many rules and precise calculations, gives the impression that it is an accurate as well as a precise tool. This is true of math itself; however, the numbers that are used for those calculations are questionable as to their degree of accuracy. Therefore, it makes sense that if the numbers are questionable, then the accuracy of the calculations is questionable. For example, when a meter is read, the meter has a tolerance and the person reading it has a tolerance. The value of resistors used in calculations often have a 10 percent tolerance.

With all the different tolerances being introduced into calculations, there must be a way of determining which digits of a number are reliable and accurate and which are questionable, or for that matter, not even needed because of their unreliability. A method of writing a number that shows the importance of each figure is called *significant figures*.

Often when writing numbers in electronics, the most commonly used number of figures is three. This is a common occurrence, not a rule.

Rules for determining significant figures

To determine the significant figures of a number, follow these guidelines:

Rule 1 Nonzero numbers (the digits 1 through 9) are always significant figures.

Rule 2 A zero used as a place holder between two nonzero digits is considered to be a significant figure.

Rule 3 A zero to the right of the last digit of a decimal is considered a significant figure.

Rule 4 A zero used only as a place holder, either on the right or the left side of the decimal point, if it is not between two nonzero digits, is not considered a significant figure.

Rule 5 When a number is greater than one, a decimal point can be placed following the number to show that all of the digits are significant, including zeros.

Examples:

1234	has four significant figures	0.0104	has three significant figures
1230	has three significant figures	0.2050	has three significant figures
1200	has two significant figures	25.0	has two significant figures
1034	has four significant figures	25.0	has three significant figures
1030	has three significant figures	53.05	has four significant figures
1030	has four significant figures	50.50	has three significant figures
0.00123	has three significant figures	005	has one significant figure
0.0123	has three significant figures	00500	has one significant figure

Rounding numbers

The process of rounding numbers is used to make working with numbers easier. You can see that the digit in a number with the least significance often has little power in changing the value of a number. For example, there is very little difference between $100 and $102.

Whenever it is necessary to round numbers, first determine how many significant figures will appeal in the final number. The next digit to the right will determine the outcome of rounding.

Rules for rounding numbers

Rule 1 Determine how many significant figures are to be used in the final number. The next number to the right determines the outcome of rounding.

Rule 2 If the number to the right is 5 or more, drop this digit and all the digits to the right. Add 1 to the last figure kept. Replace the digits dropped with zeros to maintain the place value of the figures remaining.

Rule 3 If the number to the right is less than 5, drop this digit and all the digits to the right. Replace the digits dropped with zeros to maintain the place value of the figures remaining.

Examples (rounded to three significant figures):

806956 = 807000		0.1009 = 0.101
875.5 = 876		0.0000003553 = 0.000000355
9083.2 = 9080		80.78 = 80.8
0.00387 = 0.00387		104 = 100
0.008737 = 0.00874		98 = 100

Practice problems

Round to three significant figures:

1. 5609	2. 5605
3. 5603	4. 29888
5. 359,999	6. 568.8
7. 573.09	8. 333.1
9. 12.09	10. 23.15
11. 54.51909	12. 56.99103
13. 8.2	14. 8.4300
15. 9.00199	16. 00.00333333333
17. 0.900099	18. 009315
19. 0.009999	20. 0.19999999
21. 0.6666666666	22. 0.005555555
23. 0.7447474747	24. 0.0909090909

25. 0.0125125125

Positive and negative numbers

Scientific notation expresses a number as a power of 10 in either positive or negative (*signed*) form. All numbers have a value of either positive or negative. The positive and negative signs are written in front of the number and become part of each number. For example, positive three is written +3 or simply 3; negative three is written −3. If no sign is written, assume the number to be positive.

Keep in mind fractions and decimals will still have the same relationships, even if the number is negative.

Addition and subtraction with signed numbers

Adding and subtracting positive and negative numbers is similar to adding and subtracting whole numbers. Study the rules and examples that follow.

Rules for adding and subtracting signed numbers

Rule 1 If the signs of the two numbers being added are alike, add the numbers and keep the sign.

Examples:

$$8 + 2 = 10 \qquad 4 + 3 = 7 \qquad 0.2 + 0.5 = 0.7$$

Note In the examples just shown, the signs of the numbers are positive, and the numbers are added. The results are positive. In the following three examples, the signs of the numbers are negative, and the numbers are added. The results are negative.

Examples:

$$-8 + -2 = -10 \qquad -4 + -3 = -7 \qquad -0.2 + -0.5 = -0.7$$

Rule 2 If the signs of the two numbers being added are unlike, subtract the numbers and take the sign of the larger number.

Examples:

$$-8 + 2 = -6 \qquad 8 + -2 = 6 \qquad -0.2 + 0.5 = 0.3$$

Note In the examples shown above, the signs in each of the sets of two numbers are unlike. The numbers are subtracted, lower from greater, and the result takes the sign of the greater number.

Rule 3 When subtracting two numbers, change the sign of the number being subtracted to its opposite, change the subtraction sign to addition, and then follow the rules for addition.

Examples:

$$7 - 3 = 7 + -3 = 4 \qquad -7 - 3 = -7 + -3 = -10$$
$$7 - -3 = 7 + +3 = 10 \qquad -7 - -3 = -7 + +3 = -4$$

Note When subtracting more than one number, always start at the left and work to the right.

Practice problems

Complete the operation:

1. $-6 + 4$
2. $7 + 3$
3. $8 + -2$
4. $9 - 5$
5. $-9 - -5$
6. $10 - - 6$
7. $-3 + 9$
8. $-2 + -5$
9. $-1 - 6$
10. $-1 - 1$
11. $-3 - -3$
12. $1 + 2 - 4$
13. $-3 - 5 + -6 - -3$
14. $10 - 6 + 3 - -10 - 3 + 6$
15. $0.3 - 0.5 - 0.8 + 1.5$
16. $2.5 - 3.2 + 6.8 + -8$
17. $0 + 8 - 0.01 - 0 - 0.03$
18. $-4.02 + 8.12 + 3.14 - 18.0$
19. $4.5 - 10.6 - 3.02 + 0.01$
20. $-\frac{3}{4} - \frac{5}{8}$

Multiplication and division with signed numbers

When multiplying and dividing signed numbers, disregard the signs and multiply or divide. Then follow the rules shown following:

Rule 1 Disregard the signs and multiply (or divide) the numbers. If the signs are alike, the product (or quotient) is positive.

Examples:

$$4 \times 2 = 8 \qquad -4 \times -2 = 8$$
$$8 \div 2 = 4 \qquad -8 \div -4 = 2$$

Rule 2 If the signs are unlike, the product (or quotient) is negative.

Examples:

$$4 \times -2 = -8 \qquad -4 \times 2 = -8$$
$$8 + -2 = -4 \qquad -8 \div 2 = -4$$

Practice problems

Note Parentheses written together ( ) ( ) signify multiplication. Division is represented by the use of the division symbol ÷ or the fraction bar /.

1. $(2)(3)$
2. -4×-5
3. 5×-3
4. -6×3
5. $(-2)(3)$
6. $(-1)(-2)$
7. $0 \times 3 \times -5$
8. 0.4×0.5
9. 0.3×-0.6
10. -2^2
11. -2^3
12. $20 \div 4$
13. $25 \div -5$
14. $-30 \div 6$
15. $-45 \div -9$
16. $\frac{-8}{2}$
17. $\frac{2}{-12}$
18. $\frac{2 \times -3}{-1}$
19. $-\frac{3 \times -4}{-2 \times -2}$
20. $\frac{(-3)(-3)(-3)}{(-1)(-1)(-1)}$

Scientific notation

Scientific notation is a mathematical tool to make numbers much easier to work with. Even with the use of a calculator, it is still helpful to use scientific notation when dealing with the types of numbers found in electronics.

Scientific notation is writing a number as a number between 1 and 10, times a power of 10 (see Table 1-1). First it is necessary to see how numbers that are multiples of ten are expressed as powers of ten.

Note the following key points:

- Numbers that are greater than 1 have a positive power of 10.
- Numbers that are less than 1 have a negative power of 10.
- 10 to zero power (10^0) equals 1. Any number to the zero power equals 1.

Table 1-1. Powers of 10.

Number	Power of 10	Read as
1,000,000	10^6	ten to the sixth power
100,000	10^5	ten to the fifth power
10,000	10^4	ten to the fourth power
1,000	10^3	ten to the third power
100	10^2	ten to the second power
10	10^1	ten to the first power
1	10^0	ten to the zero power
.1	10^{-1}	ten to the negative first power
.01	10^{-2}	ten to the negative second power
.001	10^{-3}	ten to the negative third power
.0001	10^{-4}	ten to the negative fourth power
.00001	10^{-5}	ten to the negative fifth power
.000001	10^{-6}	ten to the negative sixth power

Rules for writing numbers in scientific notation

Rule 1 When a number is greater than 1, express it in scientific notation by moving the decimal point to the left enough places so the number is written between 1 and 10. Count the number of decimal places the point was moved, and that is the positive power of 10.

Examples:

$86{,}500 = 8.65 \times 10^4$ $\quad$ $563{,}000{,}000 = 5.63 \times 10^8$

$230 = 2.30 \times 10^2$ $\quad$ $7{,}602 = 7.602 \times 10^3$

$7.3 = 7.3 \times 10^0$ $\quad$ $20 = 2.0 \times 10^1$

Rule 2 When a number is less than 1, express it in scientific notation by moving the decimal point to the right enough places so the number is written between 1 and 10. Count the number of decimal places the point was moved to determine the negative power of 10.

Examples:

$0.1 = 1.0 \times 10^{-1}$ $\quad$ $0.023 = 2.3 \times {}^{-2}$

$0.000356 = 3.56 \times 10^{-4}$ $\quad$ $0.03004 = 3.004 \times 10^{-2}$

$0.0000098 = 9.8 \times 10^{-5}$ $\quad$ $0.00000030 = 3.0 \times 10^{-7}$

Sometimes numbers are already written as a number times some power of 10. If the number is not expressed between 1 and 10, rewrite the number to follow the form of scientific notation.

Example:

$$365 \times 10^4 = 3.65 \times 10^6$$

The easiest way to solve this type of problem is to first write the number in its proper form, with a power of 10 corresponding to the initial amount of the decimal move. The next step is to add the exponents according to the rules of signed numbers.

Step 1 $365 = 3.65 \times 10^2$ (write the number in proper form)
Step 2 $10^2 + 10^4 = 10^6$ (add the powers of 10)
Step 3 3.65×10^6 (combine steps 1 and 2 for the final answer)

More examples:

$$560{,}000 \times 10^{-5} = 5.6 \times 10^0$$

Step 1 $560{,}000 = 5.6 \times 10^5$
Step 2 $10^5 + 10^{-5} = 10^0$

$$890{,}000{,}000 \times 10^{-3} = 8.9 \times 10^5$$

Step 1 $890{,}000{,}000 = 8.9 \times 10^8$
Step 2 $10^8 + 10^{-3} = 10^5$

$$0.00035 \times 10^8 = 3.5 \times 10^4$$

Step 1 $0.00035 = 3.5 \times 10^{-4}$
Step 2 $10^{-4} + 10^8 = 10^4$

$$0.00000057 \times 10^{-4} = 5.7 \times 10^{-11}$$

Step 1 $0.00000057 = 5.7 \times 10^{-7}$
Step 2 $10^{-7} + 10^{-4} = 10^{-11}$

Sometimes it is necessary to remove the power of 10 and return to the original number, without scientific notation. In these cases, there are two basic rules to follow:

Rule 1 If the power of 10 is positive, move the decimal to the right. Move it the number of places stated by the power of 10. This moves the decimal in the negative direction to remove the positive exponent.

$$8.6 \times 10^6 = 8{,}600{,}000$$
$$56.7 \times 10^3 = 56{,}700$$

Rule 2 If the power of 10 is negative, move the decimal to the left. Move it the number of places stated by the power of 10. This moves the decimal in the positive direction to remove the negative exponent.

$$0.0034 \times 10^{-5} = 0.00000034$$
$$56.3 \times 10^{-1} = 5.63$$

Practice problems

Write each in the form of scientific notation:

1. 876,000
2. 1,030,000,000
3. 32,000
4. 25
5. 5.8
6. 0.03
7. 0.00056
8. 00.00405

9. 0.0000001000
10. 0.200
11. 13,000 × 10^5
12. 520 × 10^4
13. 0.0045 × 10^7
14. 0.00000039 × 10^3
15. 0.000000056 × 10^5
16. 52,000 × 10^{-3}
17. 3,200 × 10^{-6}
18. 0.00046 × $^{-5}$
19. 0.00705 × 10^{-2}
20. 0.0000004 × 10^{-7}

Practice problems

Write each as a standard numeral:

1. 4.8 × 10^5
2. 7.85 × 10^2
3. 8.9 × 10^0
4. 34.6 × 10^1
5. 457 × 10^3
6. 1.0 × 10^{-1}
7. 2.01 × 10^{-3}
8. 00.003 × 10^{-5}
9. 100 × 10^{-2}
10. 100,000 × 10^{-6}
11. 0.00035 × 10^5
12. 0.000000400 × 10^6
13. .709 × 10^{-3}
14. 5600 × 10^3
15. 0.00000065 × 10^8
16. 9.8 × 10
17. 1.09 × 10^{-4}
18. 0.00000078 × 10^0
19. 10 × 10^0
20. 1 × 10^1

Engineering notation

Engineering notation is a variation of scientific notation. In electronics and many other scientific and technical fields, there are certain powers of 10 that are used more often than others. Names have been given to certain powers of 10 to make them even easier to use.

The names of the powers are given in multiples of three. For example: 10^3, 10^6, 10^9, 10^{-3}, 10^{-6}, 10^{-9}, etc. Refer to Table 1-2. Notice how the powers of 10 that are positive also have corresponding powers that are negative.

The engineering notation names have prefixes that are written in front of and attached to the unit name. For example, when dealing with volts as the unit of measure, engineering notation prefixes could be millivolts, microvolts, kilovolts, or megavolts. Refer to Table 1-2 for a list of names and powers of 10 commonly used in electronics.

Writing numbers using engineering notation

If you needed to convert 48,000 W (watts) to engineering notation, the first step would be to write the number in modified scientific notation. That is, use a power of 10 that is a multiple of 3. Then replace the power of 10 with the multiplier name. Refer to Table 1-2.

Step 1 48,000 W = 48 × 10^3 W
Step 2 10^3 = kilo . . . therefore . . . 48 kW (kilowatts)

Convert 305,000,000 Hz (hertz) to engineering notation.

Table 1-2. Engineering notation.

Multiply by	Power of 10	Multiplier name	Symbol
1,000,000,000,000	10^{12}	tera	T
1,000,000,000	10^{9}	giga	G
1,000,000	10^{6}	mega	M
1,000	10^{3}	kilo	k
1	10^{0}	basic unit	(no multiplier)
0.001	10^{-3}	milli	m
0.000 001	10^{-6}	micro	μ
0.000 000 001	10^{-9}	nano	n
0.000 000 000 001	10^{-12}	pico	p

Step 1 305×10^{6} Hz or 0.305×10^{9} Hz (writing the number with a power of 10).
Step 2 305 (megahertz) or 0.305 GHz (gigahertz).

Note When it is not specified which multiplier name to use, select either, depending on the application in the problem.

Convert 0.0025 A (amperes) to engineering notation.

Step 1 2.5×10^{-3} A
Step 2 2.5 mA (milliamperes)

Sometimes it is necessary to take a number that is already written in scientific notation and change the multiplier name to make it more convenient to use. The easiest way to do this is by removing the multiplier given and replacing it with the proper power of 10. At this point, it is best to rewrite the number, without use of scientific notation. In other words, return the number back to the basic unit. Once it is returned to the basic unit, it is simply a matter of moving the decimal the correct number of places to have a power of 10 equal to the multiplier desired.

Examples:

Convert 2500 mA to A.

Step 1 2500 mA = 2500×10^{-3} A
Step 2 $2500 \times 10^{-3} = 2.5$ A

Convert 475 kV (kilovolts) to MV (megavolts).

Step 1 475 kV = 475×10^{3} V
Step 2 475×10^{3} V = 475,000 V
Step 3 0.475×10^{6} V = 0.475 MV (megavolts)

Convert 0.255 μA (microamperes) to mA.

Step 1 0.255 μA = 0.255×10^{-6} A
Step 2 $0.255 \times 10^{-6} = 0.000000255$ A

Step 3 0.000000255 = 0.000255×10^{-3} A
Step 4 0.000255 mA

Convert 0.001 pF (picofarads) to μF (microfarads).

Step 1 0.001 pF = 0.001×10^{-12} F (farads)
Step 2 0.001×10^{-12} = 0.000 000 000 000 001 F
Step 3 0.000 000 001 $\times 10^{-6}$ F
Step 4 0.000 000 001 μF

Note the following key point:

- When you are converting from one multiplier to another, the actual value of the number does not change, even though the decimal place moves.

Practice problems

Change each to the units shown:

1. 2,400,000 Ω	________	kΩ	________	MΩ	(ohms)
2. 353,000 Hz	________	kHz	________	MHz	(hertz)
3. 2500 W	________	mW	________	kW	(watts)
4. 25 V	________	mV	________	kV	(volts)
5. 0.01 H	________	μH	________	mH	(henries)
6. 1.5 kΩ	________	Ω	________	MΩ	(ohms)
7. 25 MHz	________	kHz	________	GHz	(hertz)
8. 0.03 GHz	________	MHz	________	Hz	(hertz)
9. 56 MV	________	kV	________	V	(volts)
10. 75 kW	________	W	________	MW	(watts)
11. 25 mA	________	μA	________	A	(amps)
12. 1500 μA	________	A	________	mA	(amps)
13. 1000 μF	________	pF	________	F	(farads)
14. 0.001 pF	________	μF	________	nF	(farads)
15. 0.025 mV	________	μV	________	V	(volts)
16. 7500 mW	________	W	________	μW	(watts)
17. 500 V	________	kV	________	mV	(volts)
18. 2,400 mV	________	V	________	kV	(volts)
19. 1 A	________	kA	________	mA	(amps)
20. 10 V	________	mV	________	kV	(volts)

Multiplication and division with engineering notation

In electronics, there are a great deal of calculations, most of which involve the use of powers of 10. Calculations can be performed through the use of scientific notation or through the use of engineering notation with multiplier names. Using the multiplier names is much more common and therefore, the method that is concentrated on in this book.

There are a few basic rules to learn, as in any form of mathematics.

Rules for multiplication and division

Rule 1 To multiply numbers that contain powers of 10, multiply the numbers and add the powers of 10 exponents.

Examples:

Multiply 20 kilo × 5 milli

Step 1 Multiply the numbers 20 × 5 = 100
Step 2 Add the powers of 10 exponents

$$\text{kilo} = 10^3$$
$$\text{milli} = 10^{-3}$$
$$10^3 + 10^{-3} = 10^0 = 1$$

Step 3 Combine step 1 (the number) with step 2 (the power of 10) final answer = 100

- When a positive multiplier is multiplied by a negative multiplier of the same size (kilo and milli) the result is canceled unit multipliers. The final answer will be in basic units.

Multiply 100 × 25 milli

Step 1 Multiply the numbers 100 × 25 = 2500
Step 2 Add the powers of 10 exponents

$$\text{100 is in basic units} = 10^0$$
$$\text{milli} = 10^{-3}$$
$$10^0 + 10^{-3} = 10^{-3}$$

Step 3 Combine the number with the power of 10

2500 milli

Step 4 Adjust the decimal in the number to better fit the multiplier

2500 milli = 2.5 (basic unit)

Rule 2 To divide numbers that contain powers of 10; divide the numbers and subtract the powers of 10 exponents.

Example:

$$\text{Divide } \frac{100}{25 \text{ milli}}$$

Step 1 Divide the numbers 100 ÷ 25 = 4
Step 2 Subtract the powers of 10 exponents

$$\text{100 is in basic units} = 10^0$$
$$\text{milli} = 10^{-3}$$
$$10^0 - 10^{-3} = 10^3$$

Always subtract the denominator (bottom) from the numerator (top).

Step 3 Combine step 1 (the number) with step 2 (the power of 10)

$$\text{final answer} = 4 \text{ kilo}$$

- When a positive multiplier is divided into basic units, the result is a negative multiplier of equal size (kilo divided into basic units results in milli). Also, basic divided by negative gives a positive.

Rule 3 Whenever the power of 10 is moved from the numerator to the denominator or the denominator to the numerator, move the power of 10 by changing the sign of the exponent.

Example (to show moving of power only):

$$\frac{1}{\text{micro}} = \frac{1}{10^{-6}} = \frac{10^{6}}{1} \text{ mega}$$

$$\text{kilo} = 10^{3} = \frac{1}{10^{-3}} = \frac{1}{\text{milli}}$$

Note Although at this time it may be difficult to see a use for rule 3 above, it often comes in quite useful. The same operation can be performed by using the rules for division by allowing the 1 shown in the examples to be 10^0.

When working with multiplier names, keep in mind that there are units to deal with. Rather than dealing with the units in this chapter, you will have better success with the multiplier names if that is all you have to deal with. In electronics, multiplication or division of two units often leads to a new unit altogether. An excellent example of this is Ohm's Law:

$$E(\text{volts}) = I(\text{amperes}) \times R(\text{ohms})$$

Practice problems

Complete the operation. Write the answer in engineering notation with the most convenient multiplier and show the correct power of 10.

1. 100 kilo × 1 milli
2. 20 mega × 3 milli
3. 10 × 1.5 kilo
4. 4 kilo × 2 micro
5. 5 × 6 nano
6. $\frac{10}{2 \text{ kilo}}$
7. $\frac{50}{25 \text{ milli}}$
8. $\frac{20 \text{ micro}}{5}$
9. $\frac{25 \text{ giga}}{5 \text{ mega}}$
10. 2 × 3.14 × 1 kilo × 0.01 milli
11. 2 × 3.14 × 10 mega × 10 milli
12. $\frac{1}{2 \times 3.14 \times 1 \text{ mega} \times 1 \text{ micro}}$
13. $\frac{1}{2 \times 3.14 \times 1 \text{ kilo} \times 10 \text{ pico}}$
14. $\frac{2 \text{ giga} \times 3 \text{ kilo}}{6 \text{ mega} \times 1 \text{ milli}}$
15. $\frac{5 \text{ milli} \times 4 \text{ micro}}{2 \text{ nano} \times 2 \text{ kilo}}$

Addition and subtraction with engineering notation

Addition and subtraction with numbers having powers of 10 is an arithmetic function found quite often in the study of electronics.

There are two common ways of adding and subtracting: by keeping the power of 10 for both numbers the same, or by converting both of the numbers to basic units and using regular arithmetic. Learn both methods because they are both used at different times.

Rules for adding and subtracting with engineering notation

Rule 1 When the powers of 10 are the same, add or subtract the numbers, keeping the same power of 10.

Examples:

2 kilo + 3 kilo = 5 kilo
4 milli + 6 milli = 10 milli
300 + 5 kilo = 0.3 kilo + 5 kilo
= 5.3 kilo (it is necessary to make the units the same)

Rule 2 If the powers of 10 are not the same; it is often easier to convert the numbers to basic units and use regular arithmetic.

Examples:

100 milli + 1 = 0.1 + 1 = 1.1 basic units
1.5 kilo + 0.06 mega = 1500 + 60,000 = 61,500 = 61.5 kilo
2.7 mega + 10 = 2,700,000 + 10 = 2,700,010 = 2.7 mega

Note In this last example, the very small basic unit of 10 was dropped. Using the rules for rounding to three significant figures, this is acceptable.

Practice problems

Complete the operation. Write the answer in engineering notation with the most convenient multiplier. Show the correct power of 10. Round to three significant figures if needed.

1. 3 kilo + 41 kilo
2. 305 + 609
3. 75 kilo + 1200
4. 0.85 mega + 150 kilo
5. 0.75 mega − 25 kilo
6. 2.2 kilo − 850
7. 10 + 2500 milli
8. 25 kilo + 10 milli
9. 85 milli + 1000 micro
10. 0.90 milli + 150 micro
11. 8 milli − 2 milli
12. 10 milli − 10 micro
13. 0.1 milli − 90 micro
14. 250 micro + 1500 nano
15. 25 nano + 100 pico
16. 0.01 micro + 0.001 micro
17. 15 pico + 25 pico
18. 0.01 micro + 1500 pico
19. 1.0 + 0.1 milli
20. 201 + 20.1

Chapter summary

This first chapter covers writing numbers in either scientific notation or engineering notation. Keep in mind that although engineering notation is used most of the time in electronics and other technical areas, the multiplier names are used to replace powers of 10.

Note the following key points:

- Numbers that are greater than 1 have a positive power of 10.
- Numbers that are less than 1 have a negative power of 10.
- 10 to the zero power (10^0) equals 1. Any number to the zero power equals 1.
- When converting from one multiplier to another, the actual value of the number does not change even though the decimal place moves.
- When a positive number is multiplied by a negative multiplier of the same size (kilo and milli) the result is canceled unit multipliers. The final answer will be in basic units.
- When a positive multiplier is divided into basic units, the result is a negative multiplier of equal size (kilo divided into basic units results in milli). Also, basic divided by negative gives a positive.

2
Circuit math: dc

DIRECT CURRENT CIRCUIT MATH IS CONSIDERED THE MOST BASIC OF ALL mathematics in electronics. However, many students progress to the more difficult subject areas and find themselves with an inadequate background in basic circuit math. Because of the learning backgrounds of most students, memorizing new material is the most widely used method for learning. The process of memorizing is helpful in the beginning, but in the field of electronics, it is very important for the technician to have a thorough understanding of the subject material.

Resistive circuits are used in every aspect of electronic circuits. Resistive circuits are probably the most important of all subject matter to understand fully.

Ohm's Law

Ohm's Law states that voltage equals current times resistance. Many of the formulas in electronics have two or more units multiplied or divided to produce a completely different unit.

- Stated by Ohm's Law: voltage = amperes × ohms

$$V = I \times \Omega$$

All the rules of multipliers apply whenever units are used. Review the rules covered in chapter 1 if necessary.

Exactly what the formula means and how it is used is discussed later in this chapter. This section of the chapter helps you become proficient in performing the arithmetic of the formula and rearranging it to suit the needs of the particular application.

Because the formula has three letters, called unknowns, the formula can be used to solve for any of the three unknowns. In order to use the equation, however, you must know two of the three values. Then, with two values given, you find the third value using the formula.

Start with the basic formula: $E = I \times R$. Use algebra to change the formula as needed. To find I:

Step 1 $E = I \times R$ formula (2-1)

Step 2 $\frac{E}{R} = \frac{I \times R}{R}$ divide both sides by R

Step 3 $\frac{E}{R} = I$ cancel the R on the right side

$$I = \frac{E}{R} \quad \text{(2-1A)}$$

To find R:

Step 1 $E = I \times R$ formula (2-1)

Step 2 $\frac{E}{I} = \frac{I \times R}{I}$ divide both sides by I

Step 3 $\frac{E}{I} = R$ cancel the I on the right side

$$R = \frac{E}{I} \quad \text{(2-1B)}$$

Summaries of Ohm's Law formulas

$$E = I \times R \quad \text{(2-1)}$$

Use this formula when current (I) and resistance (R) are known; solve for voltage (E).

$$I = \frac{E}{R} \quad \text{(2-1A)}$$

Use this formula when voltage (E) and resistance (R) are known; solve for current (I).

$$R = \frac{E}{I} \quad \text{(2-1B)}$$

Use this formula when voltage (E) and current (I) are known; solve for resistance (R).

Note In each of these formulas, the letter E is used to represent voltage. The letter V may be used rather than the letter E.

The practice problems are designed to help you grasp full use of the Ohm's Law formulas. Spend time learning the formulas before trying to work the practice problems. The Ohm's Law formulas will be used repeatedly in the electrical and electronics fields.

Later in the chapter, a fuller understanding of how the formulas work to solve electrical circuits will be developed.

Learn the letter symbols below:

Voltage: *V* for volts, *E* for EMF (electromotive force)
Current: *I* for current, *A* for amperes, mA for milliamperes, μA for microamperes
Resistance: Ohms or Ω (Greek letter omega), kΩ for kilohms, MΩ for megohms, *R* for resistance.

Examples with Ohm's Law formulas

$$E = I \times R \tag{2-1}$$

Given: $I = 5$ A, $R = 20\ \Omega$. Find E.

$$\begin{aligned} E &= I \times R \\ &= 5\text{ A} \times 20\ \Omega \\ &= 100\text{ V} \end{aligned}$$

$$I = \frac{E}{R} \tag{2-1A}$$

Given: $E = 250$ V, $R = 2.5$ kΩ. Find I.

$$\begin{aligned} I &= \frac{E}{R} \\ &= \frac{250\text{ V}}{2.5\text{ k}\Omega} \\ &= 100\text{ mA} \end{aligned}$$

$$R = \frac{E}{I} \tag{2-1B}$$

Given: $E = 10$ V, $I = 5$ mA. Find R.

$$\begin{aligned} R &= \frac{E}{I} \\ &= \frac{10\text{ V}}{5\text{ mA}} \\ &= 2\text{ k}\Omega \end{aligned}$$

Practice problems

Use Ohm's Law to solve for the unknown quantity. Show the answers with the proper units and use engineering notation, whenever possible. Round answers to three significant figures.

1. I = 2 A, R = 10 Ω, find E:
2. R = 100 Ω, I = 0.5 A, find E:
3. R = 1 kΩ, I = 20 mA, find E:
4. I = 100 mA, R – 3 kΩ, find E:
5. R = 1.2 MΩ, I = 10 μA, find E:
6. E = 10 V, R = 100 Ω, find I:
7. R = 50 Ω, E = 50 V, find I:
8. R = 1 kΩ, E = 10 V, find I:
9. E = 12 V, R = 1.2 kΩ, find I:
10. E = 0.5 V, R = 50 Ω, find I:
11. E = 10 V, I = 1 amp, find R:
12. I = 5 A, E = 50 V, find R:
13. I = 10 mA, E = 20 V, find R:
14. E = 30 V, I = 15 μA, find R:
15. V = 50 V, I = 10 mA, find R:
16. V = 40 V, R = 400 Ω, find I:
17. I = 15 A, R = 2 Ω, find E:
18. R = 1500 Ω, E = 1.5 kV, find I:
19. I = 2 mA, E = 2.4 kV, find R:
20. E = 24 V, I = 24 A, find R:

Power formulas . . . $P = I \times E$

This formula is the basic power formula. It is used to calculate the power dissipated in a dc circuit when the voltage and current are known. Of course, if the voltage and current are known, then resistance can also be calculated.

- In any dc circuit there are four quantities that can be calculated; voltage, current, resistance, and power.

This section demonstrates use of formulas based on both Ohm's Law and the power formula (calculations are made of all four electrical quantities, with any two of the four given).

$$P = I \times E \tag{2-2}$$

Power (watts) = Current (amperes) × Voltage (volts)

The basic formula will allow the calculation of power if current and voltage are given. The formula can be arranged to solve for either of the other two quantities.

$$P = I \times E \qquad (2\text{-}2)$$

$$\frac{P}{E} = \frac{I \times E}{E} \quad \text{(divide both sides by } E\text{)}$$

$$P = I \times E \qquad (2\text{-}2)$$

$$\frac{P}{I} = \frac{I \times E}{I} \quad \text{(divide both sides by } I\text{)}$$

$$\frac{P}{I} = E \qquad (2\text{-}2A)$$

Combining Ohm's Law formulas with the power formulas

If Ohm's Law formulas are combined with the power formulas, it is possible to solve for any of the four quantities (P, I, R, or E), regardless of which two values are given.

To find P, substitute $I \times R$ for E.

$P = I \times E$ and $E = I \times R$
$= I \times (I \times R)$ substitute the $I \times R$ of Ohm's Law into the power formula for E
$= I^2 \times R$ combine like terms (2-3)

Formula 2-3 shown here can be used to calculate power when current and resistance are known. Rearranging this formula makes it possible to solve for either current or resistance when the other two are known.

Find I:

$$P = I^2 \times R$$

$$\frac{P}{R} = \frac{I^2 \times R}{R} \quad \text{divide both sides by } R$$

$$= I^2 \quad \text{cancel the } R \text{ on the right side}$$

$$I = \sqrt{\frac{P}{R}} \quad \text{take the square root of both sides} \qquad (2\text{-}3A)$$

Find R:

$$P = I^2 \times R$$

$$\frac{P}{I^2} = \frac{I^2 \times R}{I^2} \quad \text{divide both sides by } I^2$$

$$= R \quad \text{cancel the } I^2 \text{ on the right side} \qquad (2\text{-}3B)$$

Find P substitute $\frac{E}{R}$ for I:

$$P = I \times E \text{ and } I = \frac{E}{R}$$

$$= \frac{E}{R} \times E \quad \text{substitute the } E/R \text{ for } I$$

$$= \frac{E^2}{R} \quad \text{combine like terms} \tag{2-4}$$

Formula 2-4 can be used to calculate the power when the voltage and resistance are given. Rearranging the formula makes it possible to solve for either voltage or resistance when power is known.

Find E:

$$P = \frac{E^2}{R} \tag{2-4}$$

$$P \times R = \frac{E^2 \times R}{R} \quad \text{multiply both sides by } R$$

$$= E^2 \quad \text{cancel the } R \text{ on the right side}$$

$$E = \sqrt{P \times R} \quad \text{take the square root of both sides} \tag{2-4A}$$

Find R:

$$P = \frac{E^2}{R}$$

$$= \frac{E^2 \times R}{R} \quad \text{multiply both sides by } R$$

$$P \times R = E^2 \quad \text{cancel } R \text{ on right side}$$

$$\frac{P \times R}{P} = \frac{E^2}{P} \quad \text{divide both sides by } P$$

$$R = \frac{E^2}{P} \quad \text{cancel the } P \text{ on the left side} \tag{2-4B}$$

With Ohm's Law and the power formulas, there are 12 different combinations (see Table 2-1). Notice in each combination there are only three variables; therefore, each variation on the original formula needs only two known values.

Practice problems

Use Ohm's Law and the power formulas to complete the following table. Round answers to three significant figures.

Table 2-1. Ohm's Law and power formulas.

Given	Find	Formula	Alternate formulas
I, R	E	$E = I \times R$	
I, R	P	$P = I^2 R$	$E = I \times R$ and $P = I \times E$
E, R	I	$I = \frac{E}{R}$	
E, R	P	$P = \frac{E^2}{R}$	$I = \frac{E}{R}$ and $P = I \times E$
I, E	R	$R = \frac{E}{I}$	
I, E	P	$P = I \times E$	
P, E	I	$I = \frac{P}{E}$	
P, E	R	$R = \frac{E^2}{P}$	$I = \frac{P}{E}$ and $R = \frac{E}{I}$
P, I	E	$\mathrm{E} = \frac{\mathrm{P}}{\mathrm{I}}$	
P, I	R	$R = \frac{P}{I^2}$	$E = \frac{P}{I}$ and $R = \frac{E}{I}$
P, R	I	$I = \sqrt{\frac{P}{R}}$	
P, R	E	$E = \sqrt{P \times R}$	$I = \sqrt{\frac{P}{R}}$ and $I \times R$

	Voltage	Current	Resistance	Power
1.	$E =$ ______	$I =$ 15 amps	$R =$ 2 ohms	$P =$ ______
2.	$E =$ ______	$I =$ 2 amps	$R =$ ______	$P =$ 5 watts
3.	$E =$ 100 volts	$I =$ .01 amps	$R =$ ______	$P =$ ______
4.	$E =$ 250 volts	$I =$ ______	$R =$ 5 ohms	$P =$ ______
5.	$E =$ 15 volts	$I =$ ______	$R =$ ______	$P =$ 100 mW
6.	$E =$ ______	$I =$ ______	$R =$ 1 kΩ	$P =$ 10 mW
7.	$E =$ ______	$I =$ 25 mA	$R =$ 10 kΩ	$P =$ ______
8.	$E =$ ______	$I =$ 2 mA	$R =$ ______	$P =$ 25 mW
9.	$E =$ 50 volts	$I =$ 5 mA	$R =$ ______	$P =$ ______
10.	$E =$ 5 volts	$I =$ ______	$R =$ 2500 ohms	$P =$ ______
11.	$E =$ 0 volts	$I =$ ______	$R =$ 100 ohms	$P =$ ______
12.	$E =$ 10 volts	$I =$ ______	$R =$ ______	$P =$ 10 watts
13.	$E =$ ______	$I =$ 10 μa	$R =$ 10 kΩ	$P =$ ______
14.	$E =$ ______	$I =$ 100 mA	$R =$ ______	$P =$ 1 watt
15.	$E =$ 25 volts	$I =$ 1 amp	$R =$ ______	$P =$ ______
16.	$E =$ 20 volts	$I =$ ______	$R =$ 200 ohms	$P =$ ______
17.	$E =$ 1 volt	$I =$ ______	$R =$ ______	$P =$ 1 watt
18.	$E =$ ______	$I =$ ______	$R =$ 1000 ohms	$P =$ .1 watt
19.	$E =$ ______	$I =$ 1000 mA	$R =$ 10,000 Ω	$P =$ ______
20.	$E =$ 10 volts	$I =$ 100 mA	$R =$ ______	$P =$ ______

Direct/indirect relationships in formulas

Now that you have practiced using the formulas, it is time to look at how the units of voltage, current, resistance, and power react with each other. In other words, with any of the Ohm's Law or power formulas, what happens to the third quantity if one is held constant and the other changed?

There are two types of relationships that are of particular concern, the direct relationship and the indirect relationship.

Direct relationships

A direct relationship can be defined as follows: when one quantity is increased, the other also increases; when one is decreased, the other decreases.

Example:

$$E = I \times R \tag{2-1}$$

Relationships are always compared on opposite sides of the equals sign. If R is held constant when there is an increase in current, there will be an increase in voltage. If I is held constant when there is an increase in resistance there will be an increase in voltage.

$$P = I \times E \tag{2-2}$$

If E is held constant, there will be an increase in P when I is increased. If I is held constant, there will be an increase in P when E is increased.

By these two examples you can establish some key points:

- E is directly related to I
- E is directly related to R
- P is directly related to R
- P is directly related to I

These key points help to show the effect on different quantities in the event it is necessary to change or compare the values of different quantities in a circuit.

Indirect relationships

An indirect relationship can be defined as follows: when a quantity is increased, the other decreases; when one is decreased, the other increases. Note that with the direct relationships, the two quantities that are directly related were both in the numerator of the equation. It stands to reason that, with indirect relationships, one quantity will be in the numerator and the other in the denominator.

Example:

$$I = \frac{E}{R} \tag{2-1A}$$

Relationships are always compared on opposite sides of the equals sign. If E is held constant, an increase in R will cause the current to decrease. A decrease in R will cause the current to increase.

Notice with the example shown here the *E* and the *I* are both in the numerator. Their relationship does not change.

- *I* is indirectly related to *R*
- *P* is indirectly related to *R*

Series circuits

A series circuit is defined as having only one path for current to flow. In order to have any current flow, in any circuit, there must be a complete path for the current to return to the same power supply. That is, for each electron leaving a power source, one must return. A series circuit has four key points:

- In a series circuit, there is only one current path. Current is the same throughout a series circuit. This is the definition of a series circuit.
- Voltage drops at each resistance is directly related to the size of the resistance.
- The sum of the voltage drops in a series circuit must be equal to the supply voltage.
- The sum of the powers dissipated in a series circuit must be equal to the total power.

Refer to Fig. 2-1. This is a series circuit because current has only one path to follow. There can be any number of resistors. Provided there is only one path for current to flow, it will still be a series circuit. Notice in the drawing that there is an arrow indicating a direction to current flow. The direction is from minus to plus, starting from the minus, or negative side of the battery (smaller of the two battery terminals), flowing through the load, returning to the positive (plus) side of the battery. As long as current can return to the power source, it is considered a complete circuit.

One word about direction of current flow. This book, and most modern textbooks, refer to current flow as being from negative to positive. This is called electron current flow. Technicians need to be aware that current can be described as

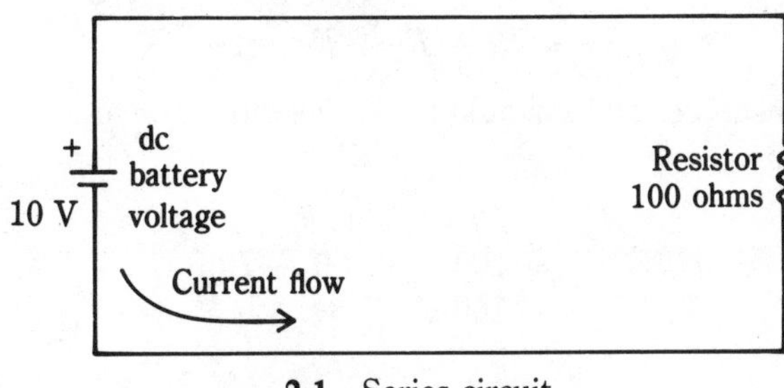

2-1 Series circuit.

flowing in the opposite direction, from plus to minus. This is called conventional current flow.

- Electron current flow is from negative to positive. Conventional is from positive to negative.

Calculations

Calculate current, often referred to as total current. See Fig. 2-1.

$$I = \frac{E}{R}$$

$$= \frac{10\ \text{V}}{100\ \Omega}$$

$$= 0.1\ \text{A or } 100\ \text{mA}$$

Calculate power. True power is only dissipated in a resistance. In a later chapter, apparent power is discussed. Power is the work done, or in the case of electricity, the heat produced.

$$P = I \times E$$

$$= 0.1\ \text{A} \times 100\ \Omega$$

$$= 10\ \text{W}$$

Note In the power calculations above, the value of current used is the result of a previous calculation. If a mistake had been made in the current calculation, power would also be wrong. Be careful or use given values in all calculations.

Figure 2-2 shows three resistors in series connected to a single power supply. The direction of current flow is from negative to positive, and current must flow through each resistor to return to the power source. The current is labeled I_T to represent total current. Keep in mind current is the same throughout a series circuit.

The circuit in Fig. 2-2 or any circuit that contains more than one resistor in series, requires more calculations than just the use of Ohm's Law. It is first necessary to determine the total resistance of the circuit. Resistors in series are added, giving the formula:

$$R_T = R_1 + R_2 + R_3 + \ldots \tag{2-5}$$

Using the resistance total formula for series circuits, calculate the total resistance of Fig. 2-2.

$$R_T = R_1 + R_2 + R_3$$

$$= 10\ \Omega + 20\ \Omega + 70\ \Omega$$

$$= 100\ \Omega$$

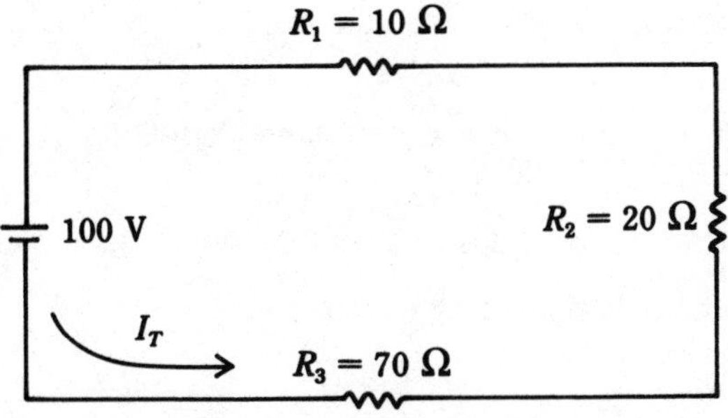

2-2 Series circuit with three resistors.

Now that the circuit resistance is known, you can use Ohm's Law to find the circuit current.

$$I = \frac{E}{R}$$

$$= \frac{100\ \text{V}}{100\ \Omega}$$

$$I_T = 1\ \text{A}$$

Using the current through the resistor and the resistor value, you can calculate the voltage drop across each resistor, often called the *IR* drop. The voltage drop is the voltage that would be measured with a voltmeter connected directly across the resistor.

$$E = I \times R \text{ (formula to be used)}$$

$$E_{R1} = I_T \times R_1$$
$$= 1\ \text{A} \times 10\ \Omega$$
$$= 10\ \text{V}$$

$$E_{R2} = I_T \times R_2$$
$$= 1\ \text{A} \times 20\ \Omega$$
$$= 20\ \text{V}$$

$$E_{R3} = I_T \times R_3$$
$$= 1\ \text{A} \times 70\ \Omega$$
$$= 70\ \text{V}$$

Summary of voltage drops:

$$E_{R1} = 10\ \text{V}$$
$$E_{R2} = 20\ \text{V}$$
$$E_{R3} = 70\ \text{V}$$

If all the voltage drops were calculated correctly, they should add to the applied voltage.

$$E_{R1} + E_{R2} + E_{R3} = \text{applied voltage, } E_T$$
$$10 + 20 + 70 = 100 \text{ V}$$

- The sum of the voltage drops in a series circuit equals the applied voltage.

Calculate the power dissipated by each resistor.

$$P = I \times E \tag{2-2}$$

$$\begin{aligned} P_{R1} &= I_T \times E_{R1} \\ &= 1 \text{ A} \times 10 \text{ V} \\ &= 10 \text{ W} \end{aligned}$$

$$\begin{aligned} P_{R2} &= I_T \times E_{R2} \\ &= 1 \text{ A} \times 20 \text{ V} \\ &= 20 \text{ W} \end{aligned}$$

$$\begin{aligned} P_{R3} &= I_T \times E_{R3} \\ &= 1 \text{ A} \times 70 \text{ V} \\ &= 70 \text{ W} \end{aligned}$$

Summary of power dissipation:

$$\begin{aligned} P_{R1} &= 10 \text{ W} \\ P_{R2} &= 20 \text{ W} \\ P_{R3} &= 70 \text{ W} \end{aligned}$$

If all the powers have been calculated correctly, the power dissipated by the resistors should be equal to the power supplied by the power supply.

$$P_{R1} + P_{R2} + P_{R3} = P_T$$
$$10 + 20 + 70 = 100 \text{ W}$$

The sum of the powers can be checked by using the power formula with total current and applied voltage.

$$P = I \times E \quad \text{formula to be used} \tag{2-2}$$

$$\begin{aligned} P_T &= I_T \times E_T \\ &= 1 \text{ A} \times 100 \text{ V} \\ &= 100 \text{ W} \end{aligned}$$

The total power dissipated in a series circuit is equal to the sum of the individual powers.

There are other observations that should be made about the series circuit and the calculations just made. Note the following key points.

- In a series circuit, the largest resistance will drop the most voltage.
- In a series circuit, the largest resistance will dissipate the most power.

There are two other conditions that can happen to a circuit, a short circuit and an open circuit. Both of these conditions are usually considered defects or faults.

- A short circuit has a resistance of zero. This condition has unlimited current, zero voltage drop, and zero power.
- An open circuit has a resistance of infinity. This condition causes zero current, applied voltage dropped at open, and zero power.

General guidelines for solving series circuits

The following general procedure is intended as guidelines for solving the series circuit. It is not intended to consider every possible condition of known and unknown values in a circuit.

Step 1 Find R_T. Finding the total resistance first is almost always the best way to proceed. If one value of resistance is not known, then the current must be known, or it must be possible to calculate from some other information.

Step 2 Calculate the total current, I_T. Keep in mind, current is the same throughout a series circuit.

Step 3 Find the individual voltage drops across each resistor. Use I_T and the resistance values.

Step 4 Check the calculated voltage drops by adding them together to see if they equal the applied voltage.

Step 5 Calculate the power dissipated in each resistor. Use the current, I_T, and the individual voltage drops.

Step 6 The individual powers should be checked by adding and see if they equal the total power. Total power is calculated using total current and applied voltage.

Note Once the total circuit resistance is found, all the other calculations will fall into place. Sometimes, one of the resistor values is not known and the circuit will have to be worked differently.

Practice problems

Use the schematic diagrams shown to find the unknown quantities. Round answers to three significant figures.

1. Find: I_T, R_T, E_{R1}, E_{R2}, P_{R1}, P_{R2}

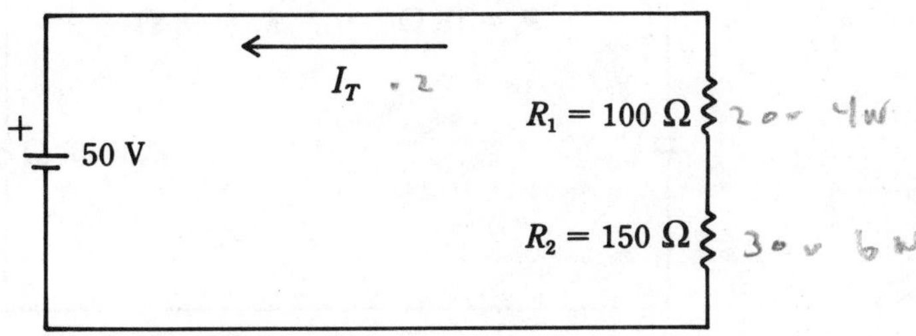

2. Find: I_T, R_T, E_{R1}, E_{R2}, E_{R3}, P_{R1}, P_{R2}, P_{R3}

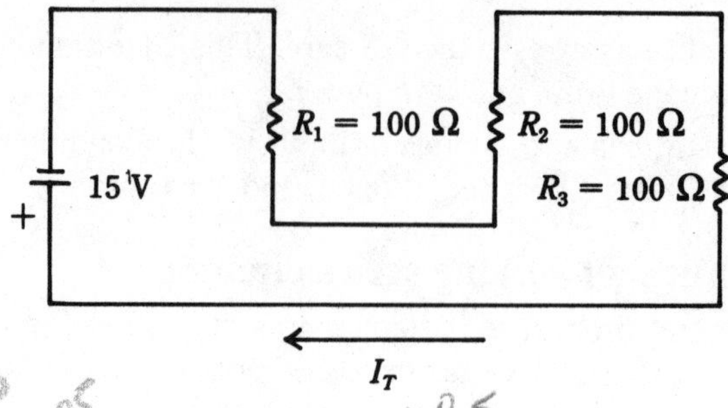

3. Find: R_T, I_T

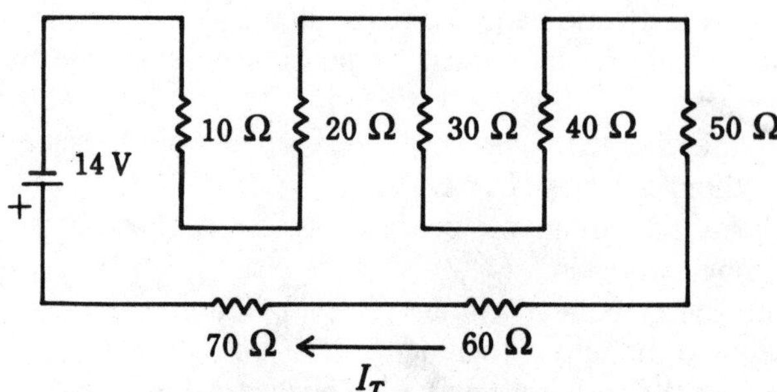

4. Given: E_{R2} = 90 V. Find: I_T, V

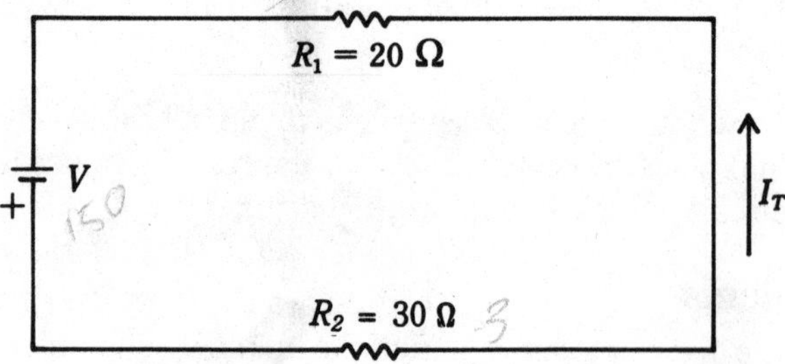

5. Given: E_{R3} = 60 V. Find: R_T, I_T, V

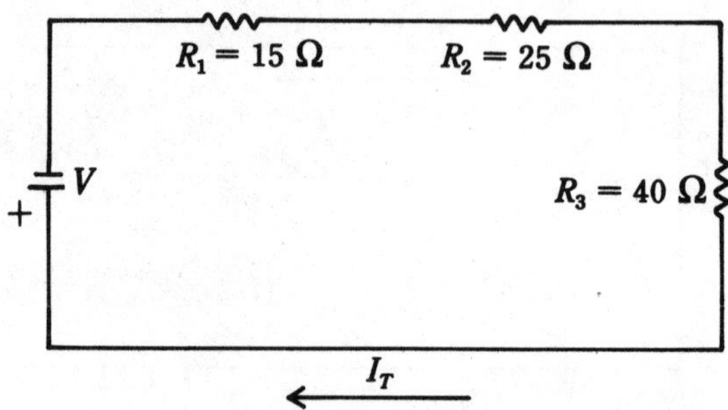

6. Given: I_T = 2 mA. Find: R_T, R_1

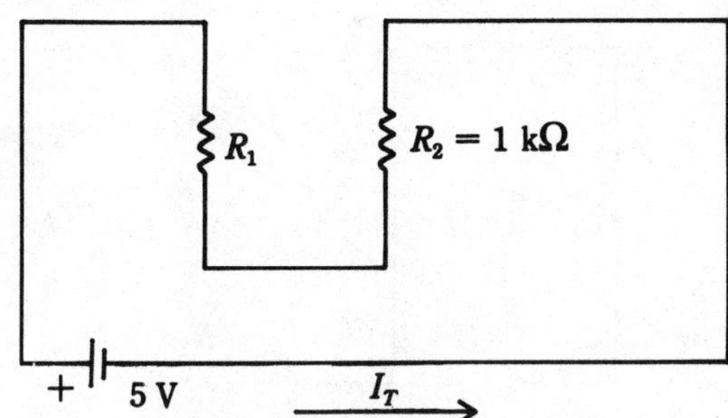

7. Given: I_T = 100 mA, E_{R3} = 10 V. Find: R_T, R_3, V

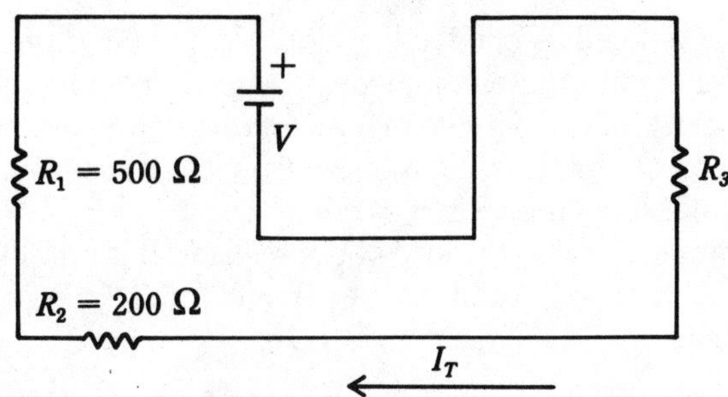

8. Given: P_{R3} = 200 mW. Find: I_T, R_2, E_{R1}

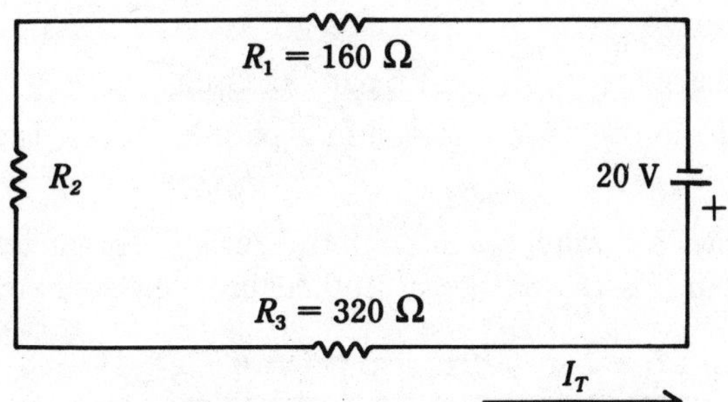

9. Given: P_T = 12.8 W. Find: V, I_T

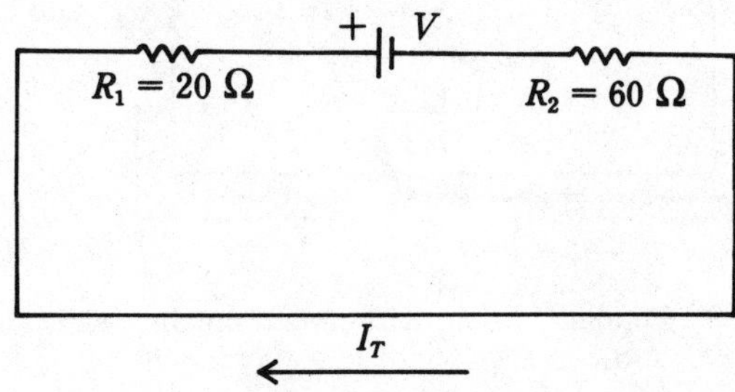

10. Given: $P_T = 240$ W, $P_{R1} = 40$ W: Find: V, I_T, R_2

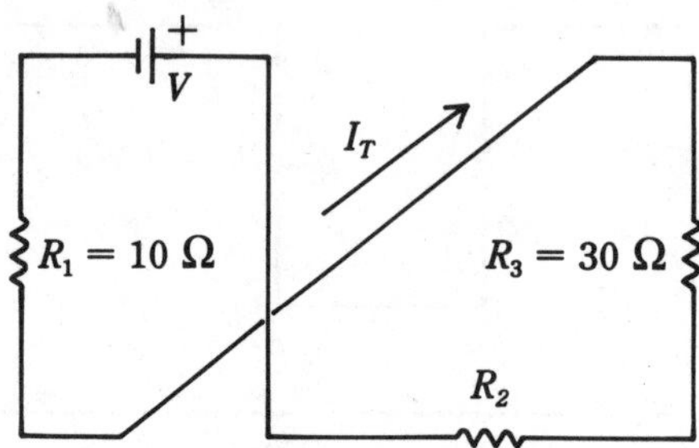

Parallel circuits

A parallel circuit is defined as one having more than one path for current to flow. In a simple parallel circuit, there is a piece of wire connecting all the resistors on one side and another piece of wire connecting the resistors on the other side.

Because a piece of wire has no voltage dropped across it, voltage to all the resistors is the same. Remember, in a series circuit, each resistor has a different voltage, directly proportional to the size of the resistor. In a parallel circuit, it is the current that will be divided to each resistor branch, indirectly proportional to the size of the resistance.

- In a parallel circuit, the current divides to the individual branches indirectly proportional to the size of the resistors.
- Total current in a parallel circuit is equal to the sum of the individual branch currents.
- Voltage is the same throughout a parallel circuit.
- The sum of the powers dissipated in a parallel circuit is equal to the total power.

Figure 2-3 shows a parallel circuit with two resistive branches. Notice how the arrows in the drawing that show the current leaving the power supply is the sum of

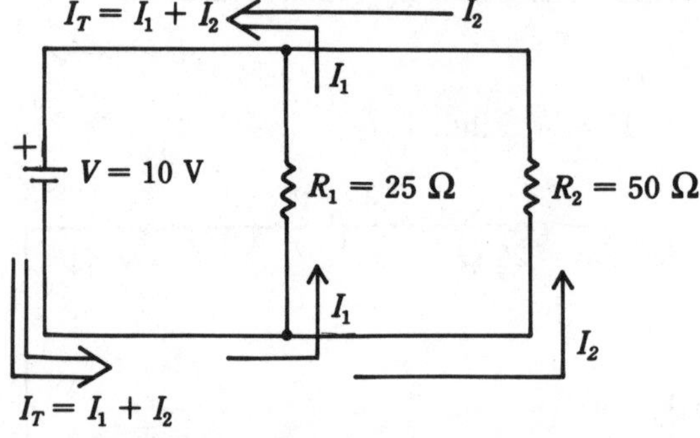

2-3 Parallel circuit with two branches showing how current divides to each branch.

the individual branch currents. The branch currents are calculated using the voltage across the resistor and the resistance value, keeping in mind the voltage is the same across both branches.

Calculating current in a parallel circuit

Each individual branch can be treated as an individual series circuit. Calculate the branch current as you would in any series circuit. Refer to Fig. 2-3 for the following example.

R_1 branch current.

$$I_1 = \frac{E}{R_1}$$

$$= \frac{10\ \text{V}}{25\ \Omega}$$

$$= 0.4\ \text{A}$$

R_2 branch current.

$$I_2 = \frac{E}{R_2}$$

$$= \frac{10\ \text{V}}{50\Omega}$$

$$= 0.2\ \text{A}$$

$$I_T = I_1 + I_2$$

$$= 0.4\ \text{A} + 0.2\ \text{A}$$

$$= 0.6\ \text{A}$$

Calculating total resistance in a parallel circuit

Calculating the resistance in a simple parallel circuit is fairly easy using Ohm's Law. To calculate the total resistance of the simple circuit shown in Fig. 2-3, use total current and applied voltage.

R_T is the total resistance of the circuit.

$$R_T = \frac{E}{I_T}$$

$$= \frac{10\ \text{V}}{0.6\ \text{A}}$$

$$= 16.7\ \Omega$$

Figure 2-4 shows an equivalent circuit from the calculated values to replace the original circuit shown in Fig. 2-3.

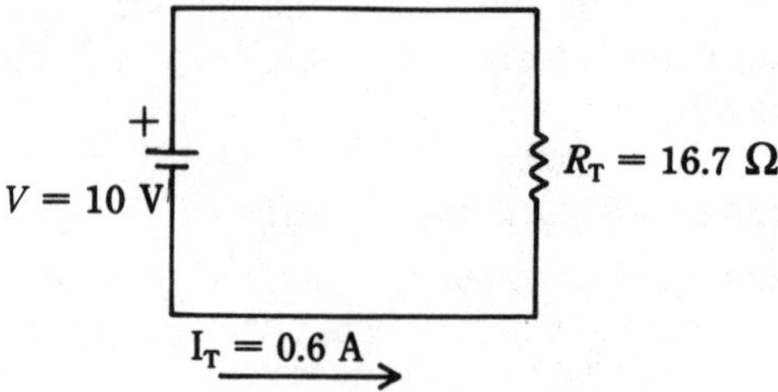

2-4 The equivalent circuit of Fig. 2-3.

Total resistance can also be called the equivalent resistance of the parallel circuit. When the cicuit is connected to the power supply, the total current will be exactly equal to the original circuit. Equivalent circuits are very helpful for circuits that are much more complicated.

Power calculations are performed exactly the same as they are done in series circuits.

The method used to calculate total resistance, shown above, is fine if total current can be calculated first. In a more complex circuit, it would be impossible to first calculate total current. Therefore, it is necessary to have a way of calculating the total resistance of the parallel circuit independently of current. That leads to a new formula.

$$\frac{1}{R_T} = \frac{1}{R_1} + \frac{1}{R_2} + \frac{1}{R_3} + \ldots \quad \text{resistors in parallel} \qquad (2\text{-}6)$$

This formula is called the reciprocal formula, and it can be used regardless of how many resistors are connected in parallel.

R_T using the reciprocal formula for Fig. 2-3.

$$\frac{1}{R_T} = \frac{1}{R_1} + \frac{1}{R_2} \qquad (2\text{-}6)$$

$$= \frac{1}{25} + \frac{1}{50}$$

$$= 0.04 + 0.02$$

$$= 0.06 \ (R_T \text{ is still in reciprocal form})$$

$$= 16.7\ \Omega$$

The reciprocal method shown above arrives at the same answer as using the total current. The reciprocal formula must be used with complicated circuits. If there are only two resistors in the circuit, a shortcut formula can be used.

Shortcut formula for two resistors in parallel:

$$R_T = \frac{R_1 \times R_2}{R_1 + R_2} \qquad (2\text{-}7)$$

Example from Fig. 2-3:

$$R_T = \frac{25 \times 50}{25 + 50}$$

$$= \frac{1250}{75}$$

$$= 16.7\ \Omega$$

Calculations for parallel circuits must be done separately from the calculations for a series circuit.

The total resistance of a parallel circuit is almost always the first thing that will be found, especially in a more complex circuit. Often, the R_T will be used to find other calculations. If a mistake is made in R_T all other calculations will also be wrong. One quick check is, the resistance of each branch is higher than the total resistance.

- The total resistance of a parallel circuit is smaller than the smallest resistance of any branch.

Figure 2-5 shows three parallel branches. This circuit will be used to demonstrate the effect of having equal resistance in each of the parallel branches. Notice in branch B there are two resistors. This branch is actually a series circuit, by itself. Two resistors in series simply add, so the middle branch has a resistance of 60 Ω, as do the other two branches.

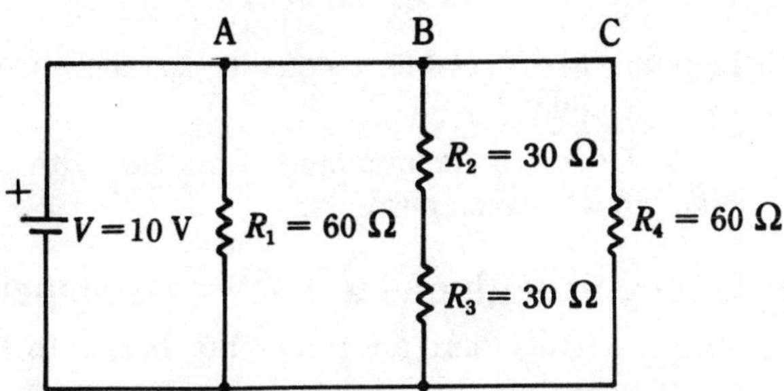

2-5 Parallel branches with equal resistance.

Calculate R_T for Fig. 2-5.

$$\frac{1}{R_T} = \frac{1}{R_A} + \frac{1}{R_B} + \frac{1}{R_{RC}}$$

$$= \frac{1}{60} + \frac{1}{60} + \frac{1}{60}$$

(branch B is a series circuit equal to 30 + 30 = 60 Ω)

$$= 0.01667 + 0.01667 + 0.01667$$

(change fractions to decimals using the calculator)

$$= 0.05\ \Omega$$

$$= 20\ \Omega$$

Notice from the calculations, the value of R_T is one-third the value of one resistor. This problem could have been solved by dividing the resistance of one branch by the number of branches having equal resistance.

- To find the total resistance of equal resistances in parallel, divide the resistance of one branch by the number of branches with equal resistance.

$$R_T = \frac{R}{n}$$

$$= \frac{60}{3}$$

$$= 20\ \Omega$$

Calculate the value of total current.

$$I = \frac{E}{R}$$

$$= \frac{10\ \text{V}}{60\ \Omega}$$

$$= 0.1667\ \text{A}$$

- In a parallel circuit, if branches have equal resistance, the current through each branch will be equal.
- The total current divided by the number of branches with equal resistances will give the individual branch currents.

Solving a parallel circuit with one unknown resistance

Refer to Fig. 2-6. In this circuit, there are three branches, with the resistance of only two known. The total current and the applied voltage is given, which means

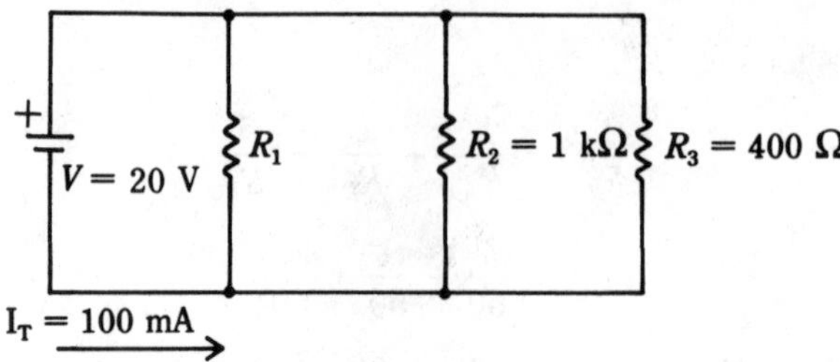

2-6 Solving a parallel circuit when one resistance is unknown.

the first step is to calculate the total resistance. From there, the unknown branch resistance can be calculated.

Calculate the total resistance of the parallel circuit, Fig. 2-6.

$$R_T = \frac{V}{I_T}$$

$$= \frac{20 \text{ V}}{100 \text{ mA}}$$

$$= 200\ \Omega$$

Using the R_T just calculated, find R_1.

$$\frac{1}{R_T} = \frac{1}{R_1} + \frac{1}{R_2} + \frac{1}{R_3} \qquad (2\text{-}6)$$

$$\frac{1}{200} = \frac{1}{R_1} + \frac{1}{1000} + \frac{1}{400}$$

$$0.005 = \frac{1}{R_1} + 0.001 + 0.0025$$

$$\frac{1}{R_1} = 0.005 - (0.001 + 0.0025)$$

$$= 0.0015$$

$$R_1 = 667\ \Omega$$

General guidelines for solving parallel circuits

The following is intended to be general guidelines for solving parallel circuits. You will need to work each circuit according to what is given and what is to be found. Having a good working knowledge of the formula is the best way to solve circuits.

Step 1 Find R_T. It is almost always best to solve a circuit by first finding the total resistance of the circuit. Keep in mind that the total resistance of a parallel circuit is always smaller than the smallest branch resistance.

Step 2 Find I_T. Use the applied voltage and the total resistance. Total current can also be found by adding each of the branch currents.

Step 3 Find branch currents. Keep in mind that voltage is the same across all branches of a parallel circuit, but the current will divide to the individual branches, indirectly proportional to the branch resistance. The largest resistance has the smallest current.

Step 4 Power is the same as with series circuits. The power is the current through a resistor times the voltage across a resistor. Total power is the sum of the individual powers.

Practice problems

Use the schematic diagrams shown to find the unknowns. Round answers to 3 significant figures.

1. Find: R_T, I_T, I_A, I_B, I_C (branch currents)

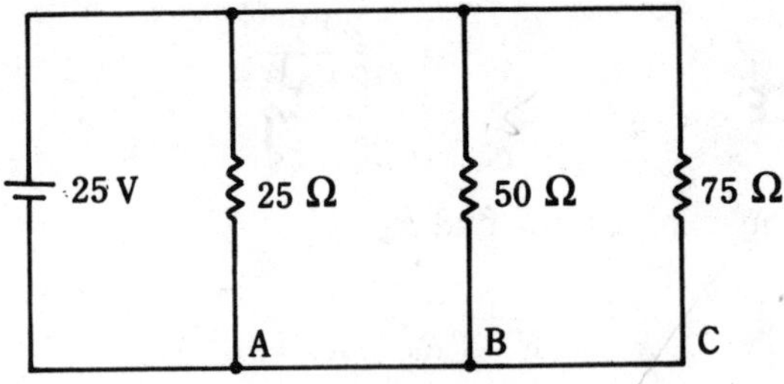

2. Find: R_T, I_T

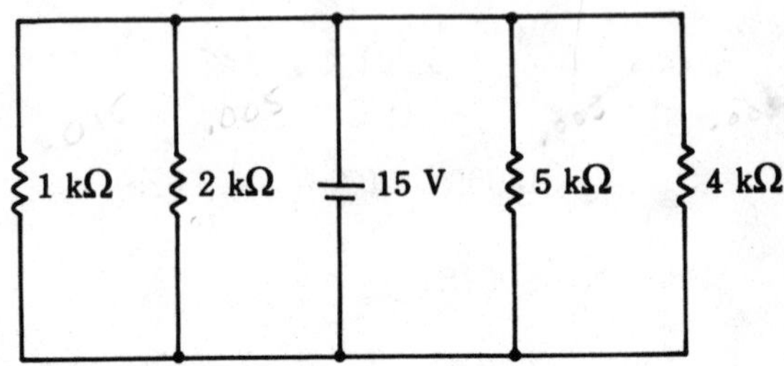

3. Find: R_T

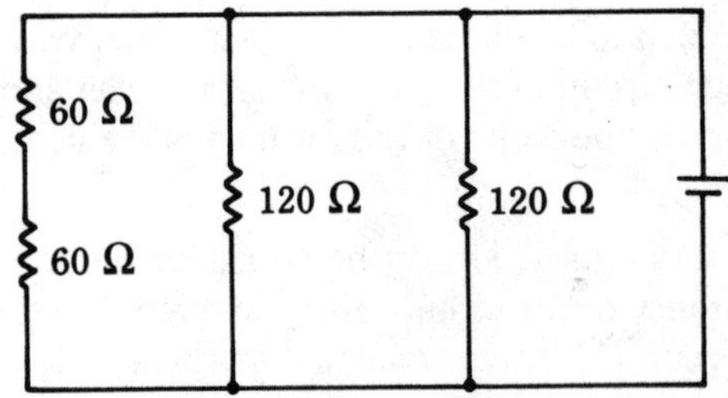

4. Find: R_T

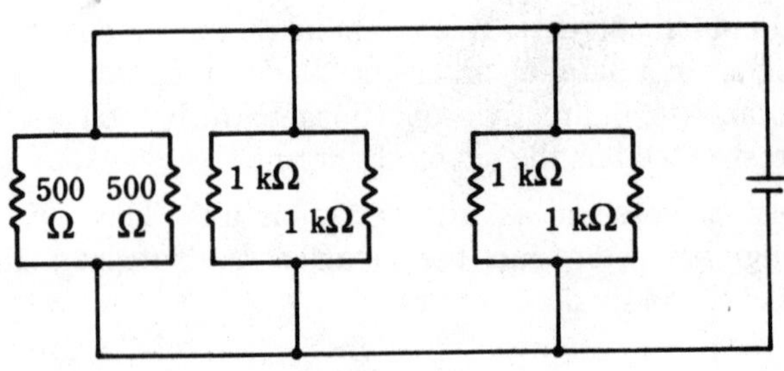

5. Given: $R_T = 150\ \Omega$. Find: R_3

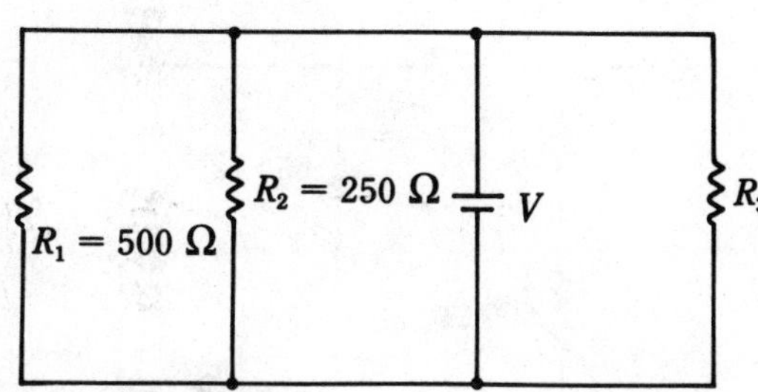

6. Given: $R_T = 5\ \Omega$. Find: R_4

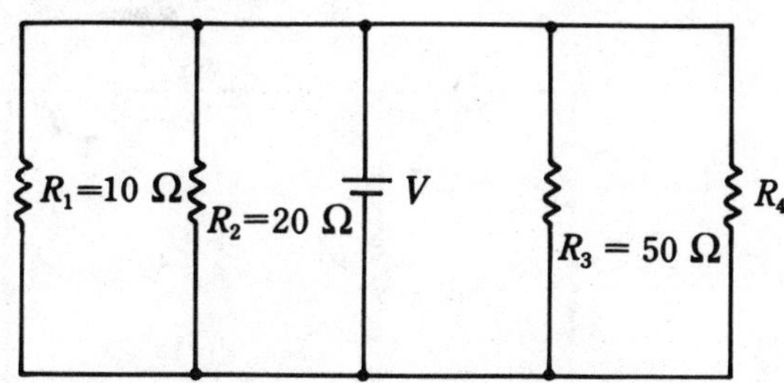

7. Find: R_3, I_T, R_T

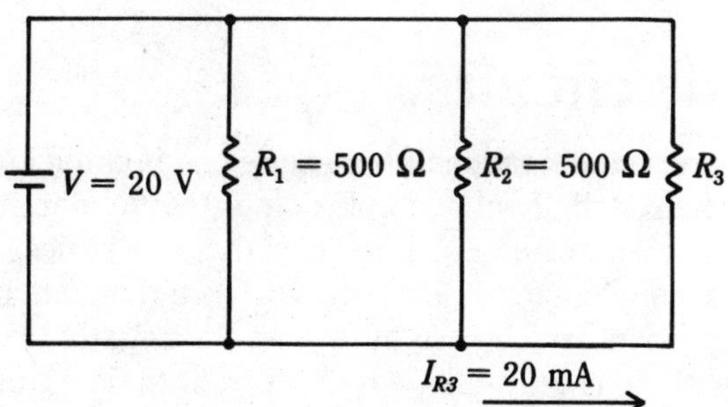

8. Find: R_T, I_A, I_B, I_C

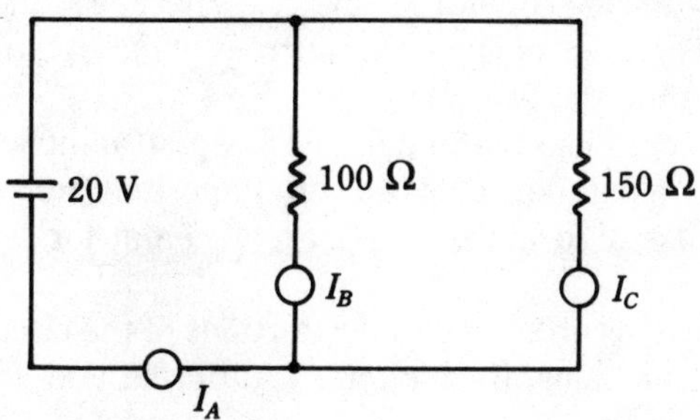

9. Find: R_T, I_A, I_B, I_C, I_D

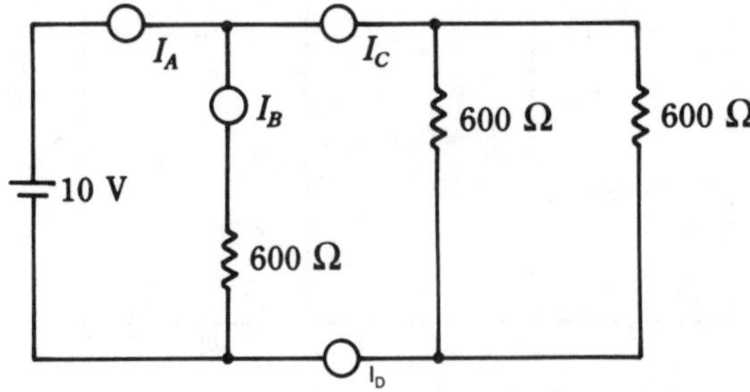

10. Find: R_T, I_A, I_B, I_C, I_D, I_E, I_F, I_G

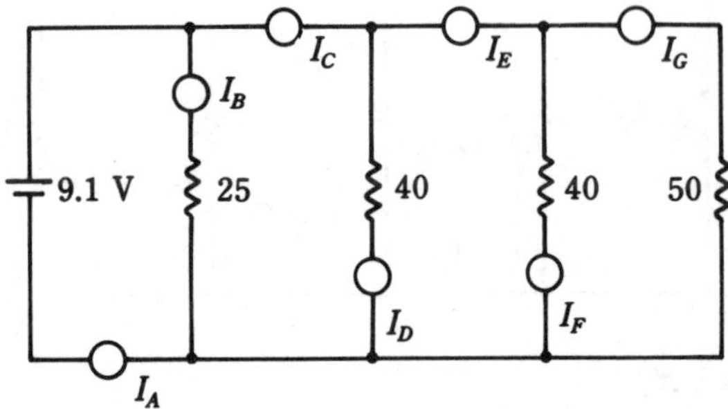

Complex dc circuits

Complex circuits are a combination of both series and parallel resistors. The most important thing when dealing with complex circuits is to identify which are series and which are parallel branches. Keep in mind that in a series circuit, current has only one path. In a parallel circuit, current has more than one path to flow.

Refer to Fig. 2-7. Notice the current leaving the negative side of the battery is labeled I_T because there is only one path for current to flow from the battery. The current then comes to R_R, which is a series resistor because there is only one path for the current to flow. Therefore, I_T flows through the series resistor, R_4. The current then comes to a place where the circuit has two paths, labeled point b. Here the current divides, flowing to both R_2 and R_3. Therefore, these two resistors are parallel resistors. Then, at point a the current joins again to form I_T to flow through R_1 and return to the battery.

The thought process used here helps to keep straight which resistors are in series and which are in parallel. This is a very important process because solving a circuit requires the use of formulas which are different for series and parallel circuits.

It is generally best to solve the parallel portions of the circuit and then combine them with the series portions. Each circuit is different and will have to be considered on its own.

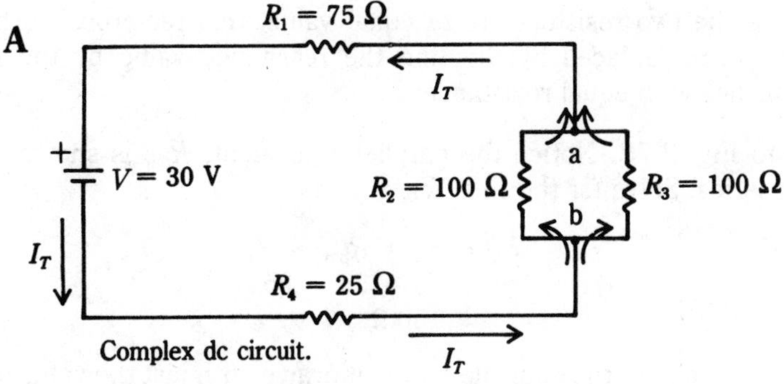

Complex dc circuit.

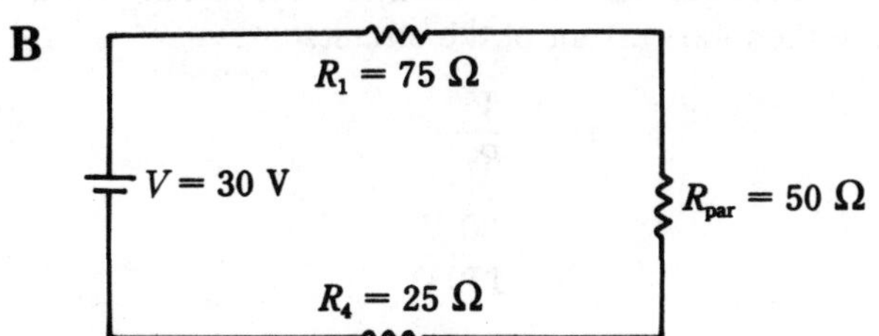

One step in solving a complex circuit

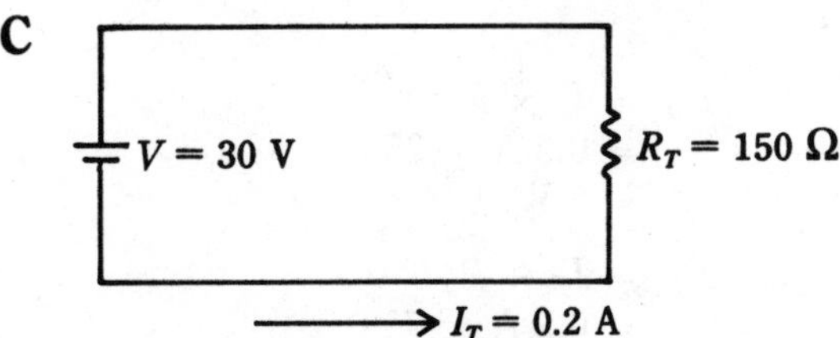

Equivalent circuit showing R_T

2-7 Steps in solving a complex dc circuit.

Solving a complex circuit

Refer to Fig. 2-7. It has been identified that R_1 and R_4 are series resistors and R_2 and R_3 are parallel. The first objective, and usually the hardest, is to solve for R_T. In a complex circuit, this will involve several steps.

Find the equivalent resistance of the parallel branches, Fig. 2-7A.

$$\frac{1}{R_{par}} = \frac{1}{R_2} + \frac{1}{R_3}$$

$$= \frac{1}{100} + \frac{1}{100}$$

$$= 0.01 + 0.01$$

$$= 0.02$$

$$R_{par} = 50\ \Omega$$

Note Since the two resistors are of equal value, this reciprocal formula could have been replaced by dividing the resistance value by the number of branches with equal resistance.

Refer to Fig. 2-7B. Notice the parallel equivalent, R_{par} is shown as a single resistor in series. Solve for the final R_T.

$$\begin{aligned} R_T &= R_1 + R_{par} + R_4 \\ &= 75 + 50 + 25 \\ &= 150\ \Omega \end{aligned}$$

Figure 2-7C shows the equivalent circuit drawn with just the value of R_T. From here we can calculate the total current of the circuit.

$$\begin{aligned} I_T &= \frac{V}{R_T} \\ &= \frac{30\ \text{V}}{150\ \Omega} \\ &= 0.2\ \text{A} \end{aligned}$$

To find branch currents and voltage drops, first find the voltage drops of the series resistors, Fig. 2-7A.

$$\begin{aligned} E_{R4} &= I_T \times R_R \\ &= 0.2\ \text{A} \times 25\ \Omega \\ &= 5\ \text{V} \end{aligned}$$

$$\begin{aligned} E_{R1} &= I_T \times R_1 \\ &= 0.2\ \text{A} \times 75\ \Omega \\ &= 15\ \text{V} \end{aligned}$$

With the voltage drops of the series resistors calculated, the voltage remaining in the circuit is dropped across the parallel circuit, Fig. 2-7A.

$$\begin{aligned} V_{applied} &= E_{R1} + E_{R4} + E_{par} \\ 30\ \text{V} &= 15\ \text{V} + 5\ \text{V} + E_{par} \\ E_{par} &= 10\ \text{V} \end{aligned}$$

Except for power, the calculations remaining in this circuit are to find branch currents. This is done using the voltage across the parallel circuit and the resistance of each branch, Fig. 2-7A.

$$I = \frac{E}{R}$$

$$I_{R2} = \frac{10\ \text{V}}{100\ \Omega}$$

$$= 0.1 \text{ A}$$

$$I_{R3} = I_{R2} \quad \text{(because this circuit has equal values of resistance in the parallel branches)}$$

It is now possible to find the power dissipated by each resistor and also find the total power. Below is a sample power calculation.

Power calculation for resistor R_2:

$$\begin{aligned} P &= I \times E \\ &= 0.1 \text{ A} \times 10 \text{ V} \\ &= 1 \text{ W} \end{aligned}$$

Total power can be calculated by either adding all the powers in the circuit, or using total current.

$$\begin{aligned} P_T &= I_T \times V \\ &= 0.2 \text{ A} \times 30 \text{ V} \\ &= 6 \text{ W} \end{aligned}$$

Solving the complex circuit

Figure 2-8A shows the original circuit before any of the resistors have been combined. With this type of circuit, and with many circuits, it is often easier to start at a point farthest from the power supply. Keep in mind that series and parallel resistors cannot be combined in the same calculations. Saying that resistors are combined means to find the equivalent value of the resistors.

Find the parallel combination of R_5 and R_6. These are two equal resistors in parallel, divide the resistance value by the number of resistors.

$$\frac{10 \text{ k}\Omega}{2} = 5 \text{ k}\Omega$$

Figure 2-8B shows resistor R_4 in series with the combined R_5 and R_6, and this branch is in parallel with R_3. First solve the series combination.

$$R_4 + R_{5,6}$$

$$5 \text{ k}\Omega + 5 \text{ k}\Omega = 10 \text{ k}\Omega \quad \text{(this equivalent resistance is in parallel with } R_3\text{)}$$

Note R_3 and the equivalent resistor are both 10 kΩ. The new equivalent resistance will be 5 kΩ.

Figure 2-8C shows the equivalent resistance from the last calculation, labeled the combined resistance of R_3, R_4, R_5, R_6. The circuit shown in Fig. 2-8C is a series circuit with three resistors and when they are combined, the result will be the final total resistance of the circuit.

$$\begin{aligned} R_T &= R_1 + R_{3,4,5,6} + R_2 \\ &= 2.5 \text{ k}\Omega + 5 \text{ k}\Omega + 1.5 \text{ k}\Omega \\ &= 9 \text{ k}\Omega \end{aligned}$$

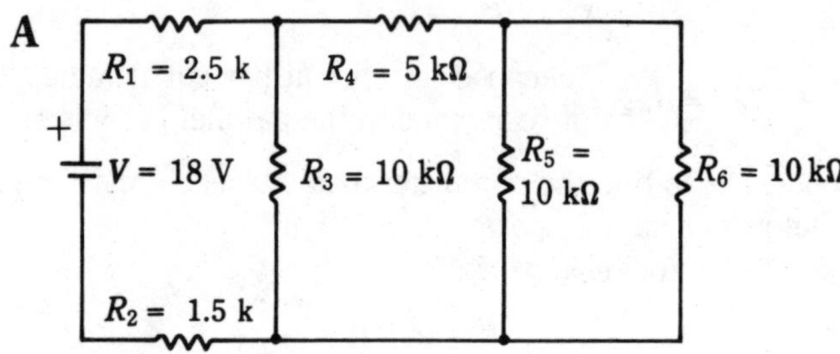

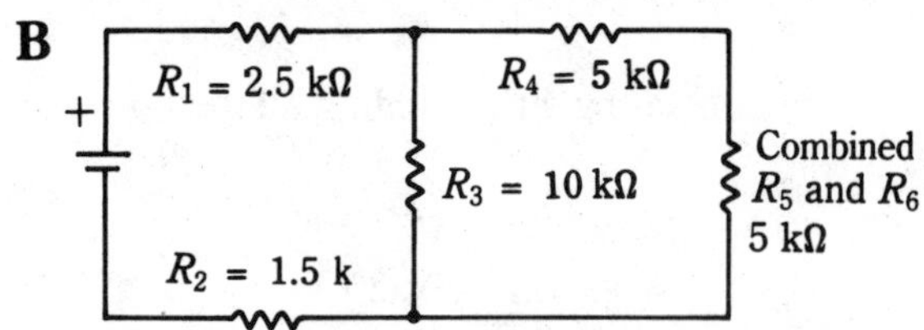

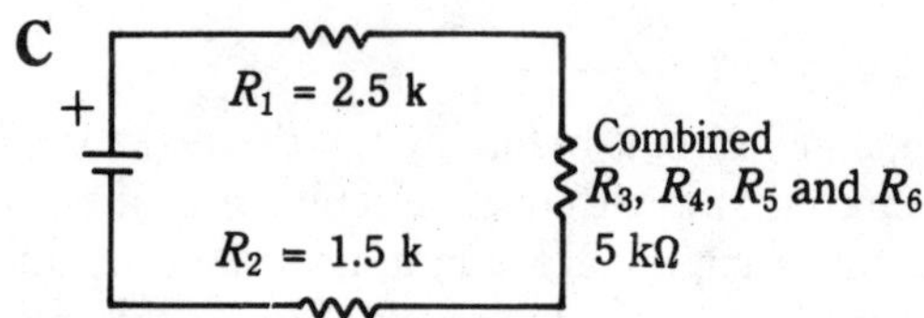

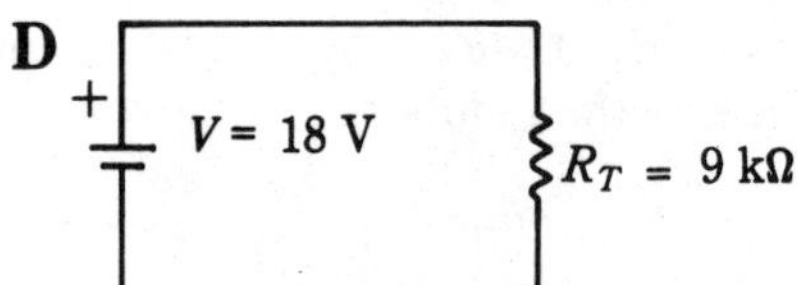

Final equivalent circuit

2-8 Using equivalent circuits to simplify a complex circuit.

The next step is always to use the final equivalent circuit to calculate the circuit total current, I_T.

$$I_T = \frac{V}{R_T} \tag{2-1A}$$

$$= \frac{18\text{ V}}{9\text{ k}\Omega}$$

$$= 2\text{ mA}$$

Using Fig. 2-8C, find the voltage drops in the series equivalent circuit.

$$E_{R1} = I_T R_1$$
$$= 2\text{ mA} \times 2.5\text{ k}\Omega$$
$$= 5\text{ V}$$

$$E_{R3,4,5,6} = I_T R$$
$$= 2 \text{ mA} \times 5 \text{ k}\Omega$$
$$= 10 \text{ V}$$

$$E_{R2} = I_T R_2$$
$$= 2 \text{ mA} \times 1.5 \text{ k}\Omega$$
$$= 3 \text{ V}$$

Check the voltage drops just calculated to be sure they add to the applied voltage. This is the only time this check can be performed easily.

$$E_{R1} + E_{R3,4,5,6} + E_{R2} = V$$
$$5 + 10 + 3 = 18 \text{ V}$$
$$V = 18 \text{ V}$$

Find the current through all the resistors.

R_1 and R_2 are both series resistors, therefore have I_T flowing through them.

$$I_T = I_{R1} = I_{R2} = 2 \text{ mA}$$

Find current in R_3. Notice the voltage dropped across $R_{3,4,5,6}$ is applied directly across R_3. Therefore, $E_{R3} = E_{R3,4,5,6}$.

$$I_{R3} = \frac{10 \text{ V}}{10 \text{ k}\Omega}$$
$$= 1 \text{ mA}$$

The current in R_3 should be one half the total current since I_T splits to two parallel branches with equal resistance. Refer to Fig. 2-8B. $I_{R4} = I_{R5,6}$ series resistors.

$$I_{R4,5,6} = \frac{E_{R3,4,5,6}}{R_4 + R_{5,6}}$$
$$= 1 \text{ mA}$$

$$I_{R4} = 1 \text{ mA}$$

This calculated current is the same as through R_3.

The current of $I_{R5,6}$ splits to go to R_5 and R_6 individually. Refer to the original circuit. In order to calculate the current in these resistors, it is first necessary to calculate the voltage drop across the series resistor R_4.

$$E_{R4} = I_{R4} \times R_4$$
$$= 1 \text{ mA} \times 5 \text{ k}\Omega$$
$$= 5 \text{ V}$$

Voltage dropped across R_5 and R_6, parallel combination will be $E_{R3,4,5,6}$ minus E_{R4}.

$$E_{R5,6} = E_{R3,4,5,6} - E_{R4}$$
$$= 10 \text{ V} - 5 \text{ V}$$
$$= 5 \text{ V}$$

Note I_{R5} and I_{R6} will have the same current because they are equal values with equal voltage. $E_{R5} = E_{R6}$ since they are in parallel.

$$I_{R5} = I_{R6} = \frac{E_{R5,6}}{R_5}$$

$$I_{R5,6} = \frac{5\text{ V}}{10\text{ k}\Omega}$$

$$= 0.5\text{ mA}$$

Summary of voltage and current for each resistor.

$$E_{R1} = 5\text{ V}, I_{R1} = 2\text{ mA}$$
$$E_{R2} = 3\text{ V}, I_{R2} = 2\text{ mA}$$
$$E_{R3} = 10\text{ V}, I_{R3} = 1\text{ mA}$$
$$E_{R4} = 5\text{ V}, I_{R4} = 1\text{ mA}$$
$$E_{R5} = 5\text{ V}, I_{R5} = 0.5\text{ mA}$$
$$E_{R6} = 5\text{ V}, I_{R6} = 0.5\text{ mA}$$

Note This circuit is a good circuit to use as an example on how to work through the different stages of a complex circuit. The values of resistors were selected to make the calculations as easy as possible. When following a sample problem, it is best to have the calculations easy enough to be done in your head, when possible.

The purpose of this sample problem is to show the steps and allow you to follow through the problem while reading the text material.

General guidelines for solving a complex circuit

Keep in mind the following is intended to be only a general guideline. Each circuit must be worked by its own individuality.

Step 1 Find R_T. When solving complex circuits, find R_T first. This will often involve many steps.

Step 2 Use equivalent circuits for each step of R_T. The equivalent circuits will be a great help when finding currents and voltages.

Step 3 Find I_T. Use the final equivalent circuit showing R_T to find the total current. Remember, total current will flow through any resistor that is in series with the power supply and the rest of the circuit.

Step 4 Calculate the voltage drop for any series resistors. Whatever voltage is not dropped across a series resistor will be available for the parallel circuit combinations.

Step 5 Using the voltage applied to the parallel combination, determine how the current will split to each branch.

Step 6 Use branch currents and resistance values to determine the voltage of each resistor.

Note When working through these steps, keep in mind that the equivalent circuits are quite easy to use to determine the relationships of different resistors, currents and voltages.

Step 7 Make a summary of the voltage drops and currents for each resistor.

Practice problems

Use the schematic diagrams shown to find the unknown quantities. Round answers to 3 significant figures.

1. Find: R_T, I_T, E_{R1}, E_{R2}, E_{R3}, I_{R2}, I_{R3}

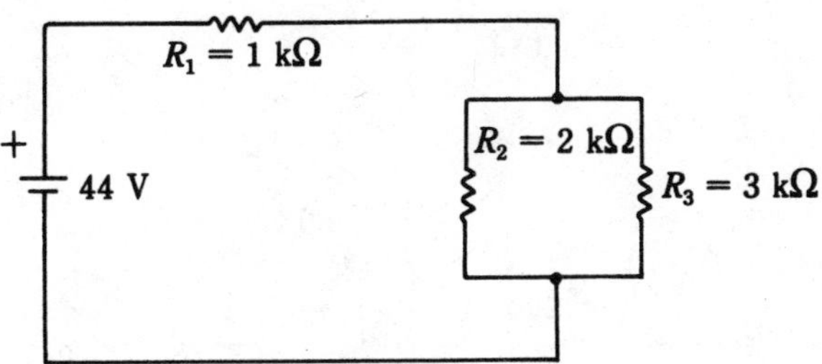

2. Find: R_T, I_T, E_{R1}, E_{R2}, E_{R3}, E_{R4}, I_{R2}, I_{R3}

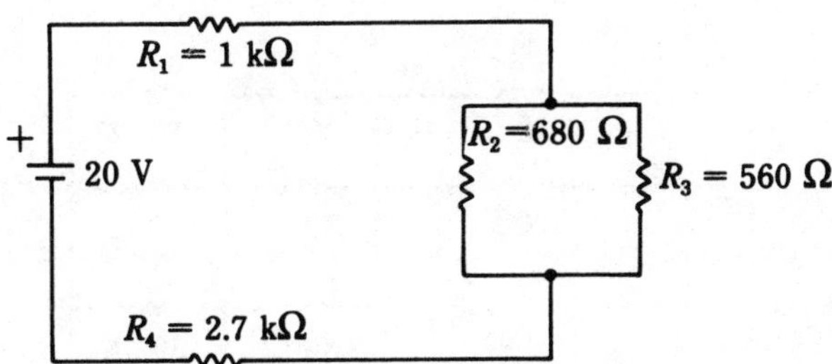

3. Find: R_T, I_T, I_{R2}, I_{R5}, voltage drops for all resistors

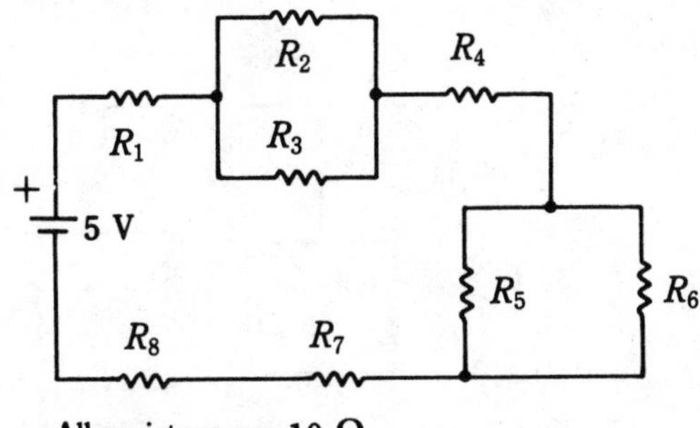

All resistors are 10 Ω

4. Find: R_T, I_T, current in all resistors and voltage drops across all resistors

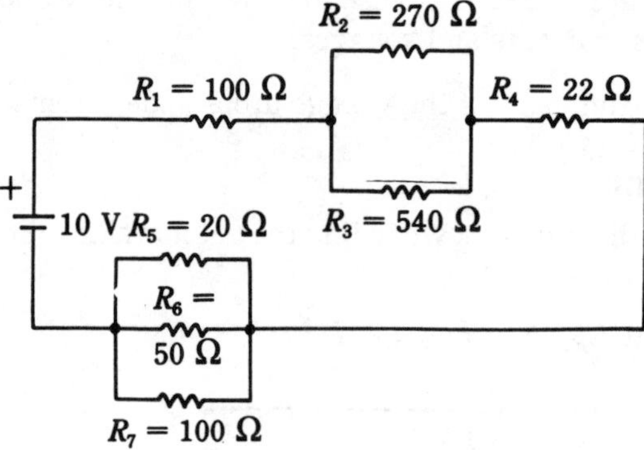

5. Find: R_T, I_T, E_{R1}, E_{R2}, I_{R2}, I_{R3}

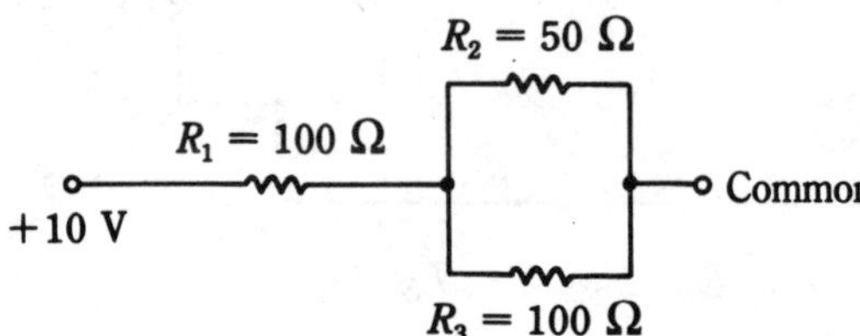

6. Find: R_T

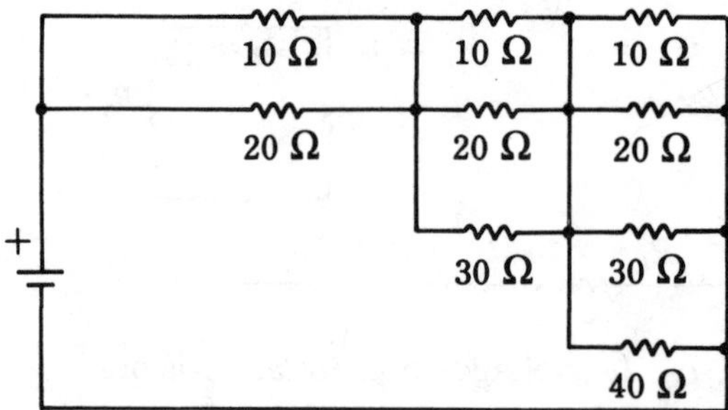

7. Find: R_T

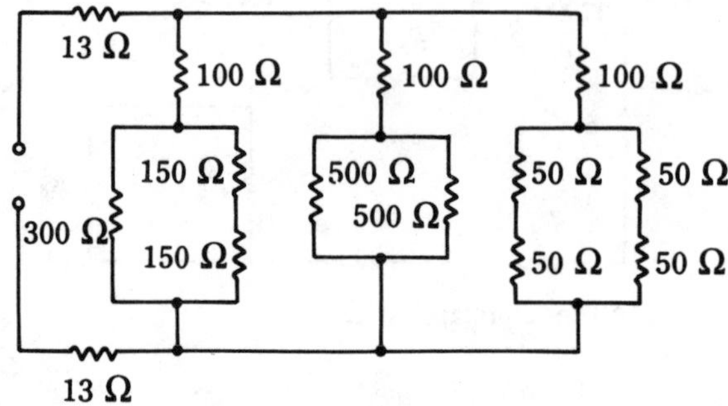

8. Find: R_T

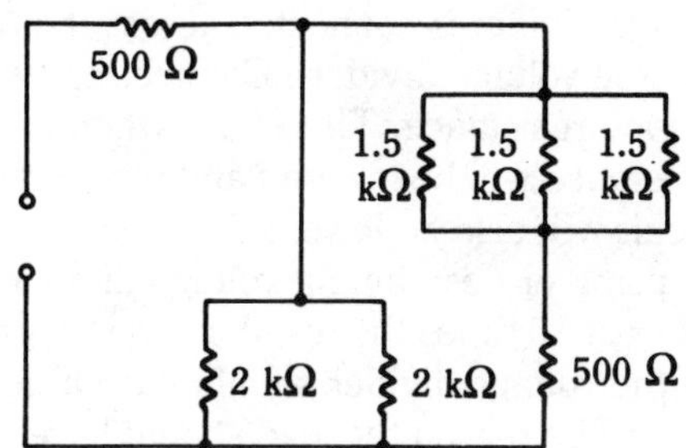

9. Find: R_T, I_T, current in all resistors and voltage drops across all resistors

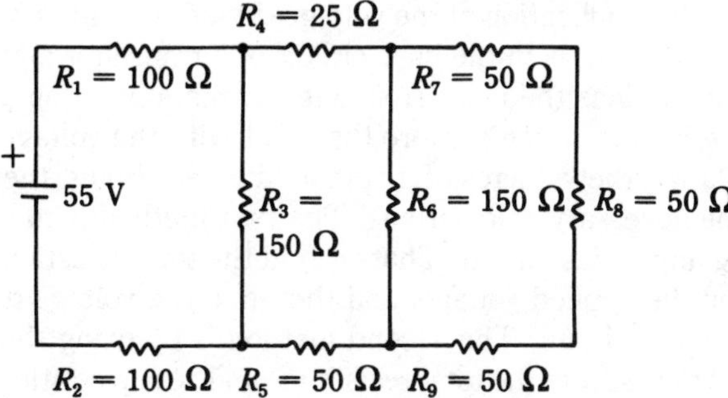

10. Find: R_T, I_T

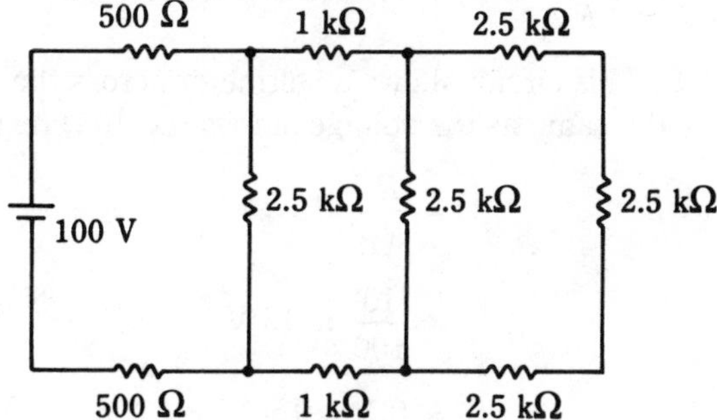

Voltage-divider circuits

A voltage-divider circuit is one application of a series circuit. It is possible using a series circuit to adjust the voltage drops across the resistors, by adjusting the resistance values, and making a circuit with several voltages available from one battery. Note that any circuit can be used as a voltage divider, provided there are at least two series components.

The major consideration for a voltage divider is how each of the voltages available will be used. If they are not used for some other part of a circuit, then there is no sense in simply making a voltage-divider circuit. The portion of the circuit that is connected to the voltage divider is connected across the resistor (in other words, directly in parallel with the voltage divider). Connecting two, or more, resistors in parallel will lower the total resistance. Therefore, when a load is connected to the voltage divider, the resistance will be lowered and the voltage drop in that point of the voltage-divider circuit will also be lowered.

A basic rule when using or designing a voltage divider circuit is that the load resistance should be at least 10 times the resistor it is in parallel with. Another way of saying this is to say the voltage divider should have at least 10 times more current than the load circuit. The reason the load should have 10 times the resistance, or 1/10 the current, is this can be considered to have so little effect on the voltage divider, it need not be considered when making the calculations. This guideline will limit the uses for the voltage divider circuit.

One very popular application of the voltage-divider circuit, when the load does not need to be considered, is in the base circuit of a transistor amplifier. The voltage divider is used to bias the transistor to its normal operating point. The accuracy needed for this application is more than met with the voltage-divider circuit.

When calculating the voltages of a voltage-divider circuit, there are two ways of performing the necessary calculations. The first method is to use the standard means of solving any series circuit. That is, find the total resistance, find the total current, based on the applied voltage, and then find the voltage drops of the individual resistors in the circuit. The second method is by using the voltage-divider formula, which is to first find the total resistance and set up a ratio of the resistor in question, divided by the total resistance, times the applied voltage.

$$V = \frac{R}{R_T} \times V_T \quad \text{(voltage divider formula)} \tag{2-8}$$

Refer to Fig. 2-9. This circuit shows a voltmeter across the 10 Ω resistor. The voltage at point a is the same as the voltage across the 10 Ω resistor, to ground.

$$\begin{aligned} V &= \frac{R}{R_T} \times V_T \\ &= \frac{10}{100} \times 15 \text{ V} \\ &= 0.1 \times 15 \\ &= 1.5 \text{ V} \end{aligned}$$

It is important to note that the voltage across a particular resistor is a ratio of the two resistors. However, the ratio is not one resistor to the other, rather, the ratio is the resistor in question to the total resistance.

Refer to Fig. 2-10. The circuit shown is also a voltage-divider circuit. This circuit shows two voltage taps, point a to ground and point b to ground. Whenever a point is shown and it is not specified where the second connection is, always assume the second connection to be on ground.

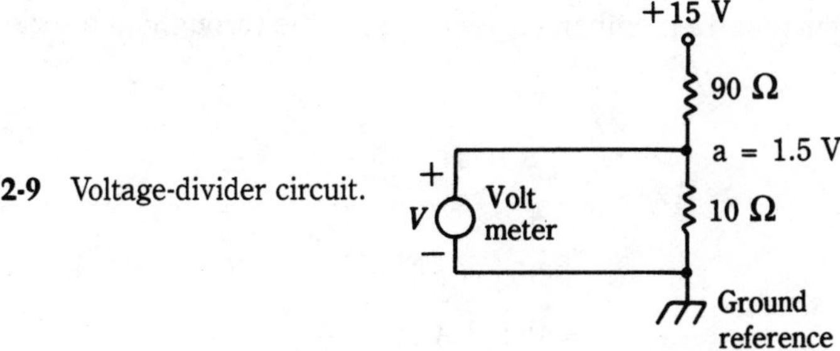

2-9 Voltage-divider circuit.

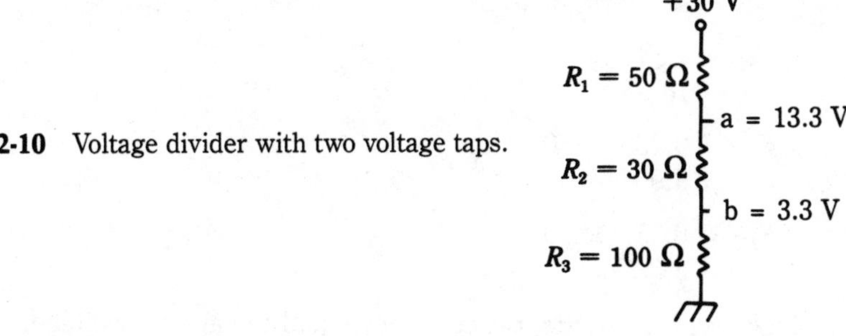

2-10 Voltage divider with two voltage taps.

Calculations

Calculate point a first.

$$V = \frac{R}{R_T} \times V_T$$

$$= \frac{40}{90} \times 30$$

$$= 1.3 \text{ V}$$

Calculate point b next.

$$V = \frac{10}{90} \times 30$$

$$= 3.33 \text{ V}$$

Again, refer to Fig. 2-10. It is possible, rather than using the voltage-divider formula, to make the calculations for the voltages at the points using Ohm's Law with regular dc circuit math.

Calculate total resistance.

$$R_T = R_1 + R_2 + R_3$$

$$= 50 + 30 + 10$$

$$= 90 \ \Omega$$

Calculate total current. Remember, current is the same throughout a series circuit.

$$I_T = \frac{E}{R_T} \qquad \text{(2-1A)}$$

$$= \frac{30}{90}$$

$$= 0.333 \text{ A}$$

Calculate the voltage across each resistor.

$$E = I_T R \text{ formula to be used for all voltage drops}$$

$$E_{R1} = 0.333 \times 50$$
$$= 16.65 \text{ V}$$

$$E_{R2} = 0.333 \times 30$$
$$= 9.99 \text{ V}$$

$$E_{R3} = 0.333 \times 10$$
$$= 3.33 \text{ V}$$

Check voltage drops by adding to see if they equal the applied voltage.

$$E_{R1} + E_{R2} + E_{R3} = V_T$$
$$16.65 + 9.99 + 3.33 = 29.97 \text{ V}$$

Voltage at point b is equal to the voltage drop across R_T.

Point b = 3.33 V checks with using the voltage divider formula

Voltage at point a is equal to the voltage drops across $R_3 + R_2$.

Point a = 3.33 + 9.99 = 13.32 V checks with using the voltage divider formula

Dual-battery circuits

Refer to Fig. 2-11. This figure shows an arrangement somewhat different from the usual dc circuit. The two power supplies represent the fact that the ground connection is actually a reference point. If all measurements are in reference to the ground connection, then that point is considered zero volts. Therefore, point a is 16 V above ground, or +6 V. Point c is 3 V below ground, or −3 V. This results in point b, using a reference or ground, having a voltage somewhere between +6 V

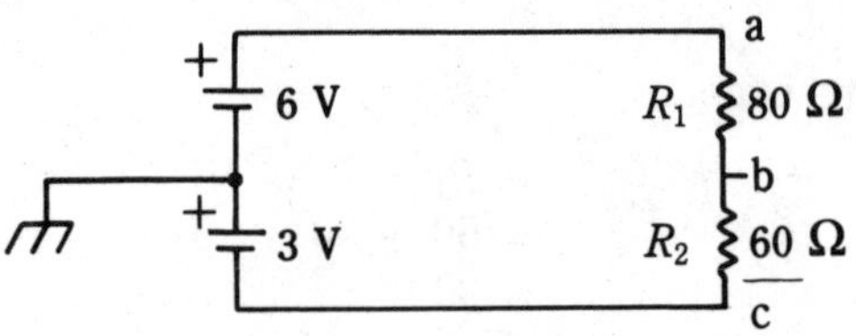

2-11 Voltage divider with two power supplies.

and −3 V. We can, therefore arrive at the conclusion that there is 9 V difference from points 1 to c. 9 V, then, is the total applied voltage to use when calculating the voltage divider.

Calculations

$$V = \frac{R}{R_T} \times V_T$$

$$= \frac{60}{140} \times 9$$

$$= 3.8 \text{ V (this is not the voltage at point b)}$$

The voltage calculated above is from point b to point c. Because it is the voltage across the 60 Ω resistor. To find the voltage at point b, to ground, add this value to the −3 V supply.

$$-3 + 3.8 = 0.8 \text{ V (at point b to ground)}$$

Using the Ohm's Law method; total resistance is 140 Ω, calculate current.

$$I = \frac{E}{R} \quad (2\text{-}1A)$$

$$= \frac{9}{140}$$

$$= 64 \text{ mA}$$

Using the total current, calculate the voltage drops across each resistor.

$$E = IR$$
$$= 0.064 \times 80$$
$$= 5.12 \text{ V}$$

$$E_{R2} = 0.064 \times 60$$
$$= 3.84 \text{ V}$$

The voltage drop across R_1 is actually from point a to point b. Therefore, point b is less than the voltage at point a by the voltage drop.

point a = 6 V
voltage drop across R_1 = 5.12 V
voltage at point b = 6 − 5.12 = 0.88 V

The voltage drop across R_2 is actually from point b to point c. Therefore, point b is higher than point c by the voltage drop across R_2.

voltage drop across R_2 = 3.84 V
voltage at point c = −3 V
voltage at point b = 3.84 + −3 = 0.84 V

The difference between these values is due to rounding of numbers in the calculations.

Wheatstone bridge

The Wheatstone bridge is the name of a particular circuit that is used for a very precise measuring instrument. The Wheatstone bridge is made up of voltage-divider circuits. The principle of the bridge circuit is that there are two voltage divider circuits formed. R_1 and R_2 form a voltage divider, and R_A and R_B form the other voltage divider. Refer to Fig. 2-12.

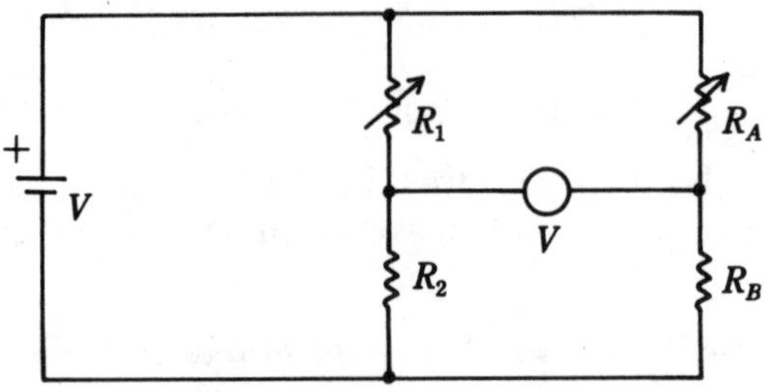

2-12 Wheatstone bridge.

The voltmeter is to be treated like an infinite resistance, or an open circuit. In other words, the right side of the circuit is completely independent of the left side of the circuit. A change in the power supply voltage will not affect the voltage divider circuits, because they are ratios.

This instrument is used for measuring resistance in a very precise manner. R_2 and R_B are fixed resistors of a very precise type. R_1 is part of the instrument and is used for balancing the bridge. R_A is shown as a variable, but is really the unknown resistance being measured. The voltmeter is center zeroed, which means it can swing either way with the center of the meter being zero. The meter is used for balancing the bridge circuit.

There are two possible conditions for the bridge circuit, the first condition is unbalanced and the second condition is balanced. If the bridge is unbalanced, that means there is a difference in the voltage divider ratios between the two sides. This will produce a difference in voltage drops across the resistors and result in the voltmeter showing a voltage. If the bridge is balanced, then the ratio of the two voltage dividers is equal. This will result in an equal voltage drop across the resistors and result in no voltage on the meter.

Sample Wheatstone bridge

$$R_1 = 1\text{ k}\Omega,\ R_2 = 2\text{ k}\Omega,\ R_A = 12\text{ k}\Omega,\ R_B = 20\text{ k}\Omega,\ V = 10\text{ V}$$

Voltage-divider formula

$$V = \frac{R}{R_T} \times V_T$$

Voltage across R_2

$$V_{R2} = \frac{2\ \text{k}\Omega}{3\ \text{k}\Omega} \times 10\ \text{V}$$

$$= 6.67\ \text{V}$$

Voltage across R_B

$$V_{RB} = \frac{20\ \text{k}\Omega}{32\ \text{k}\Omega} \times 10\ \text{V}$$

$$= 6.25\ \text{V across}$$

Notice there is a difference in the two voltages. V_{R2} is 0.42 V (6.67 − 6.25) higher than V_{RB}. This means the voltmeter will have this reading, with a positive on the left and negative on the right. This is considered an unbalanced bridge. Notice the voltage divider ratio portion of the circuit is only the $R < R_T$. This leads to a new formula for Wheatstone bridges; disregarding the supply voltage:

$$\frac{R_1}{R_2} = \frac{R_A}{R_B} \quad \text{(balancing a Wheatstone bridge)} \tag{2-9}$$

Balancing using the sample values

$R_1 = 1\ \text{k}\Omega$, $R_2 = 2\ \text{k}\Omega$, $R_A =$ unknown, $R_B = 20\ \text{k}\Omega$

$$\frac{1\ \text{k}\Omega}{2\ \text{k}\Omega} = \frac{R_A}{20\ \text{k}\Omega}$$

$$0.5 = \frac{R_A}{20\ \text{k}\Omega}$$

$$0.5 \times 20\ \text{k}\Omega = R_A$$

$$R_A = 10\ \text{k}\Omega \text{ (value of } R_A \text{ to balance the bridge)}$$

Maximum transfer of power

The maximum transfer of power occurs when the value of the load is equal to the internal resistance of the power supply.

Refer to Fig. 2-13 for a sample circuit of maximum transfer of power. Notice that this is a very simple circuit. It is easier this way to prove the point; however, any circuit can be used.

2-13 Sample for maximum transfer of power.

Calculations

For maximum transfer of power, the load must be equal to the internal resistance of the power supply, represented by r_i. This means R_{load} must be equal to 200 Ω.

Calculate power with the 200 Ω load.

$$P = \frac{E^2}{R} \tag{2-4}$$

$$= \frac{15^2}{400}$$

$$= 0.563 \text{ W (maximum power, } 1/2 \text{ to the load)}$$

$$P_{load} = 0.282 \text{ W (power to the load)}$$

To prove the power in the previous calculation is the maximum power that can be delivered to the load from a 15 V power supply with 200 Ω internal resistance, a value of load resistor above 200 Ω and one below will be tried.

Use a larger resistor. It is probably easier to calculate load power by first finding circuit current. Try 500 Ω for the load.

$$I = \frac{E}{R}$$

$$= \frac{15}{700}$$

$$= 0.021 \text{ A}$$

Calculate load power.

$$P = I^2R$$
$$= 0.021^2 \times 500$$
$$= 0.221 \text{ W}$$

Use a smaller resistor. Try 75 Ω.

$$I = \frac{15}{275}$$

$$= 0.055 \text{ A}$$

Calculate load power.

$$P = I^2R$$
$$P = 0.055^2 \times 75$$
$$P = 0.227 \text{ W}$$

Chapter summary

Direct current circuit math is one of the most important single subjects in the study of electronics. It sets the basic rules for all of electronic circuits. The following is a list of the key points in this chapter.

- Stated by Ohm's Law: voltage = amperes × ohms
- In any dc circuit there are four quantities that can be calculated; voltage, current, resistance, and power.
- P is directly related to I.
- I is directly related to E.
- E is directly related to R.
- P is directly related to E.
- I is indirectly related to R.
- P is indirectly related to R.
- In a series circuit, there is only one current path. Current is the same throughout a series circuit.
- Voltage drops at each resistance are directly related to the size of the resistance.
- The sum of the voltage drops in a series circuit must equal the supply voltage.
- The sum of the powers dissipated in a series circuit must be equal to the total power.
- Electron current flow is from negative to positive. Conventional current flow is from positive to negative.
- In a series circuit, the largest resistance will drop the most voltage.
- In a series circuit, the largest resistance will dissipate the most power.
- A short circuit has a resistance of zero. This condition has unlimited current, zero voltage drop, and zero power.
- An open circuit has a resistance of infinity. This condition causes zero current, applied voltage dropped at open, and zero power.
- In a parallel circuit, the current divides to the individual branches indirectly related to the size of the resistors.
- Total current in a parallel circuit is equal to the sum of the individual branch currents.
- Voltage is the same throughout a parallel circuit.
- The sum of the powers dissipated in a parallel circuit is equal to the total power.
- The total resistance of a parallel circuit is smaller than the smallest resistance of any branch.
- To find the total resistance of equal resistances in parallel, divide the resistance of one branch by the number of branches with equal resistance.
- In a parallel circuit, if branches have equal resistance, the current through each branch will be equal.
- The total current divided by the number of branches with equal resistances will give the individual branch currents.

Summary of formulas

$$E = I \times R \quad \text{Ohm's Law} \tag{2-1}$$

$$P = I \times E \quad \text{power formula} \tag{2-2}$$

$$P = I^2 \times R \quad \text{modified power formula} \tag{2-3}$$

$$P = \frac{E^2}{R} \quad \text{modified power formula} \tag{2-4}$$

$$R_T = R_1 + R_2 + R_3 + \ldots \quad \text{total of resistors in series} \tag{2-5}$$

$$\frac{1}{R_T} = \frac{1}{R_1} + \frac{1}{R_2} + \frac{1}{R_3} + \ldots \quad \text{total of resistors in parallel reciprocal formula} \tag{2-6}$$

$$R_T = \frac{R_1 \times R_2}{R_1 + R_2} \quad \text{two resistors in parallel shortcut formula} \tag{2-7}$$

$$V = \frac{R}{R_T} \times V_T \quad \text{voltage-divider formula} \tag{2-8}$$

$$\frac{R_1}{R_2} = \frac{R_A}{R_B} \quad \text{balancing a Wheatstone bridge} \tag{2-9}$$

Practice problems

Use the schematic diagrams for each problem to find the unknown values. It might be easier with some problems to redraw the schematic to determine which resistors are in series and which are in parallel. Round answers to 3 significant figures.

1. Find: R_T, I_T, E_{R1}, E_{R2}

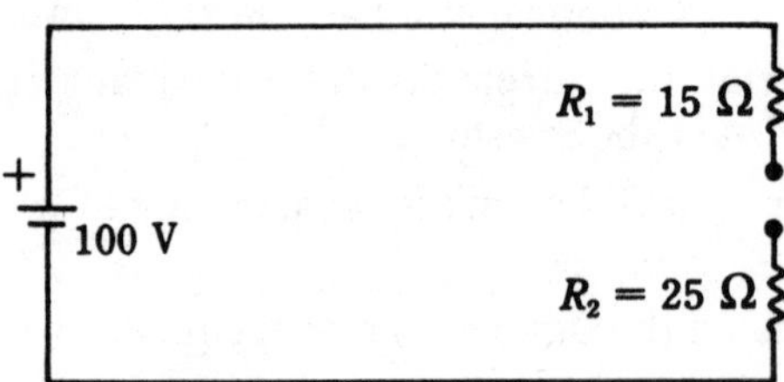

2. Find: R_T, I_T, E_{R1}, E_{R2}, E_{R3}

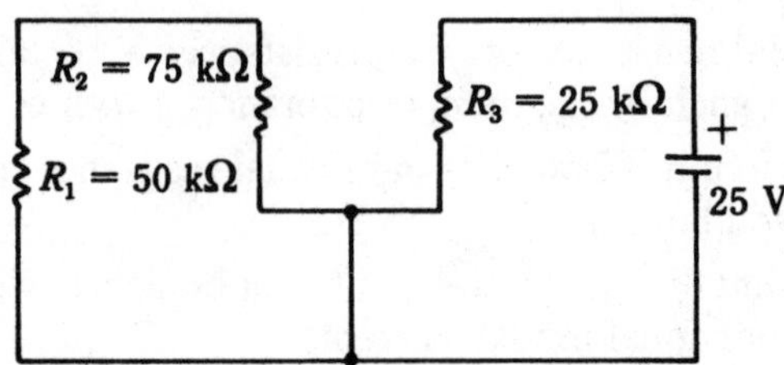

3. Find: R_T, I_T, V_a, V_b

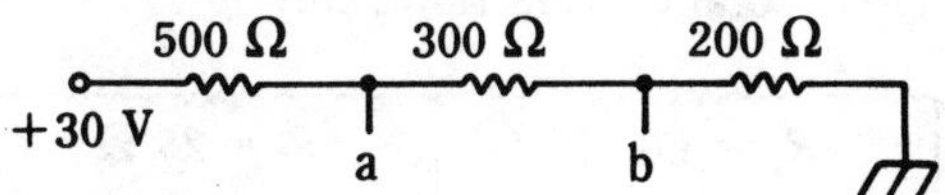

4. Find: R_T, I_T, I_{R1}, I_{R2}

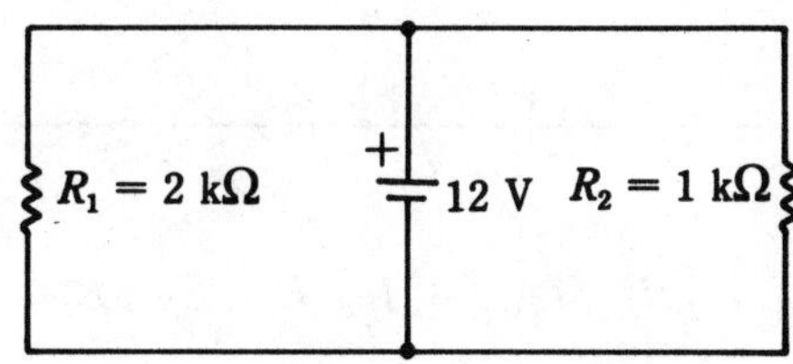

5. Find: R_T, I_T, I_{R1}, I_{R2}, I_{R3}

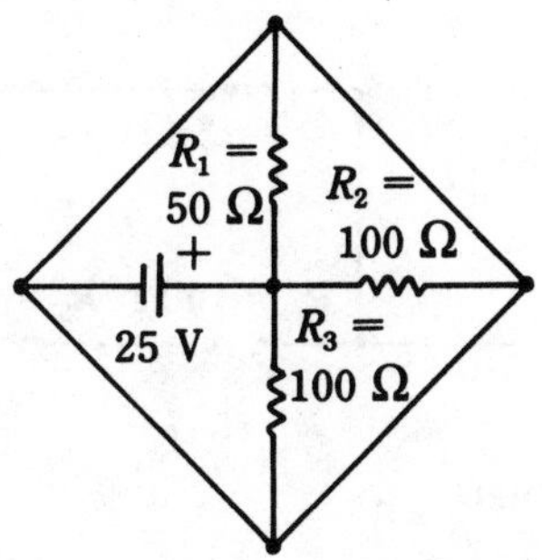

6. Find: R_T, I_T, P_T

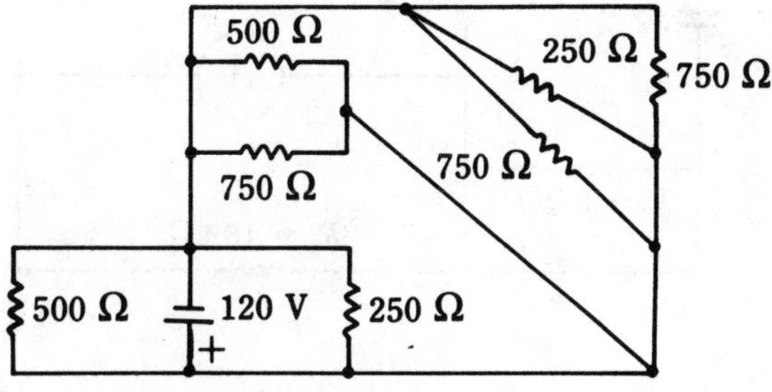

7. Find: R_T, I_T, P_T, R_1, R_2, R_3, R_4

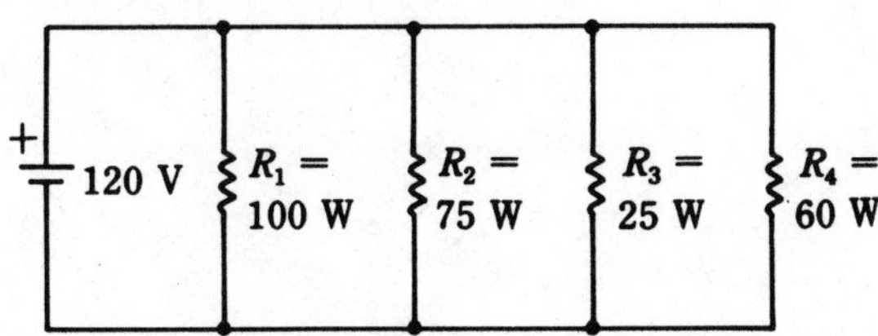

8. Find: R_T, I_T, I_{R1}, I_{R2}, I_{R3}, E_{R1}, E_{R2}, E_{R3}
What percentage of total current flows through R_2?

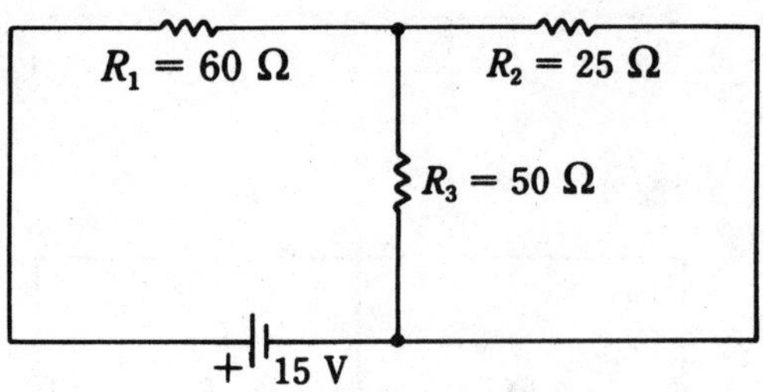

9. Find: R_T, I_T, I_{R1}, I_{R2}, I_{R3}, I_{R4}, I_{R5}, I_{R6}, E_{R1}, E_{R2}, E_{R3}, E_{R4}, E_{R5}, E_{R6}

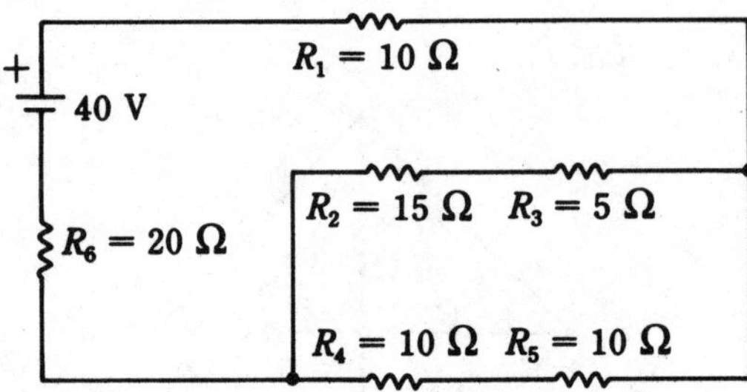

10. Find: Current in the meter

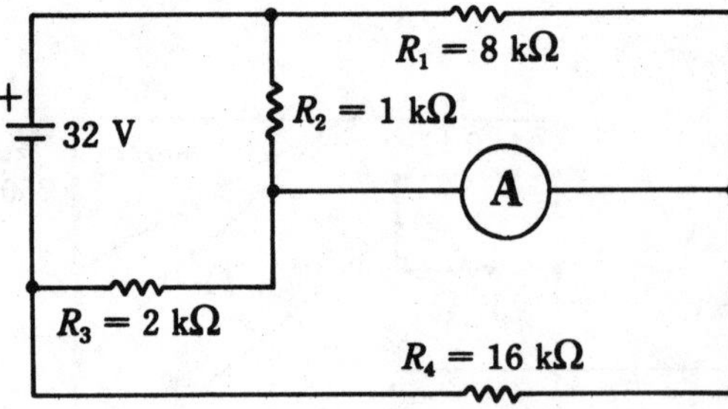

3
Reading a volt-ohm-milliammeter

THE MULTIMETERS OR VOMs (VOLT-OHM-MILLIAMMETERS) ARE USED TO measure electrical quantities such as; voltage (volts), resistance (ohms) and current (amperes). Meters are available in two types: analog, with a needle and digital, with digits. The difference between these two types of meters is similar to the difference between the two types of wristwatches. The analog meter is examined in this chapter.

Reading a multimeter

Figure 3-1 is a drawing of a meter showing the scales and range switch. This figure is used to explain how to read the meter.

Range

The dial at the bottom of the drawing has a rotary switch to select the various ranges. The ranges on this meter fall into four different categories: DCV, ACV, OHMS, DCmA. The four categories represent the type of measurement being made, but they do not change how the meter scale is read.

The range, indicated by the range switch pointer, determines the maximum for that scale (except OHMS, which is discussed later). For example: 250 V range means the scale used has a maximum reading of 250 V.

Scale (DCV, ACV, and DCmA)

The scale (the meter face) has different sets of numbers to match it to the range selected. The scale to be used when making measurements is determined by the

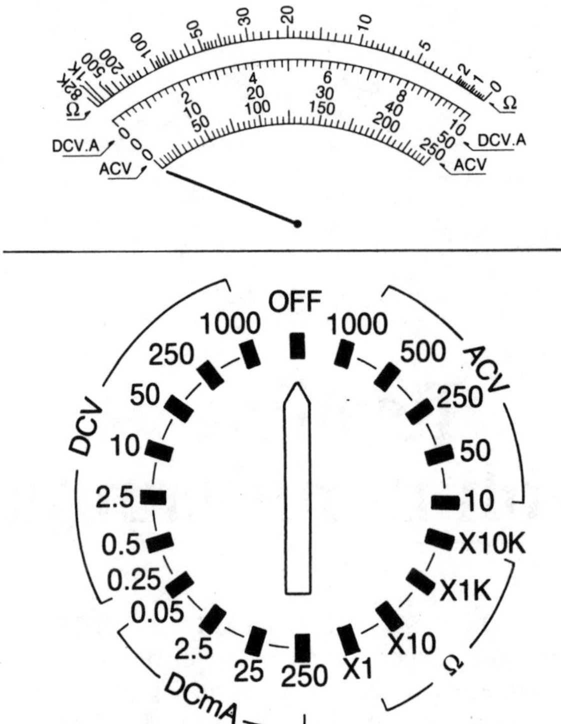

3-1 Sample VOM scale (top) and range selector (below).

numbers on the right side of the scale (maximum) corresponding to the range setting. Note: this does not apply to the ohms ranges.

More than one range can use each scale by adjusting the decimal point. Examples:

10 and 1000 both use the "10" scale
2.5, 25 and 250 use the "250" scale
0.05, 0.5, 50 and 500 use the "50" scale

Scale with decimal adjusted

When the decimal place on the scale does not correspond to the measurement desired, it can be adjusted as needed to allow the maximum reading to match the selected scale.

Move the decimal of the scale to match the range setting as follows:

1000 scale:	0 ...	200 ... 400 ... 800 ... 1000
500 scale:	0 ...	100 ... 200 ... 300 ... 400 ... 500
250 scale:	0 ...	50 ... 100 ... 150 ... 200 ... 250
50 scale:	0 ...	10 ... 20 ... 30 ... 40 ... 50
25 scale:	0 ...	5 ... 10 ... 15 ... 20 ... 25
10 scale:	0 ...	2 ... 4 ... 6 ... 8 ... 10

2.5 scale:	0 ...	0.5 ... 1.0 ... 1.5 ... 2.0 ... 2.5
0.5 scale:	0 ...	0.1 ... 0.2 ... 0.3 ... 0.4 ... 0.5
0.25 scale:	0 ...	0.05 ... 0.1 ... 0.15 ... 0.20 ... 0.25
0.05 scale:	0 ...	0.01 ... 0.02 ... 0.03 ... 0.04 ... 0.05

Value of each scale line

The values of the lines change with each range and scale. There are 10 lines between each marked number. After the scale decimal has been adjusted, divide the number closest to zero by 10. For example: range = 1000, scale first number = 200, divide 200 by 10 = 20, therefore each line = 20. This can be checked by counting each line by 20s.

1000 scale: each line = 20
500 scale: each line = 10
250 scale: each line = 5
50 scale: each line = 1
25 scale: each line = 0.5
10 scale: each line = 0.2
2.5 scale: each line = 0.05
0.5 scale: each line = 0.01
0.25 scale: each line = 0.005
0.05 scale: each line = 0.001

Scale arc (line markings)

The lines are marked on two different arcs on this meter. (Not including the ohms scale.) Both ends of the scale are labeled, with an arrow, to indicate which arc to use for which type of reading. dc volts and mA are read using the set of lines above the numbers and ac volts are read using the lines below the numbers.

Ohms

The Greek letter omega (Ω) is the symbol for ohms, which is the unit of measure of resistance. The ohms scale is numbered backwards, with zero on the right and infinity, ∞, on the left (looks like an 8 on its side). On this meter, the ohms scale is across the top.

The ohms scale is *nonlinear*, meaning that the spacing between the numbers is not equal. Each line between numbers has a different value.

The reading from the scale must be multiplied by the range switch. For example, in Fig. 3-2, the needle reads 24 on the ohms scale. The range switch is selected at × 1; therefore, the resistance is 24 Ω.

Suppose the range in Fig. 3-2 had been selected at:

×10 ... 24 × 10 = 240 Ω
×1k ... 24 × 1k = 24 kΩ
×10k ... 24 × 10k = 240 kΩ

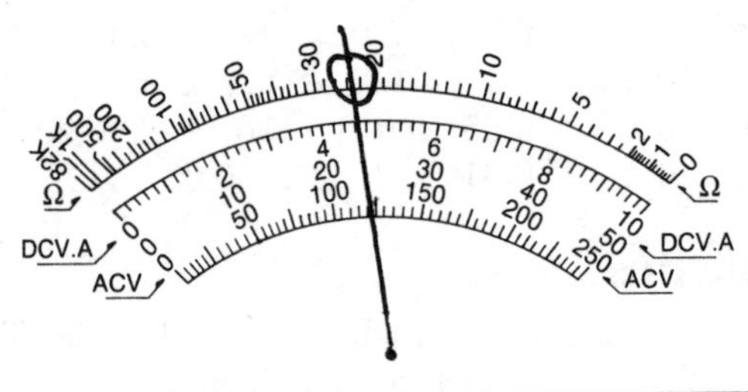

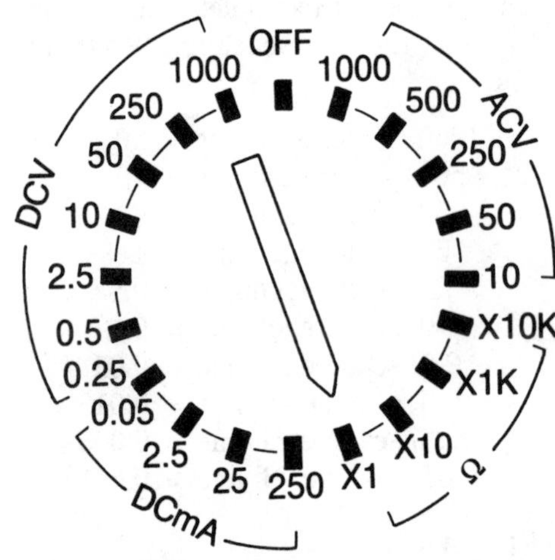

3-2 Sample Ohm scale.

Sample multimeter readings

With each of the following sample meters:

(Figure 3-2)

a. Identify the range. (This is found by the setting of the range switch.)

b. Adjust the decimal for the scale to be used and write the numbers for that range.

c. Find the value of each scale line by dividing the first number on the scale by 10.

d. Identify which arc to use and give the value that the meter is measuring.

(Figure 3-3)

a. Range switch setting = 250 ACV.

b. Set of numbers on the scale with the decimal adjusted.

0 50 100 150 200 250

c. Value of each line is found by dividing first number by 10:

$$\frac{50}{10} = 5 \text{ volts per line}$$

d. Read Vac on lower set of lines.
Measurement = 115 ACV

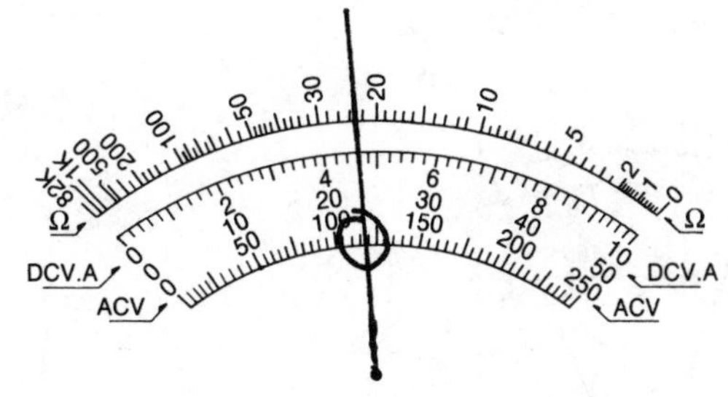

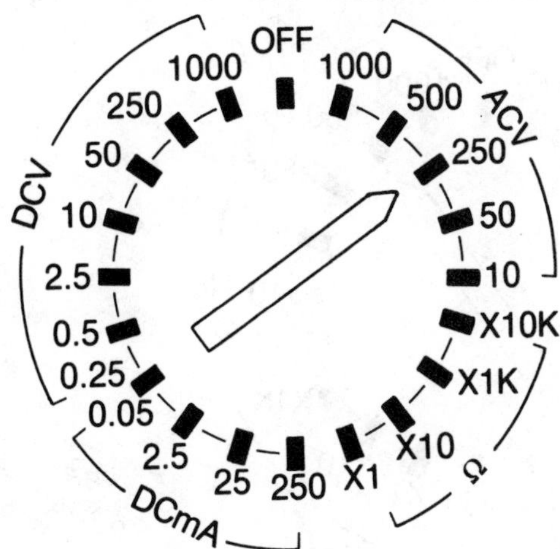

3-3 Multimeter reads 115 ACV (ac volts).

(Figure 3-4)

a. Range switch setting = 500 ACV.

b. Set of numbers on the scale with the decimal adjusted.

0 100 200 300 400 500

c. Value of each line is found by dividing first number by 10:

$$\frac{100}{10} = 10 \text{ volts per line}$$

d. Read ACV on lower set of lines.
Needle a = 150 ACV
Needle b = 380 ACV

(Figure 3-5)

a. Range switch setting = 10 DCV.

b. Set of numbers on the scale with the decimal adjusted.

0 2 4 6 8 10

c. Value of each line is found by dividing first number by 10:

$$\frac{2}{10} = 0.2 \text{ volts per line}$$

(Go to p. 65)

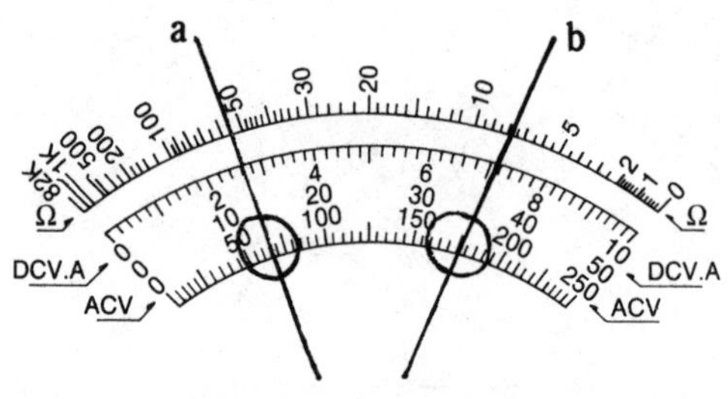

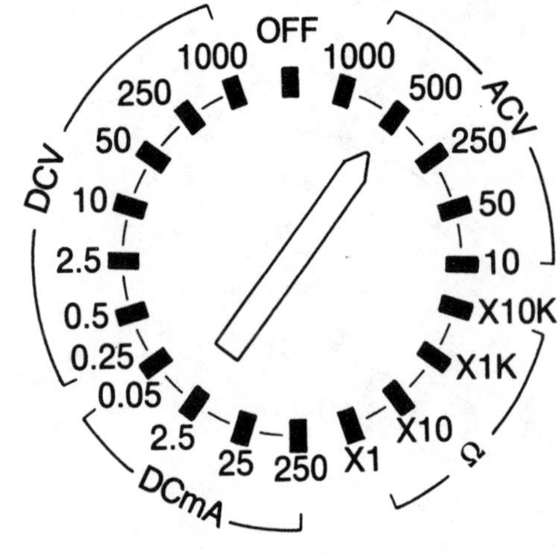

3-4 Needle a reads 150 ACV; needle b reads 380 ACV.

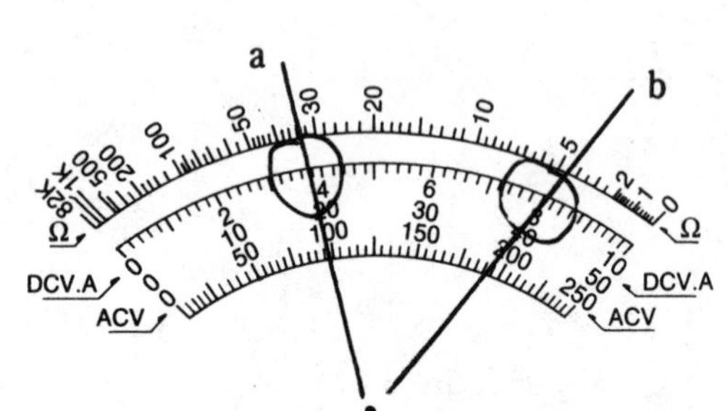

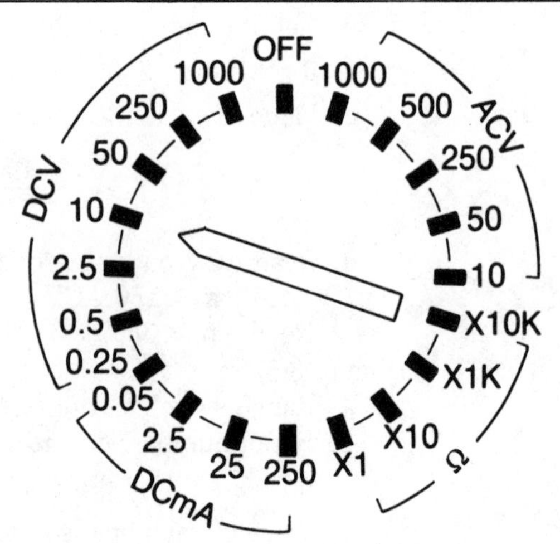

3-5 Needle a reads 3.6 DCV; needle b reads 8.2 DCV.

d. Read DCV on upper set of lines.
 Needle a = 3.6 DCV
 Needle b = 8.2 DCV

(Figure 3-6)

a. Range switch setting = ×10 Ω. All readings on the ohms scale must be multiplied by the range setting.
b. Use the set of numbers across the top of the meter.
c. Value of each line is found on an individual basis between each marked number.
d. Measurements:

Needle	Each Line	Reading	Range	Measurement
a	20	140	×10	1400
b	2	25	×10	250
c	0.5	7.5	×10	75

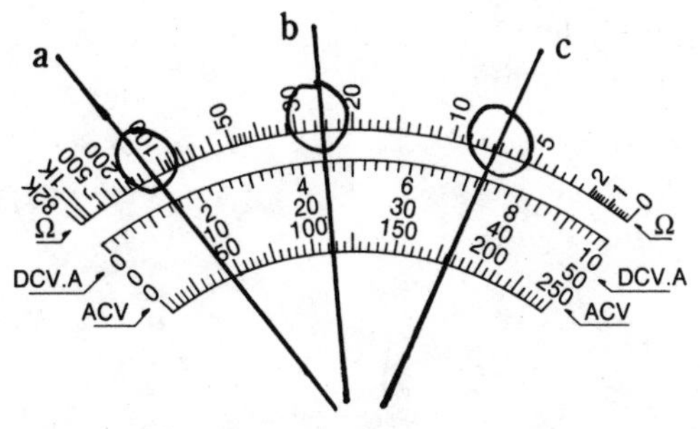

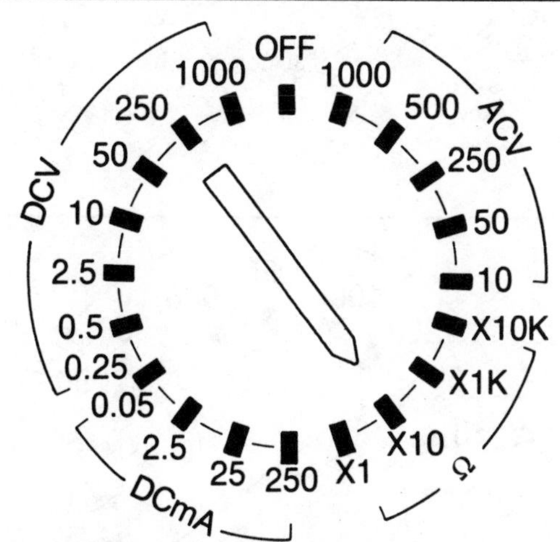

3-6 Needle a reads 1400 Ω; needle b reads 250 Ω; needle c reads 75 Ω.

Practice multimeter readings

With each of the following practice meters:

a. Identify the range (found by reading the setting of the range switch).
b. Adjust the decimal for the scale to be used and write the numbers for that range.
c. Find the value of each scale line by dividing the first number on the scale by 10.
d. Give the value that the meter is measuring.

Practice Meter 1

a. Range ....
b. Set of numbers on scale:

0 ....

c. Value of each scale line ....
d. Value the meter is reading:

Needle a = ...
Needle b = ...
Needle c = ...

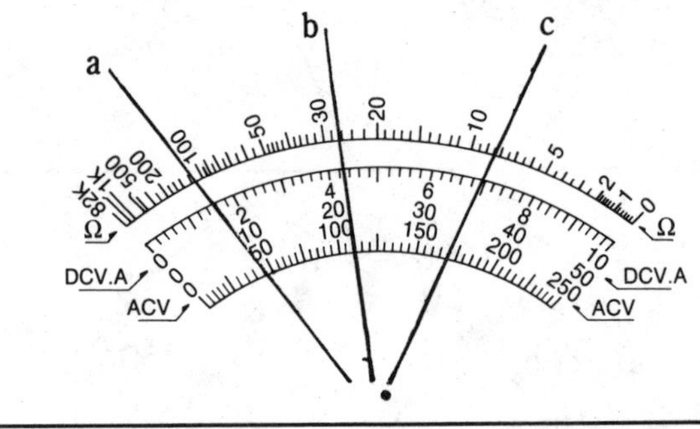

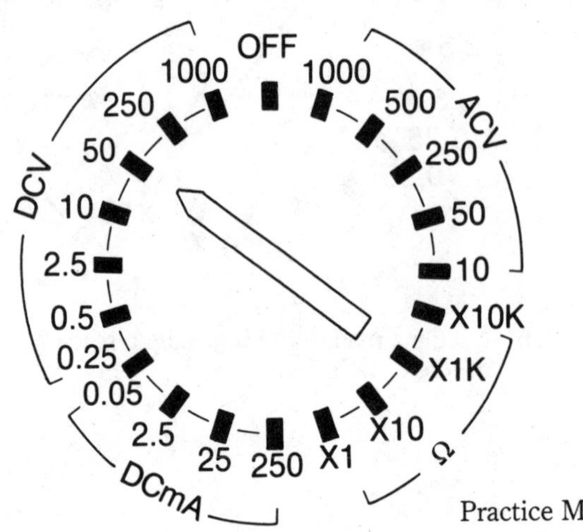

Practice Meter 1

Practice Meter 2

a. Range ....

b. Set of numbers on scale:

0 ....

c. Value of each scale line ....

d. Value the meter is reading:

Needle a = ...

Needle b = ...

Needle c = ...

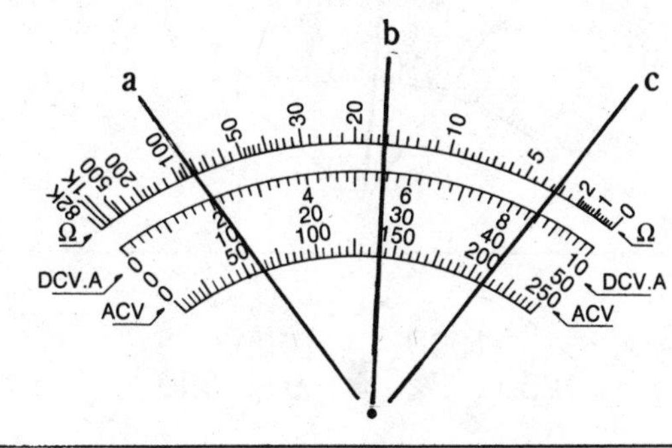

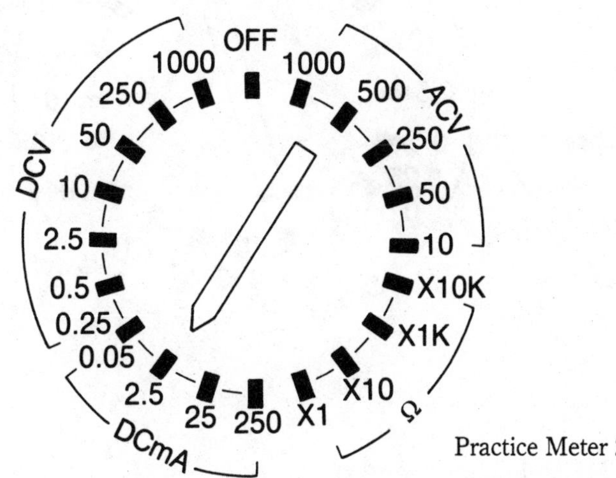

Practice Meter 2

Practice Meter 3

a. Range ....

b. Set of numbers on scale:

0 ....

c. Value of each scale line ....

d. Value the meter is reading:
 Needle a = ...
 Needle b = ...
 Needle c = ...

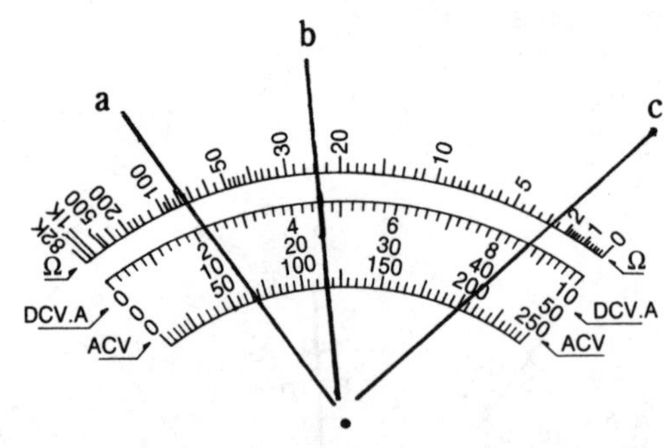

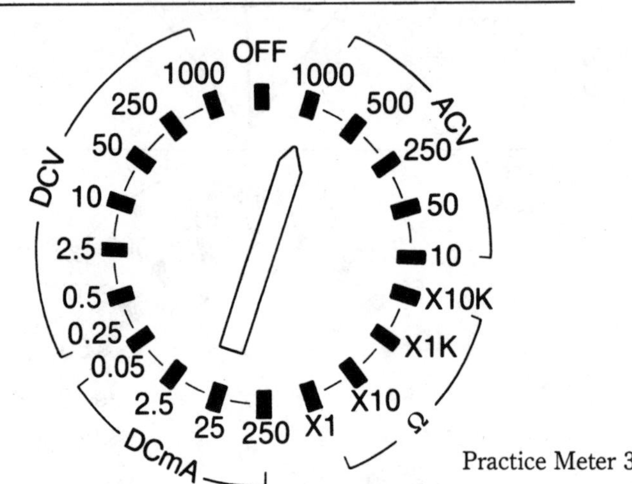

Practice Meter 3

Practice Meter 4

a. Range ....
b. Set of numbers on scale:
 0 ....
c. Value of each scale line ....
d. Value the meter is reading:
 Needle a = ...
 Needle b = ...
 Needle c = ...

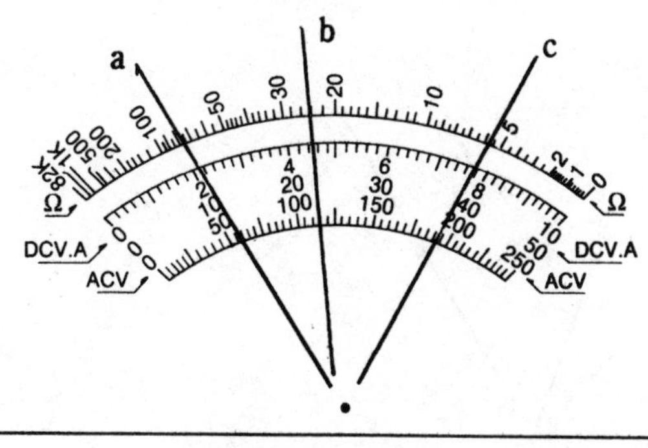

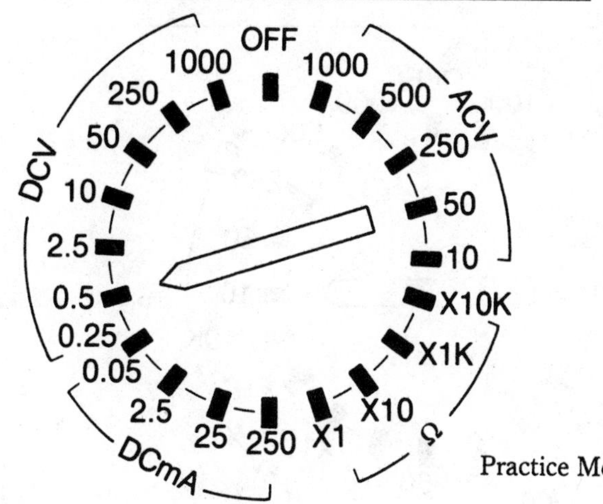

Practice Meter 4

Practice Meter 5

a. range ....

b. Set of numbers on scale:

0 ....

c. Value of each scale line ....

d. Value the meter is reading:

Needle a = ...

Needle b = ...

Needle c = ...

Practice Meter 6

a. Range ....

b. Set of numbers on scale:

0 ....

c. Value of each scale line ....

d. Value the meter is reading:

Needle a = ...

Needle b = ...

Needle c = ...

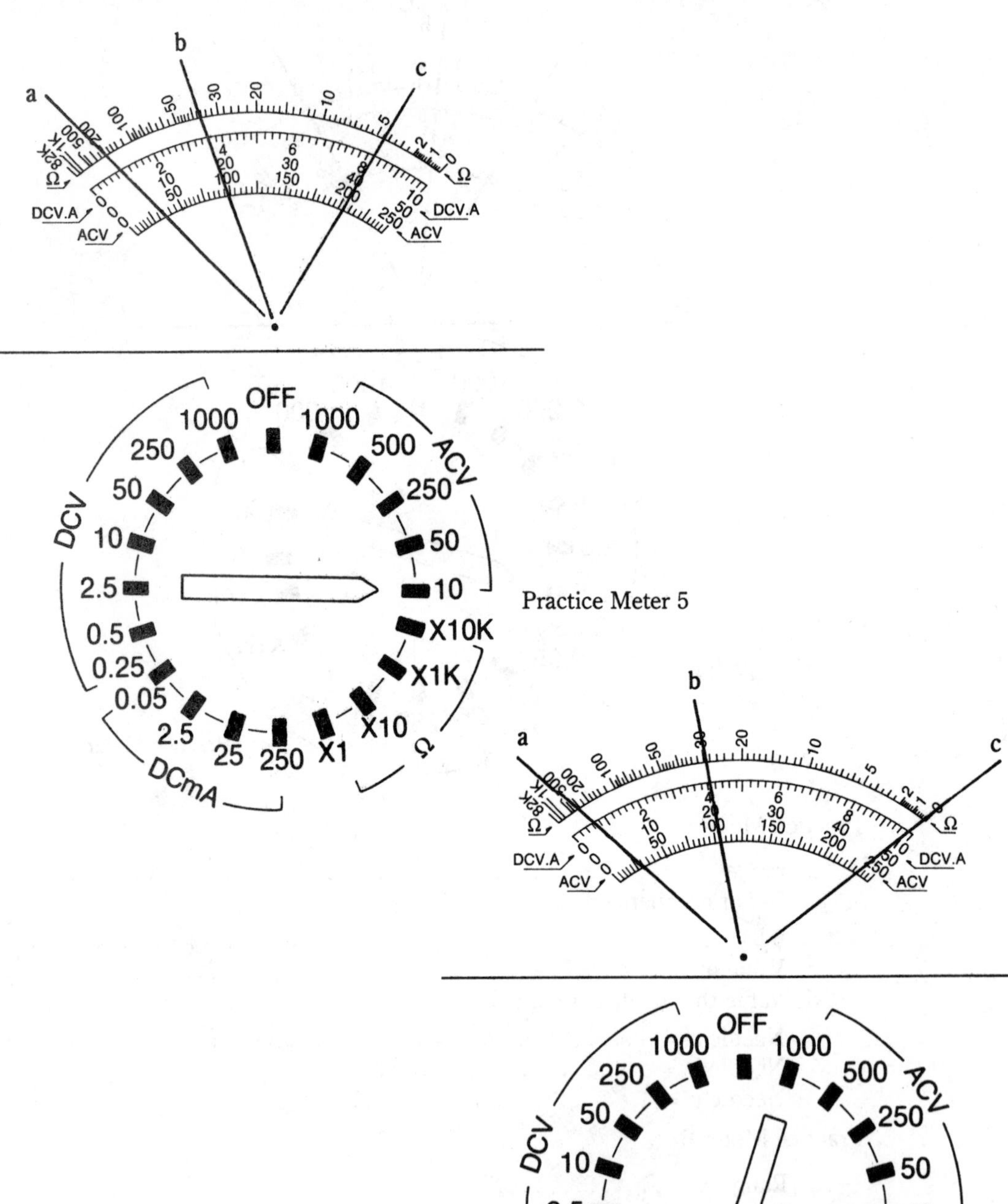

Practice Meter 5

Practice Meter 6

Practice Meter 7

a. Range ....

b. Set of numbers on scale:

0 ....

c. Value of each scale line ....

d. Value the meter is reading:

Needle a = ...

Needle b = ...

Needle c = ...

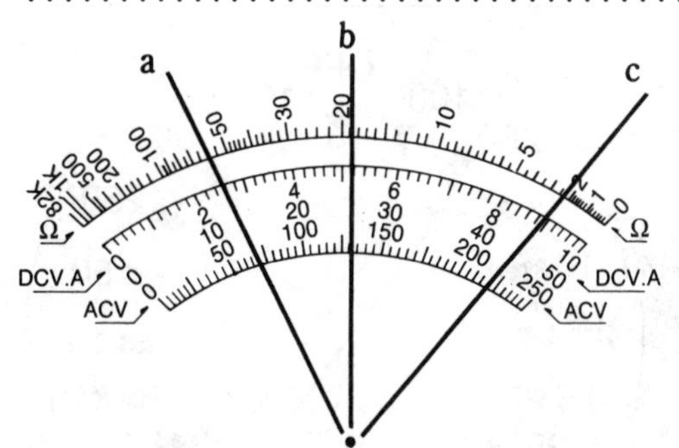

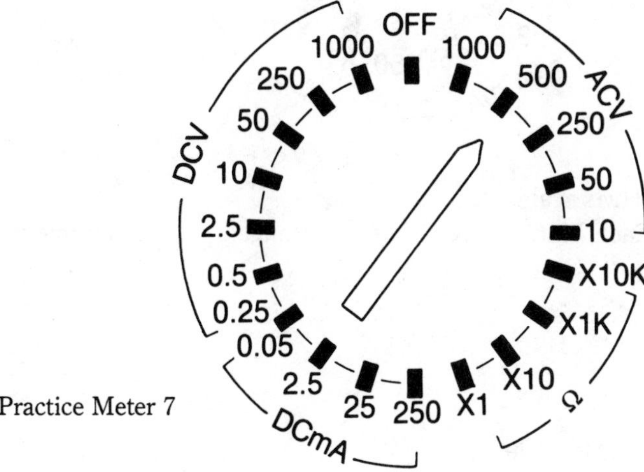

Practice Meter 7

Practice Meter 8

a. Range ....

b. Set of numbers on scale:

0 ....

c. Value of each scale line ....

d. Value the meter is reading:

Needle a = ...

Needle b = ...

Needle c = ...

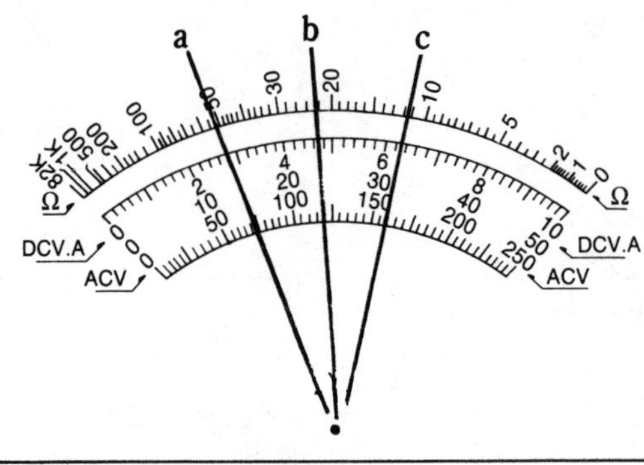

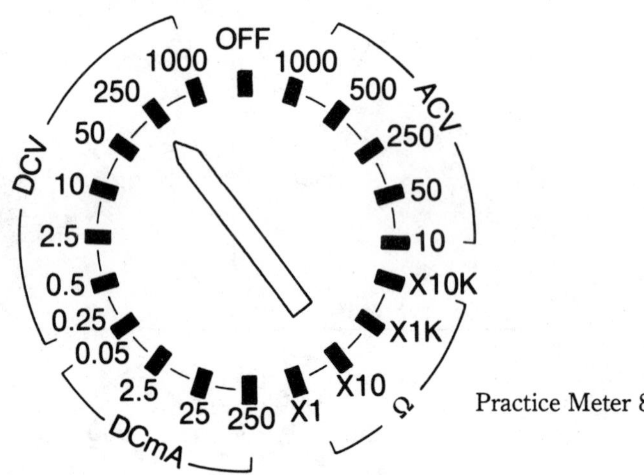

Practice Meter 8

Practice Meter 9—OHMs scale

Needle	Each Line	Reading	Range	Measurement
a	....	....	.....	....
b	....	....	.....	....
c	....	....	.....	....

Practice Meter 10—OHMs scale

Needle	Each Line	Reading	Range	Measurement
a	....	....	.....	....
b	....	....	.....	....
c	....	....	.....	....

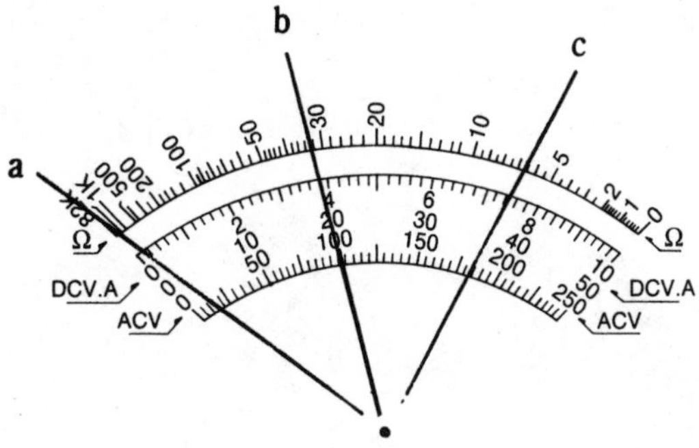

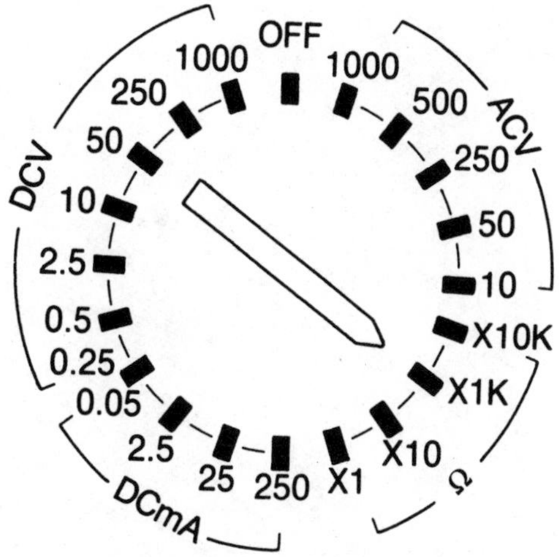

Practice Meter 9

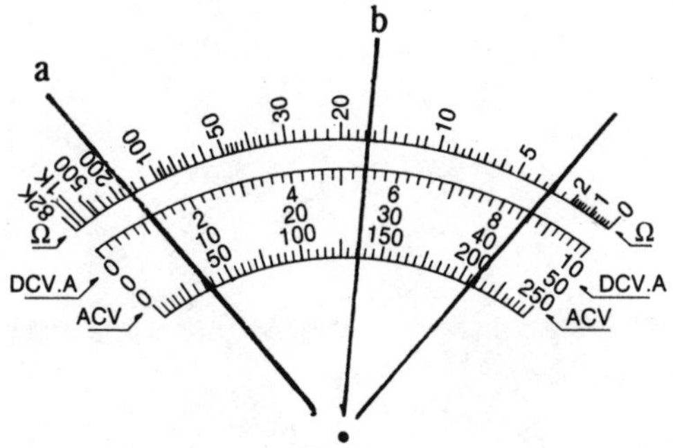

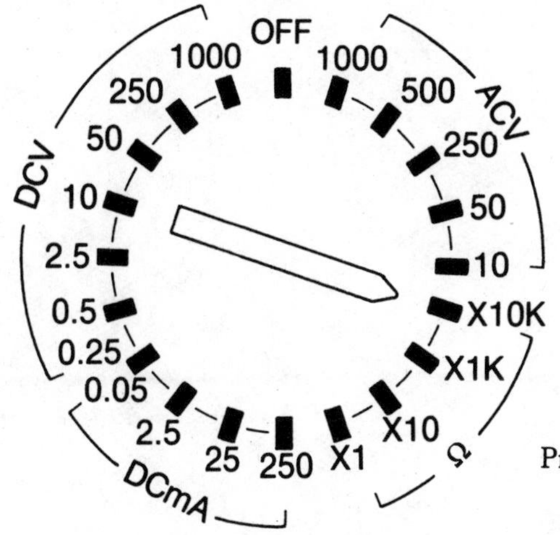

Practice Meter 10

4
Reading an oscilloscope

THE *SCOPE* (OSCILLOSCOPE) IS AN EXTREMELY IMPORTANT PIECE OF TEST equipment in electronics. It can be used to measure voltages or time periods. What makes the scope different than a voltmeter is that the scope draws a picture of the waveform. It allows you to observe the behavior pattern of the signal being tested.

The oscilloscope screen

The scope screen is graduated in blocks, called *divisions*, which are one centimeter on a standard size screen. Refer to Fig. 4-1. The center lines, both up and down, are marked with four small graduations. Each of these small lines represents two tenths (0.2) of a division. Figure 4-1 is labeled as if counting from one division to the next: 0.2, 0.4, 0.6, 0.8 parts of a division.

Read the screen with each full division representing a whole number and each subdivision representing a decimal part. If you read the screen between the subdivisions, you can make readings accurate to one-tenth (one decimal place).

Voltage measurements

Voltage measurements are made in the vertical direction. To measure voltage, count the divisions up or down. Multiply the number of divisions times the volts per division.

dc voltages

A pure dc voltage appears as a straight line on an oscilloscope. Sometimes, dc voltages contain some slight variations, which can be seen on the scope, but not with a voltmeter.

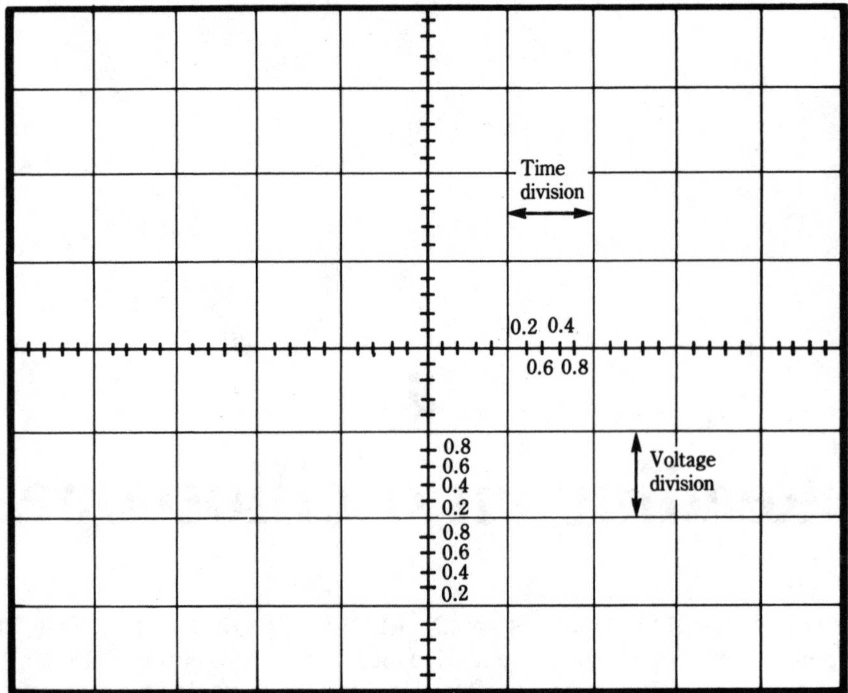

4-1 Oscilloscope screen showing how divisions are subdivided.

With a dc voltage, count up (or down) starting from the ground reference line. The scope has a switch labeled AC/GND/DC. When this switch is placed in GND, the trace on the screen will be a straight line at the ground reference point. Then, when the switch is placed in dc, the trace will jump up (or down) to the level of the dc voltage being measured.

ac voltages

An ac voltage is continually varying, usually with a repetitive pattern. Voltages are measured from the top of the wave to the bottom (or vice versa). Other measurements of ac voltages are discussed in a later chapter.

An ac waveform also has a measurement of the time of one cycle. One cycle is the unique shape of the pattern, until it repeats itself. Time is measured in the horizontal direction, from left to right. It is best to make measurements at the centerline, whenever possible, because this is the most accurate.

To make a measurement of time, multiply the number of divisions times the time per division.

Sample oscilloscope measurements

The remaining figures in this chapter are sample scope measurements. Each figure includes an explanation of how to read the measurement.

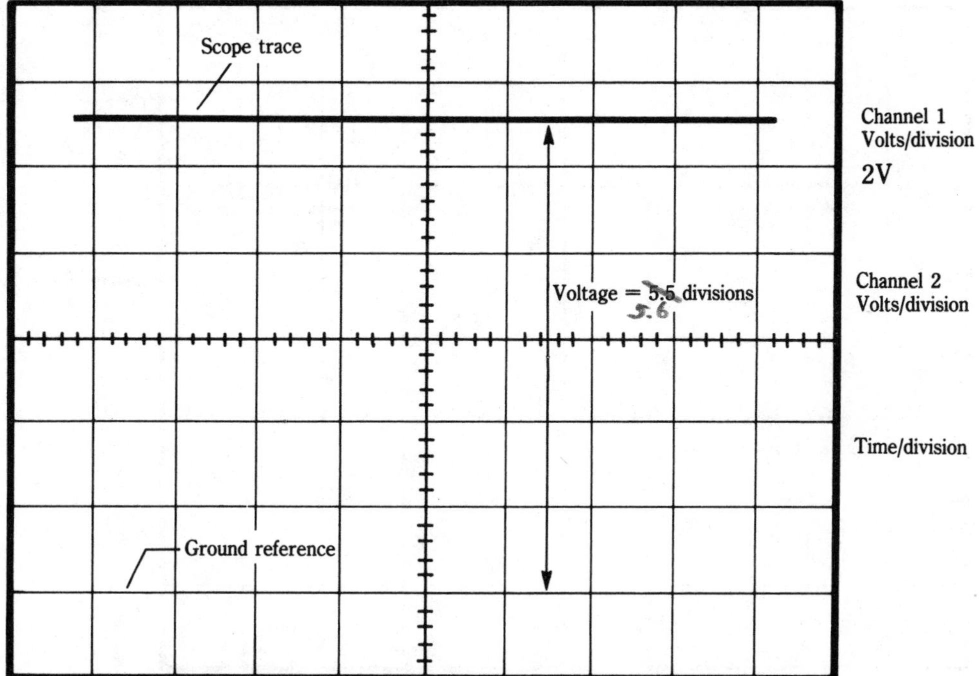

4-2 dc voltage = 11 V above ground.

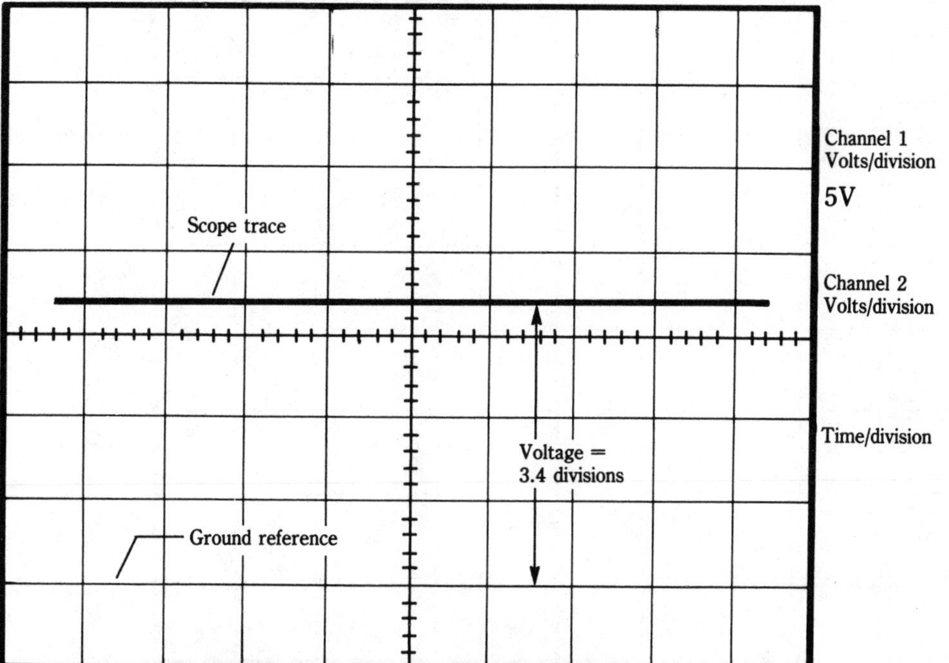

4-3 dc voltage = 17 V above ground.

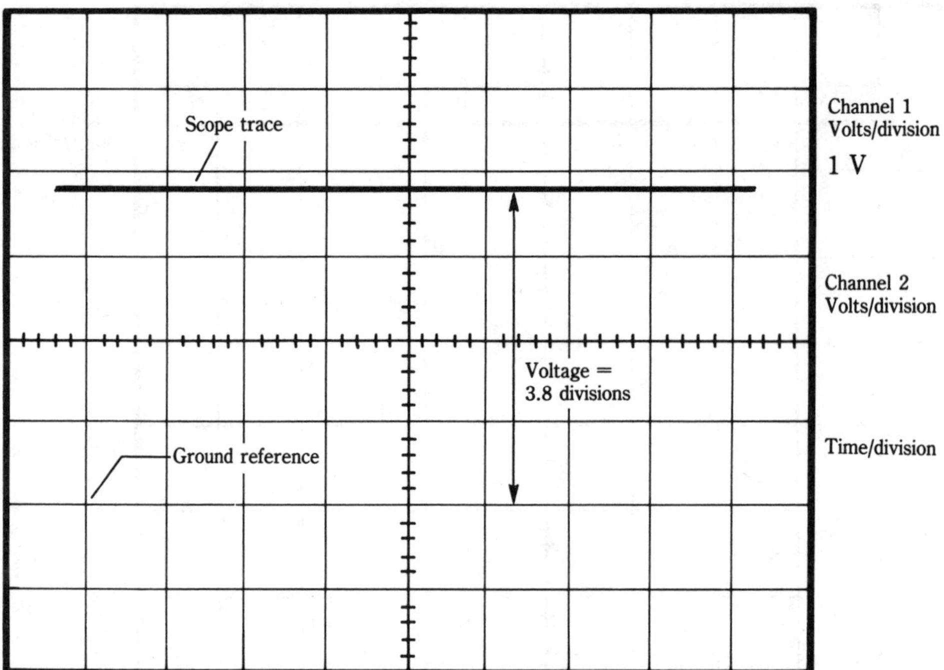

4-4 dc voltage = 3.8 V above ground.

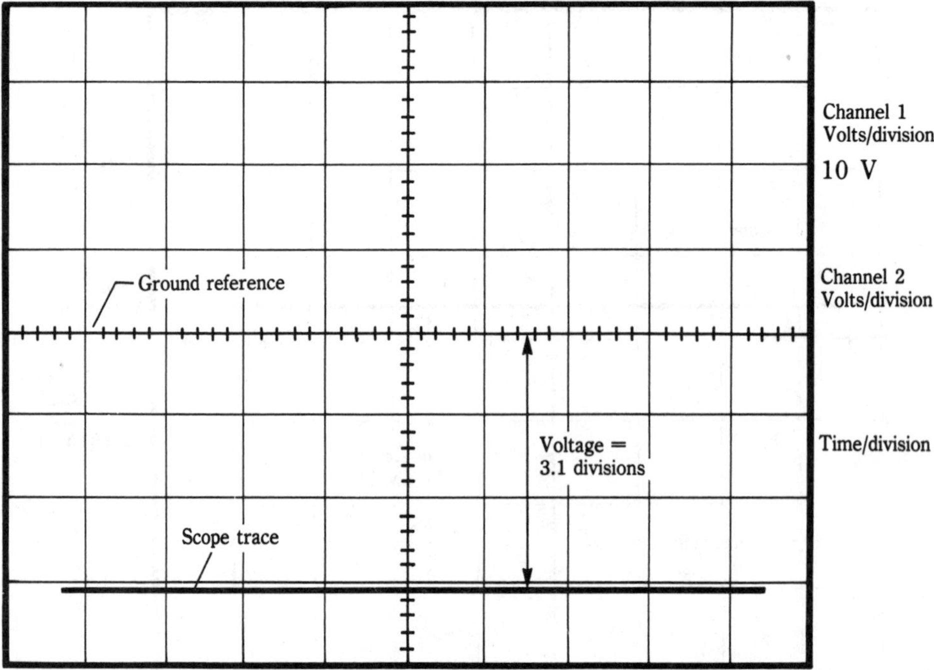

4-5 dc voltage = 31 V below ground.

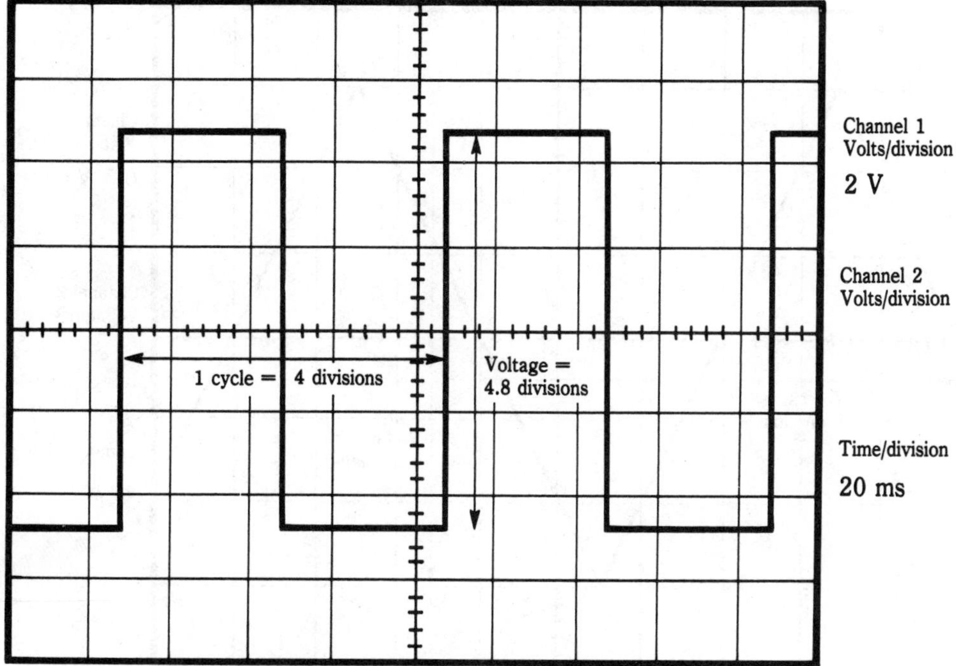

4-6 ac voltage = 9.6 V; 1 cycle = 80 ms.

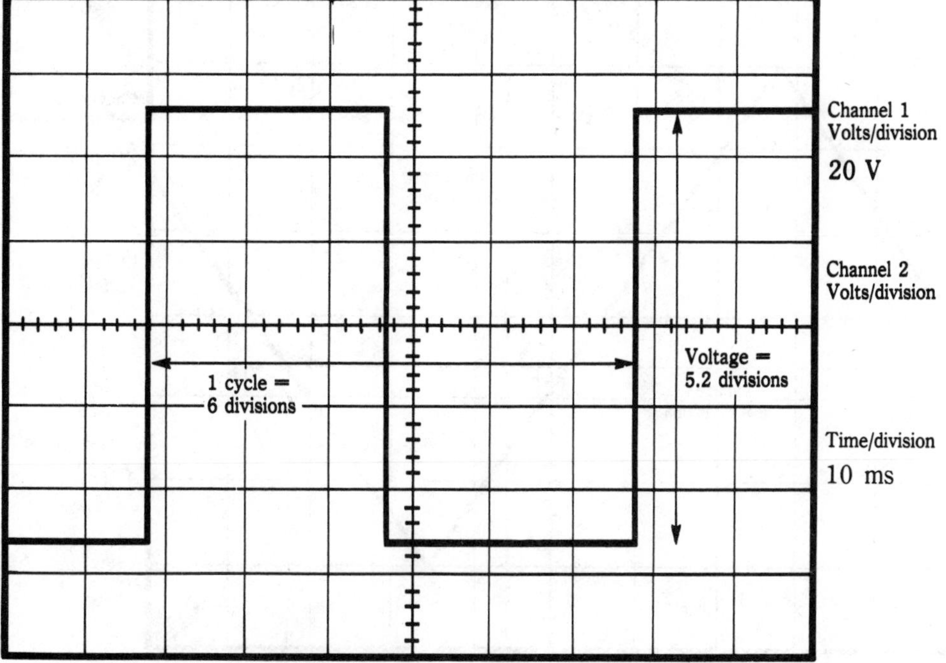

4-7 ac voltage = 104 V; 1 cycle = 60 ms.

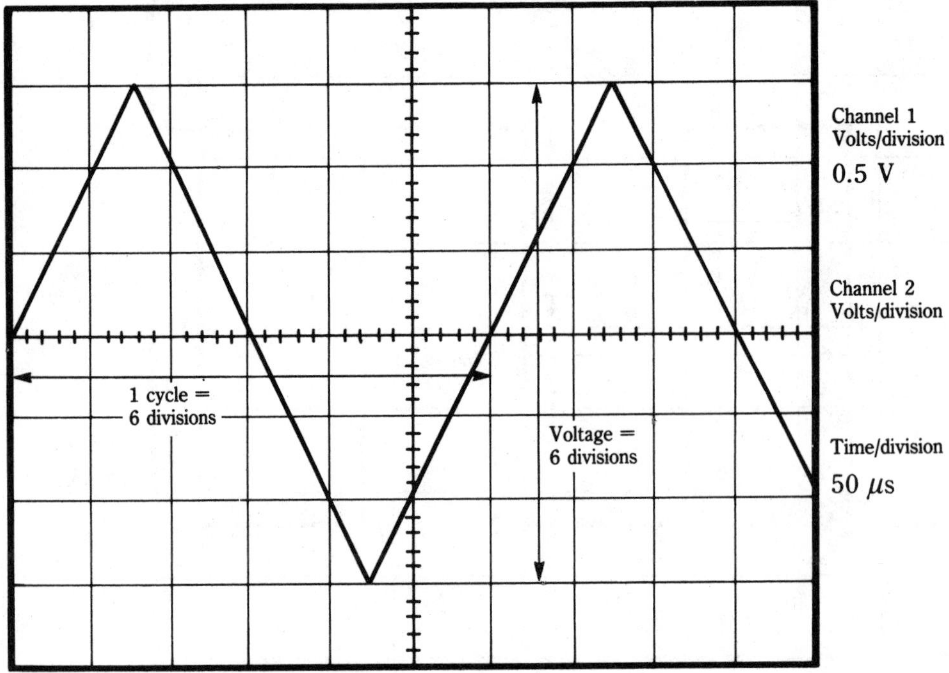

4-8 ac voltage = 3.0 V; 1 cycle = 300 μs.

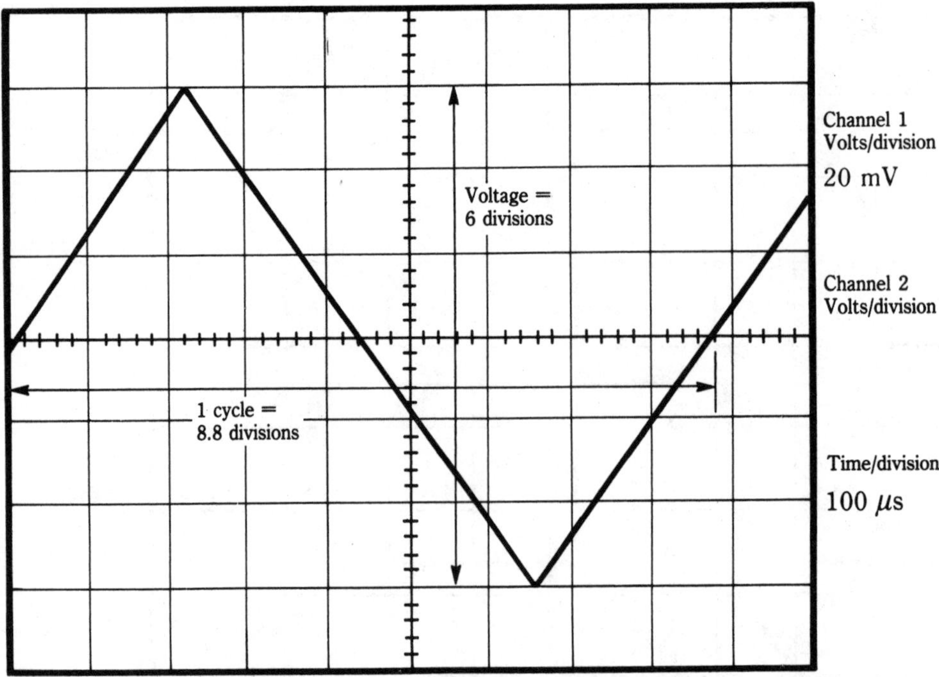

4-9 ac voltage = 120 mV; 1 cycle = 880 μs.

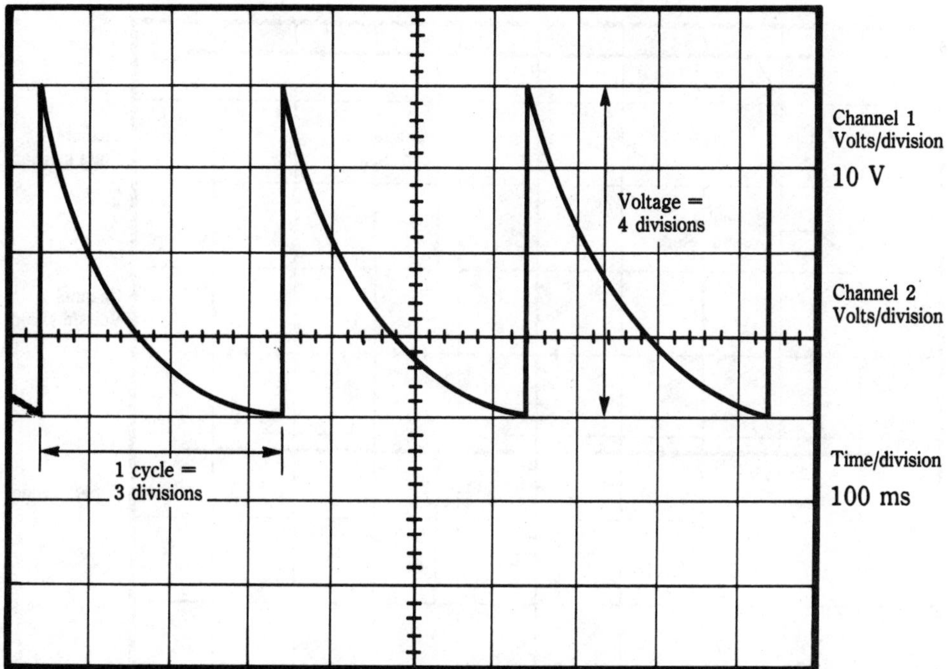

4-10 ac voltage = 40 V; 1 cycle = 300 ms.

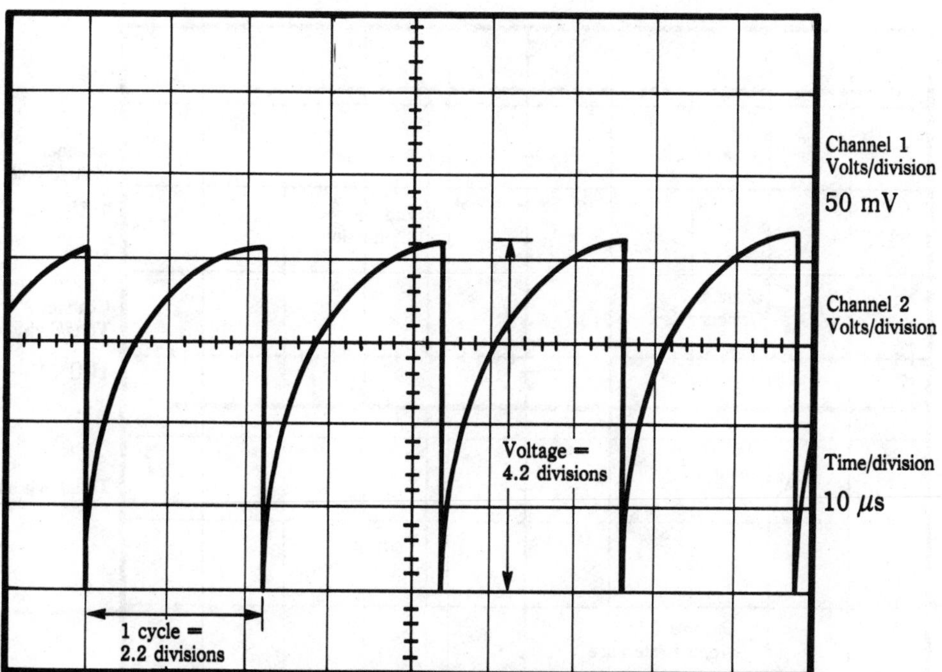

4-11 ac voltage = 210 mV; 1 cycle = 22 μs.

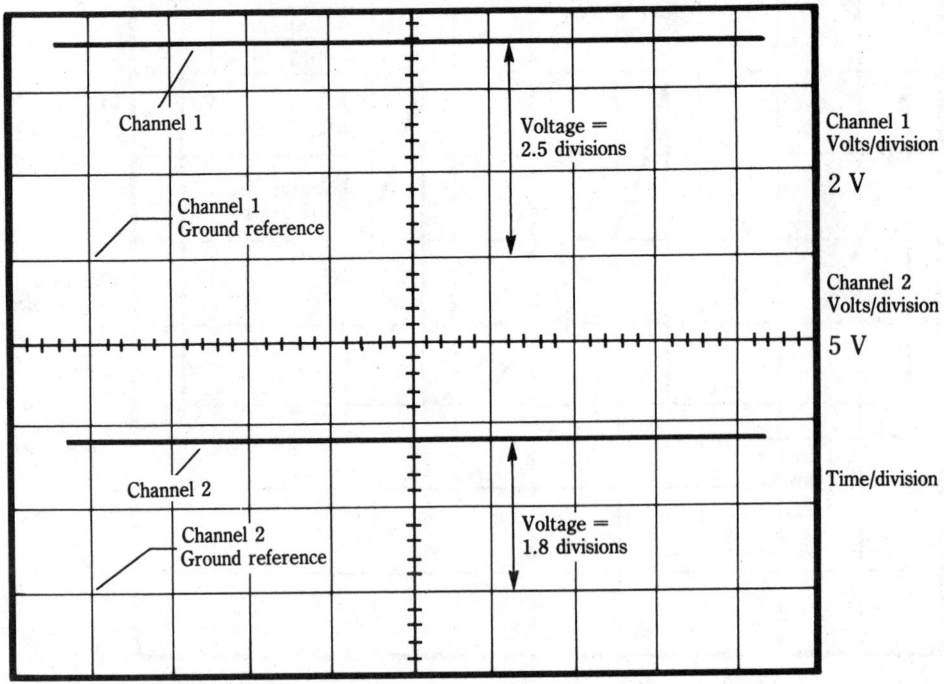

4-12 Channel 1: dc voltage = 5.0 V. Channel 2: dc voltage = 9.0 V.

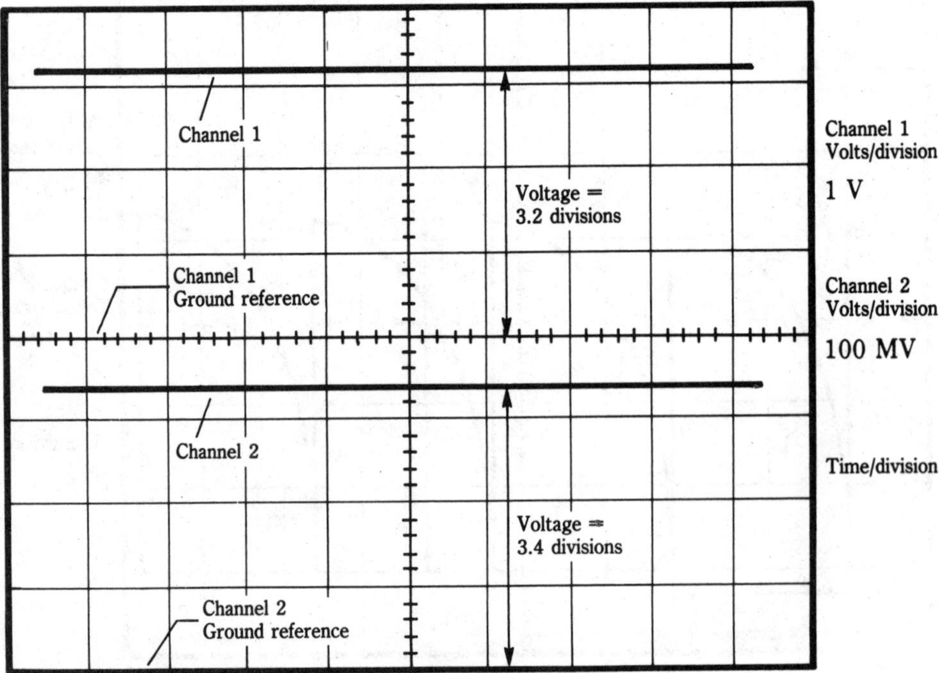

4-13 Channel 1: dc voltage = 3.2 V. Channel 2: dc voltage = 340 mV.

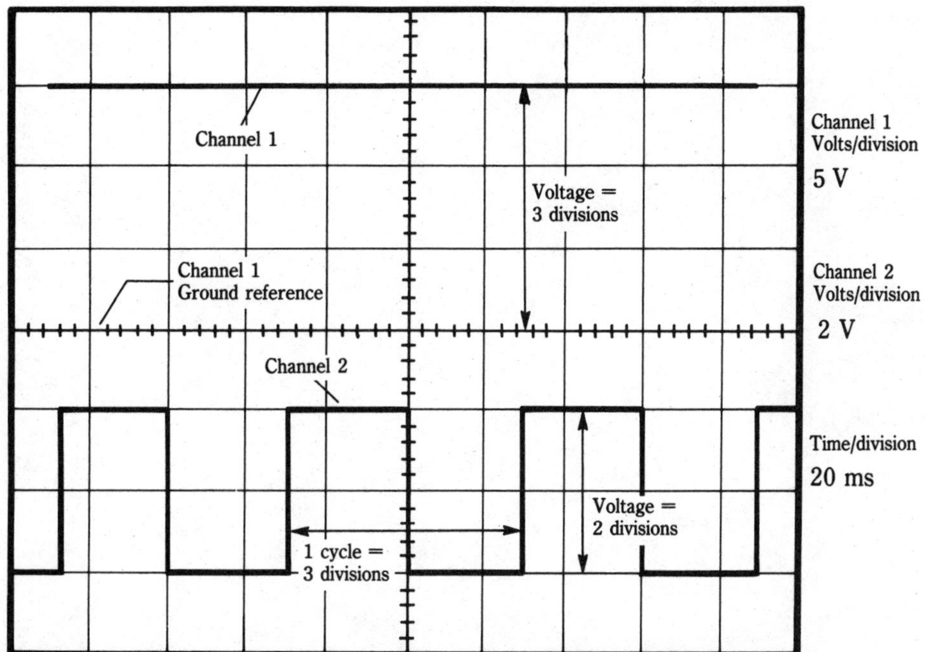

4-14 Channel 1: dc voltage = 15 V. Channel 2: ac voltage = 4.0 V; 1 cycle = 60 ms.

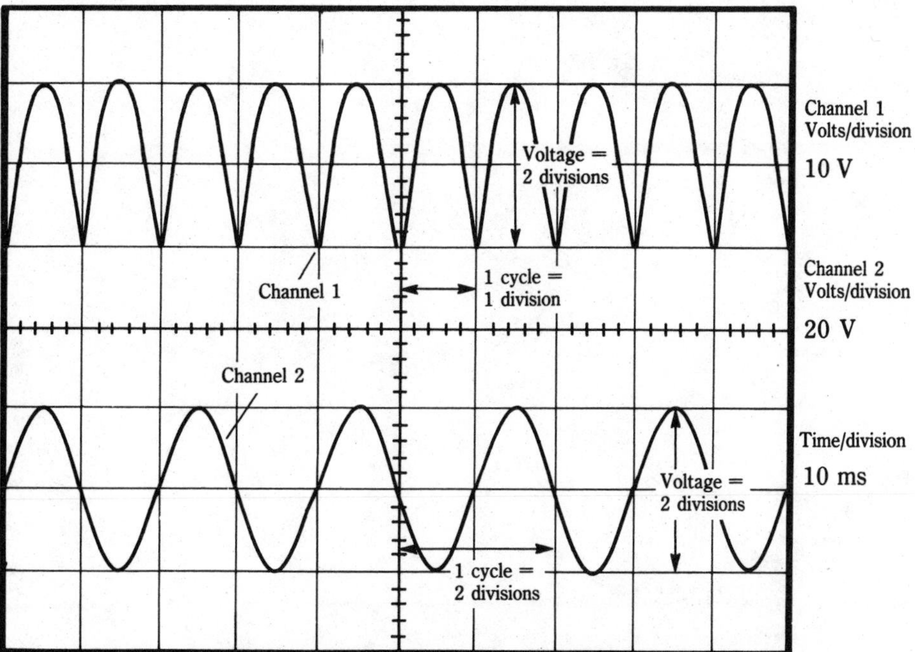

4-15 Channel 1: ac voltage = 20 V; 1 cycle = 10 ms. Channel 2: ac voltage = 40 V; 1 cycle = 20 ms.

5
The sine wave

THE SINE WAVE IS THE MOST IMPORTANT WAVEFORM IN ALL ELECTRONIC circuits and is often referred to as the primary frequency. The basic components of inductors and capacitors have ac resistance, called *reactance*, based on the sine wave.

One common example of the use of a sine wave is household electricity. Household electricity used in the United States is a 120 V, 60 Hz (60-cycle) sine wave.

Producing a sine wave

There are basically two ways to produce a sine wave. One way is with an oscillator, and the other way is with a generator.

An *oscillator* is an electronic circuit that changes a dc supply voltage into the sine wave. The circuit contains a parallel resonant circuit and an amplifier. An oscillator is usually used to produce high frequencies.

A generator is used to produce the sine wave of household electricity. Because the generator is an easy way to see the formation of a sine wave, this method will be discussed.

The alternating current generator

The *ac generator* produces electricity by rotating a wire through a magnetic field. The wire moving in the magnetic field causes a voltage to be induced, with the strongest voltage happening when the wire is perpendicular (at right angles) to the magnetic lines of force. Or, cutting across the magnetic lines produces the maximum voltage. When the wire is going in the same direction as the magnetic lines of force, no voltage is induced. All the positions of the wire, between the maximum and minimum (zero) voltage points, produce a voltage proportional to the position.

As the wire is rotated through a complete circle, the wire will have a position associated with the opposite magnetic pole. This will produce electricity having an opposite polarity.

Figure 5-1 shows a loop of wire rotating in a magnetic field produced by a two-pole generator. Position a represents 0 degrees, the starting point. At 0 degrees, the wire is in direct line with the magnetic lines of force and therefore there is no voltage produced. Position b represents one-quarter turn, or 90 degrees of rotation. At 90 degrees, the wire is cutting the lines of force at a maximum point, producing maximum voltage. This is the positive peak of the sine wave. At position c, the wire has rotated through one half of its full cycle, or 180 degrees of rotation. At this point, the wire is again in line with the magnetic lines of force except the wire is heading in the opposite direction. There is no zero voltage produced. Position d

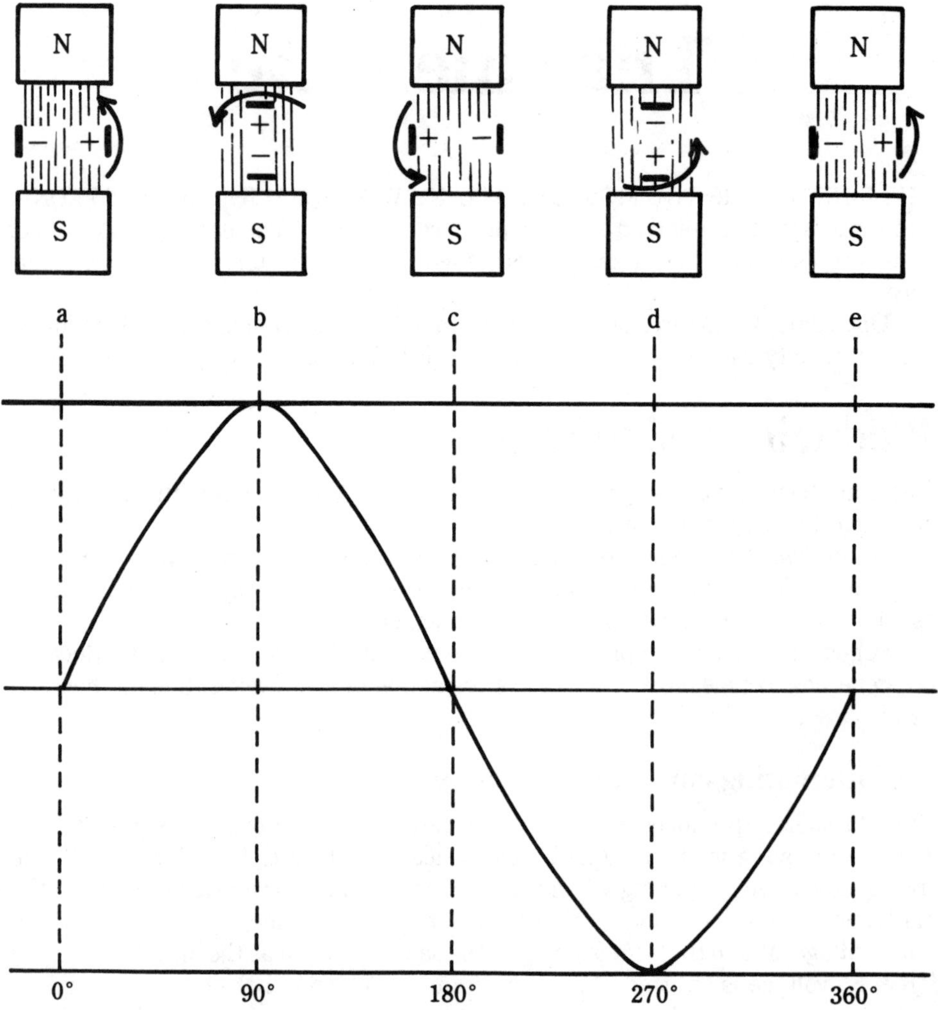

5-1 One cycle of a sine wave produced by rotating a loop of wire in a two-pole generator.

shows the wire cutting the maximum lines of force except produced by the opposite pole. This is three-quarters of a cycle, or 270 degrees, producing voltage in the opposite direction, or a negative voltage at the maximum, or peak of the sine wave. Then in position e, the wire has returned to the original starting point. At this point, it again produces zero volts, and the waveform is headed in the positive direction. This is one complete cycle or 360 degrees.

If this wire were to be rotated in this generator at a rate of 60 complete revolutions every second, the sine wave would be reproduced at a rate of 60 Hz (cycles per second).

Plotting a sine wave

The value of a sine wave can be plotted, or calculated with the use of a formula:

$$V = V_{in} \sin \Theta \text{ instantaneous voltage of a sine wave} \qquad (5\text{-}1)$$

The formula states that the instantaneous voltage (V) is equal to the maximum voltage the sine wave reaches (V_m, also called the peak value) times the sine of the angle at the particular time the calculation is made. Note *sin* is the abbreviation of sine.

Figure 5-2 shows one cycle of a sine wave plotted using the formula for instantaneous voltage. Refer to Table 5-1 for the results of the calculations. The calculations were made, for this example, every 15 degrees. The voltage selected for the maximum is 100 V. This voltage was chosen because it makes it very easy to change it to percent of full voltage.

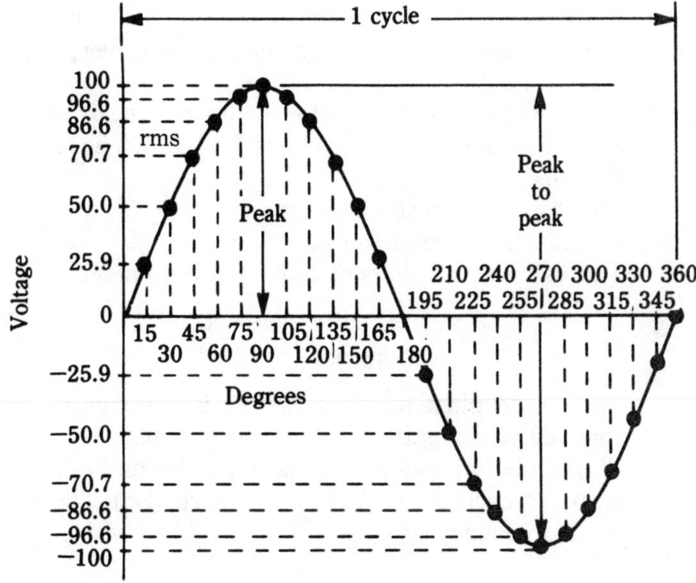

5-2 Using the formula $V = V_M \sin 0$ to plot a sine wave. Also refer to Table 5-1.

Table 5-1. $V = V_M \sin \Theta$.

Degrees	Voltage	Degrees	Voltage
0	0	180	0
15	25.9	195	− 25.9
30	50.0	210	− 50.0
45	70.7	225	− 70.7
60	86.6	240	− 86.6
75	96.6	255	− 96.6
90	100	270	−100
105	96.6	285	− 96.6
120	86.6	300	− 86.6
135	70.7	315	− 70.7
150	50.0	330	− 50.0
165	25.9	345	− 25.9
180	0	360	0

Sample calculation for Table 5-1, Fig. 5-2

$V = V_m \sin \Theta$

$V = 100 \sin 45$

sin 45 degrees = 0.707

$V = 100 \times 0.707$

$V = 70.7$ V this point is the rms—root-mean-square—value

When dealing with sine waves, there are several new words and definitions to be learned. Refer to Figs. 5-2 and 5-3.

cycle A complete cycle is when the waveform repeats itself. Figures 5-2 and 5-3a show one cycle, Fig. 5-3b is two cycles. Fig. 5-3c is $2^1/_2$ cycles, and Fig. 5-3d is four cycles.

period The length of time it takes for the wave to make one cycle. Fig. 5-3a is $^1/_2$ second, Fig. 5-3b is $^1/_4$ second, Fig. 5-3c is $^1/_6$ second, and Fig. 5-3d is $^1/_8$ second.

wavelength The distance a wave moves in a full cycle. The symbol for wavelength is the Greek letter lambda λ.

This calculation results in lambda in centimeters.

$$\lambda = \frac{3 \times 10^{10} \text{ centimeters/second}}{\text{frequency (Hz)}} \tag{5-2}$$

The wavelength of radio waves in air is an important consideration when determining the length of an antenna.

frequency The number of cycles in one second. The unit of measure for frequency is cycle per second (cps) or hertz (Hz). The formula to calculate frequency is:

$$f = \frac{1}{T} \tag{5-3}$$

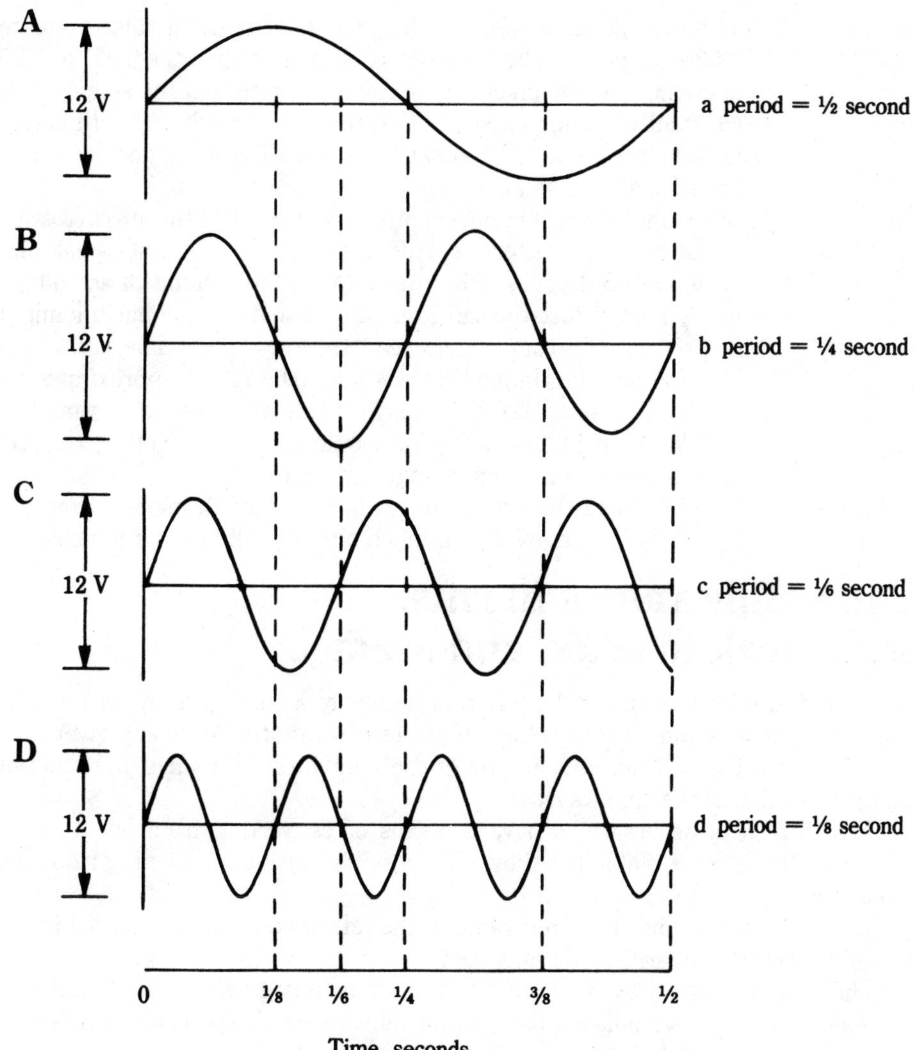

5-3 Four sine waves with different frequencies.

Frequency is the inverse of time.
Figure 5-2 has a frequency of 4 Hz. Fig. 5-3a is 2 Hz, Fig. 5-3b is 4 Hz, Fig. 5-3c is 6 Hz, and Fig. 5-3d is 8 Hz.

peak value The maximum value a sine wave reaches as measured from the zero reference line. Peak value can be either positive or negative. Figure 5-2 has a peak value of 100 V, Fig. 5-3 (all four waveforms, a – d) have a peak value of 6 V.

peak to peak The value of the sine wave from the negative peak to the positive peak. It is a value equal to two times the peak. Figure 5-2 has a peak-to-peak value of 200 V, and all four in Fig. 5-3 have a peak-to-peak value of 12 V.

average The average value of one-half the sine wave has a value equal to 0.636 × peak. The average occurs at a point equal to 39.5 degrees. Average takes into account the face that the sine wave is constantly changing. Notice that the average value of both halves of the sine would be 0. That is because the wave goes as much negative as it does positive.

rms rms stands for root mean square. It is also called the *effective* value or the *dc equivalent* value. rms is equal to 0.707 times peak and occurs at 45 degrees. The rms value is the amount of ac voltage needed to produce the same amount of heat as an equal amount of dc voltage. Most all meters read the rms value. The household electricity in the United States is said to be 120 V (approximately), 60 Hz. The stated 120 V is the rms value, the peak value would be 170 V, the peak-to-peak value would be 340 V. Notice, the frequency has no effect on voltage measurements.

amplitude The height of the waveform. This is a word developed from the use of oscilloscopes. It can mean either peak or peak-to-peak.

Converting values in rms, peak, peak-to-peak, and average

From the definitions given for the different values of a sine wave, there are four names that are all related to the voltage or current waveform. With any given sine wave, it is possible to describe it in any of the four different names and still not change the value of the sine wave.

Therefore it is necessary to have a means of converting from one form to another and vice-versa. Table 5-2 shows the possible ways of converting from one to the other.

To use the table, find the given value in the left hand column and, looking to the right, perform the arithmetic indicated.

The table is formed by modifying the relationships given in the definitions. The following list of formulas is the relationships given by the basic definitions. Notice they are all given in terms of the peak value.

$$\text{Peak to peak} = \text{peak} \times 2 \tag{5-4}$$

$$\text{Average} = \text{peak} \times 0.636 \tag{5-5}$$

$$\text{rms} = \text{peak} \times 0.707 \tag{5-6}$$

Conversion problems

With each of the following, use a peak value of 20.

Find peak to peak (p to p)

$$\begin{aligned} \text{p to p} &= 2 \times \text{peak} \\ &= 2 \times 20 \\ &= 40 \end{aligned}$$

Table 5-2. Conversion factors.

Given value	Peak	Peak to peak	rms	Average
Peak	- - -	$\times 2$	$\times 0.707$	$\times 0.636$
Peak to peak	$\times 2$	- - -	$\times \frac{0.707}{2}$	$\times \frac{0.636}{2}$
rms	$\times 1.414$ or $\times \frac{1}{0.707}$	peak $\times 2$	- - -	$\times 0.9$ or peak $\times$ 0.636
Average	$\times 1.57$ or $\times \frac{1}{0.636}$	peak $\times 2$	$\times 1.11$ or peak $\times$ 0.707	- - -

Find rms

$$\begin{aligned} \text{rms} &= 0.707 \times \text{peak} \\ &= 0.707 \times 20 \\ &= 14.14 \end{aligned}$$

Find average

$$\begin{aligned} \text{average} &= 0.636 \times \text{peak} \\ &= 0.636 \times 20 \\ &= 12.72 \end{aligned}$$

Find the peak value for each.

peak to peak = 280

$$\begin{aligned} \text{peak} &= \text{p to p} \times \frac{1}{2} \\ &= 280 \times 0.5 \\ &= 140 \end{aligned}$$

rms = 156

$$\begin{aligned} \text{peak} &= \text{rms} \times \frac{1}{0.707} \\ &= 156 \times 1.414 \\ &= 220 \end{aligned}$$

average = 63.6

$$\text{peak} = \text{average} \times \frac{1}{0.636}$$
$$= 63.6 \times 1.57$$
$$= 100$$

Practice problems

Convert each of the following. Round answers to 3 significant figures.

1. 30 volts peak; convert to: a) rms, b) p to p, c) average
2. 100 volts peak; convert to: a) p to p, b) average, c) rms
3. 80 volts p to p; convert to: a) rms, b) peak, c) average
4. 260 volts p to p; convert to: a) peak, b) average, c) rms
5. 70 volts rms; convert to: a) peak, b) p to p, c) average
6. 12 volts rms; convert to: a) average, b) p to p, c) peak
7. 16 volts average; convert to: a) rms, b) peak, c) p to p
8. 120 volts average; convert to: a) peak, b) p to p, c) rms
9. 85 volts rms; convert to: a) peak, b) p to p, c) average
10. 70 volts peak; convert to: a) p to p, b) average, c) rms

In problems 11 through 15, find: a) p to p, b) peak, c) rms, d) average

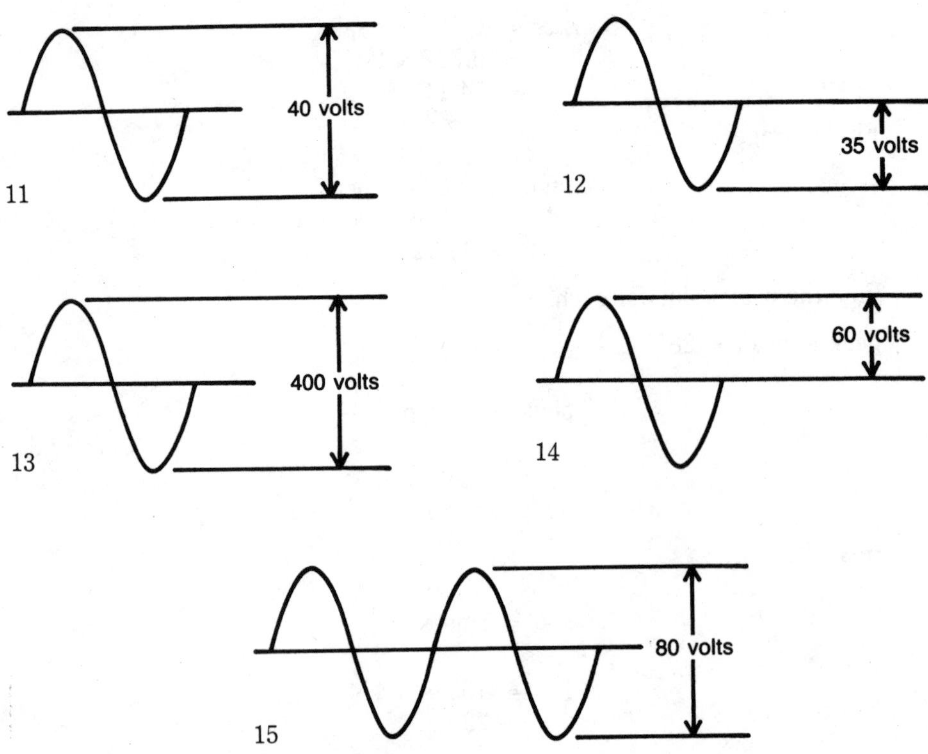

Converting values in frequency, period, and wavelength

When dealing with a sine wave, the amplitude is important for measuring the voltage. The frequency and period are also important characteristics of the sine wave. When using an oscilloscope, the amplitude is the vertical portion of the waveform and the time (frequency and period) are the horizontal portion of the waveform.

Wavelength is a measure of how long one cycle is, such as a radiowave traveling through the air. Therefore, wavelength has a unit of measure in distance.

Frequency and period

Frequency has a unit of measure in hertz. Its alternate unit of measure is cycles per second, which means frequency is very closely related to time. Period is the measure of time of one cycle. The higher the frequency, that is, the more cycles per second the shorter period of time to complete one cycle. The relationship of frequency to time is an inverse relationship.

$$\text{frequency} = \frac{1}{\text{time}} \qquad (5\text{-}3)$$

$$\text{time} = \frac{1}{\text{frequency}}$$

Frequency and time problems

The period of a sine wave is 1 ms. What is the frequency?

$$\text{frequency} = \frac{1}{\text{time}} \qquad (5\text{-}3)$$

$$= \frac{1}{1\text{ ms}}$$

$$= 100\text{ Hz}$$

What is the period (time) of the 60 Hz household line voltage?

$$\text{time} = \frac{1}{\text{frequency}}$$

$$= \frac{1}{60}$$

$$= 16.7\text{ ms}$$

Wavelength λ

When dealing with radio waves traveling through the air, it is necessary to know the wavelength of the frequency being transmitted. Radio waves are considered to travel at a velocity equal to the speed of light, which is 186,000 miles per second or 3×10^{10} centimeters per second.

$$\text{wavelength} = \frac{\text{velocity (speed of light for radiowaves)}}{\text{frequency (frequency of the transmitted signal)}}$$

$$\text{wavelength } (\lambda) = \frac{3 \times 10^{10} \text{ centimeters/second}}{\text{frequency (Hz)}} \qquad (5\text{-}2)$$

Wavelength problems

What is the wavelength of a radio station signal when the station frequency is 630 kHz (AM broadcast)?

$$\lambda = \frac{3 \times 10^{10} \text{ centimeters/second}}{\text{frequency}}$$

$$= \frac{3 \times 10^{10} \text{ centimeters/second}}{630{,}000 \text{ Hz}}$$

$$= 47{,}619 \text{ cm}$$

What is the wavelength of the same 630 kHz signal in feet? (2.54 centimeters = 1 inch and 12 inches = 1 foot.)

$$\lambda = \frac{47{,}619 \text{ centimeters}}{2.54 \text{ centimeters/inch}} = 18{,}747 \text{ inches}$$

$$= \frac{18{,}747 \text{ inches}}{12 \text{ inches/foot}} = 1562 \text{ feet}$$

Practice problems

Calculate the period of the following frequencies:

1. 10 Hz
2. 100 Hz
3. 1 kHz
4. 10 kHz
5. 100 kHz
6. 1 MHz
7. 10 MHz
8. 100 MHz
9. 1 GHz
10. 10 GHz

Calculate the frequencies whose period is:

11. 50 ms
12. 5 ms
13. 0.5 ms
14. 50 μs
15. 0.5 s

Calculate the wavelength in centimeters of the following frequencies:

16. 500 kHz
17. 5 MHz

18. 50 MHz
19. 5 GHz
20. 500 GHz

Using an oscilloscope to measure a sine wave

The oscilloscope is probably the most useful instrument a technician has for measuring any kind of waveform. Learning to make measurements on the oscilloscope with a sine wave is a very logical starting point.

The scope is capable of two types of measurement; voltage and time. Other things can be found from these, but usually require an intermediate step with calculations. For example, frequency can be found by measuring the time of one complete cycle and calculating the frequency based on the time measurement.

The front of the scope is often filled with many knobs, switches, and connectors. The controls can usually be placed into one of three categories: 1. voltage, amplitude, the up and down of the waveform; 2. time, back and forth, sideways; 3. trigger, or the means to provide a stable trace on the screen. The two categories of prime interest here are volts and time.

Measuring voltage on the oscilloscope

The primary control on the scope for use when controlling the amplitude is called volts per division. The oscilloscope screen is divided into blocks or squares called *divisions*. Each of the divisions have four small marks to divide the divisions into subdivisions (most scope screens have the subdivisions marked only on the center axis), which represent part of a division, for example: 0.2, 0.4, 0.6, 0.8, of a division. (Also refer to chapter 4 for further discussion of the oscilloscope.)

Most scopes have the volts per division control calibrated in steps such as: 0.01, 0.05, 0.1, 0.5, 1, 5. When preparing the scope display to measure voltage, it is best to adjust the volts per division for a waveform that fills the screen as full as possible (from top to bottom) without going beyond the limits of the screen. Whenever the waveform has an amplitude that is too small, it is very difficult to read the screen and as a result, the accuracy will suffer.

Figures 5-4, 5-5, 5-6, are examples of adjusting the oscilloscope, all for the same sine wave. Notice the difference in the amplitude of the waveforms as the volts per division is adjusted. Figure 5-4 shows the correct way to adjust the scope to make measurements of the voltage and the frequency.

Refer to Fig. 5-4. The easiest way to measure the voltage is to start off by adjusting the position of the waveform to a location that is convenient to work with. Note that because the voltage measurement is to be a peak-to-peak measurement, it is not necessary to have the center of the sine wave on the center line of the screen. The drawings shown here do have the center of the sine wave on the center line of the scope screen.

To make the voltage measurement, start at the peak of the wave (either positive or negative peak) and count the number of divisions and subdivisions to the opposite peak. In Fig. 5-4, starting at the negative peak, located half way between

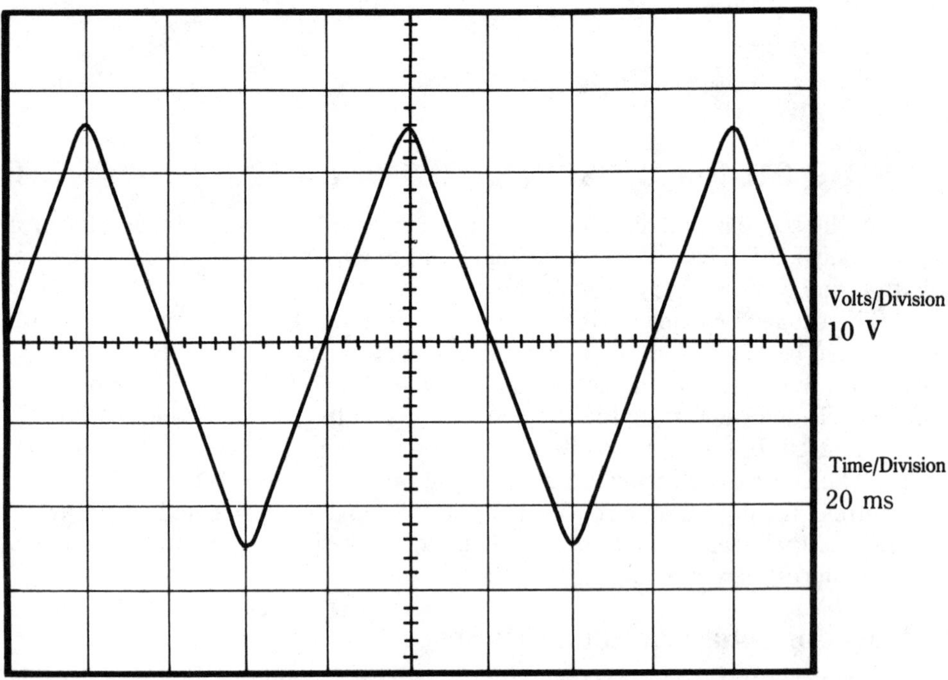

5-4 Oscilloscope display with volts and time adjusted correctly.

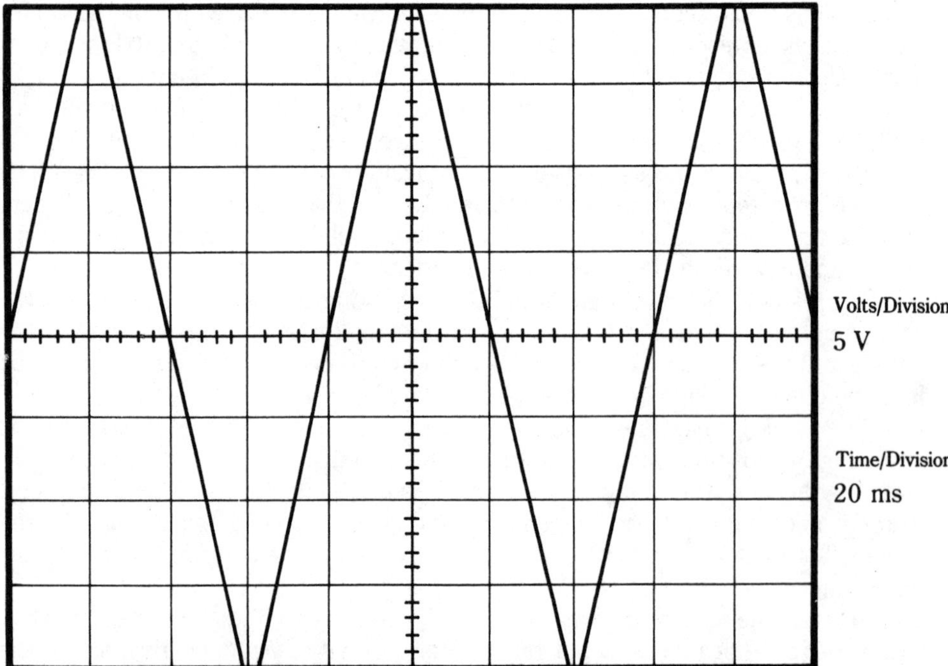

5-5 Oscilloscope display with volts per division expanded too much.

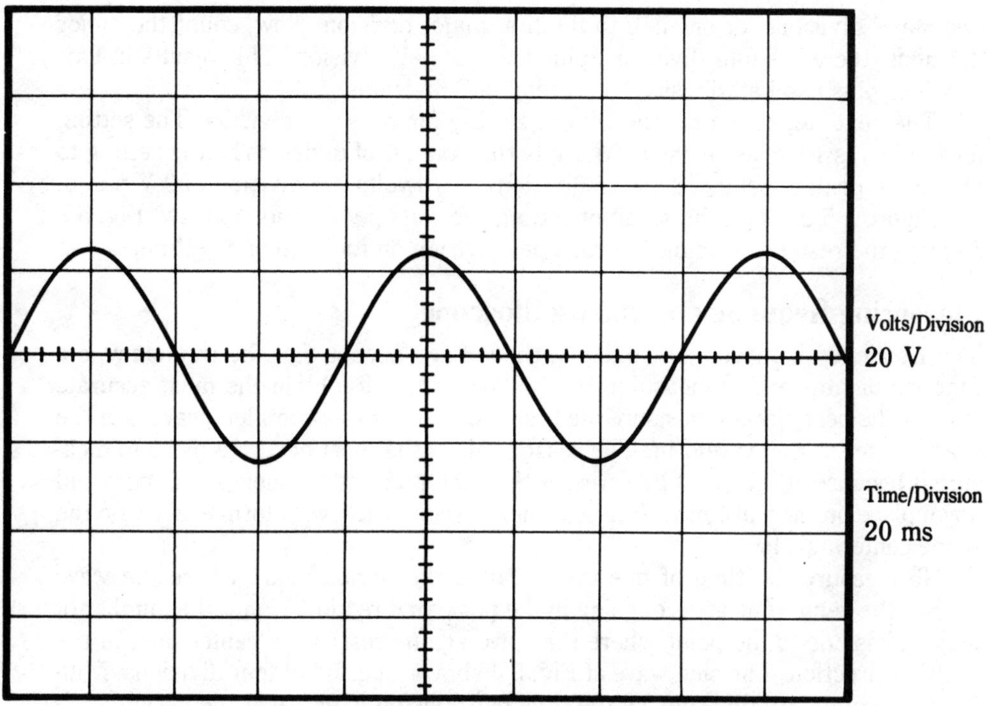

5-6 Oscilloscope display with volts per division not expanded enough.

1. Find: (a) p to p, (b) peak, (c) rms, (d) avg, (e) period, (f) frequency.
2. Find: (a) p to p, (b) peak, (c) rms, (d) avg, (e) period, (f) frequency.
3. Find: (a) p to p, (b) peak, (c) rms, (d) avg, (e) period, (f) frequency.
4. Find: (a) p to p, (b) peak, (c) rms, (d) avg, (e) period, (f) frequency.
5. Find: (a) p to p, (b) peak, (c) rms, (d) avg, (e) period, (f) frequency.
6. Find: Channel 1: (a) p to p, (b) period, (c) frequency.
   Channel 2: (d) p to p, (e) period, (f) frequency.
7. Find: Channel 1: (a) p to p, (b) period, (c) frequency.
   Channel 2: (d) p to p, (e) period, (f) frequency.
8. Find: Channel 1: (a) p to p, (b) period, (c) frequency.
   Channel 2: (d) p to p, (e) period, (f) frequency.
9. Find: Channel 1: (a) p to p, (b) period, (c) frequency.
   Channel 2: (d) p to p, (e) period, (f) frequency.
10. Find: Channel 1: (a) p to p, (b) period, (c) frequency.
    Channel 2: (d) p to p, (e) period, (f) frequency.

two major divisions, or one-half to the first major division. Now, count the major divisions: there are four divisions, plus another half division. This results in four divisions plus two half divisions, or a total of 5 divisions.

The next step is to read the control setting for volts per division. The setting for Fig. 5-4 is 10 V per division. Multiply the number of divisions by the setting to obtain the peak to peak voltage—5 divisions × 10 volts per division—50 V p to p.

Figure 5-5 displays the result of setting the volts per division too low. Fig. 5-6 displays the result of placing the volts per division on too high of a setting.

Measuring frequency on the oscilloscope

The frequency is measured similar to the voltage except that the frequency is a function of time and is measured on the horizontal. To obtain the most accurate results, the best place to measure the time required for one complete wave is at the exact center of the waveform. The reason this is the most accurate place to measure is because at this point line crosses the center axis at the most straight up and down place on the waveform. It is best then, to center the waveform exactly on the scope center axis line.

To measure the time of one cycle, called the period, start where the wave crosses the center line at zero, going in the positive direction. Count the number of major divisions to the point where the wave again crosses the center line, in the positive direction. The sine wave of Fig. 5-4 shows a period of four divisions. Multiply the number of divisions by the time per division to arrive at the period.

four divisions × 20 ms per division = 80 ms
period = 80 ms
frequency = 1/t = 1/80 ms = 12.5 hertz

Practice problems

Each of the following drawings represents an oscilloscope display of a sine wave. Find the requested values using the drawings. Round answers to 3 significant figures.

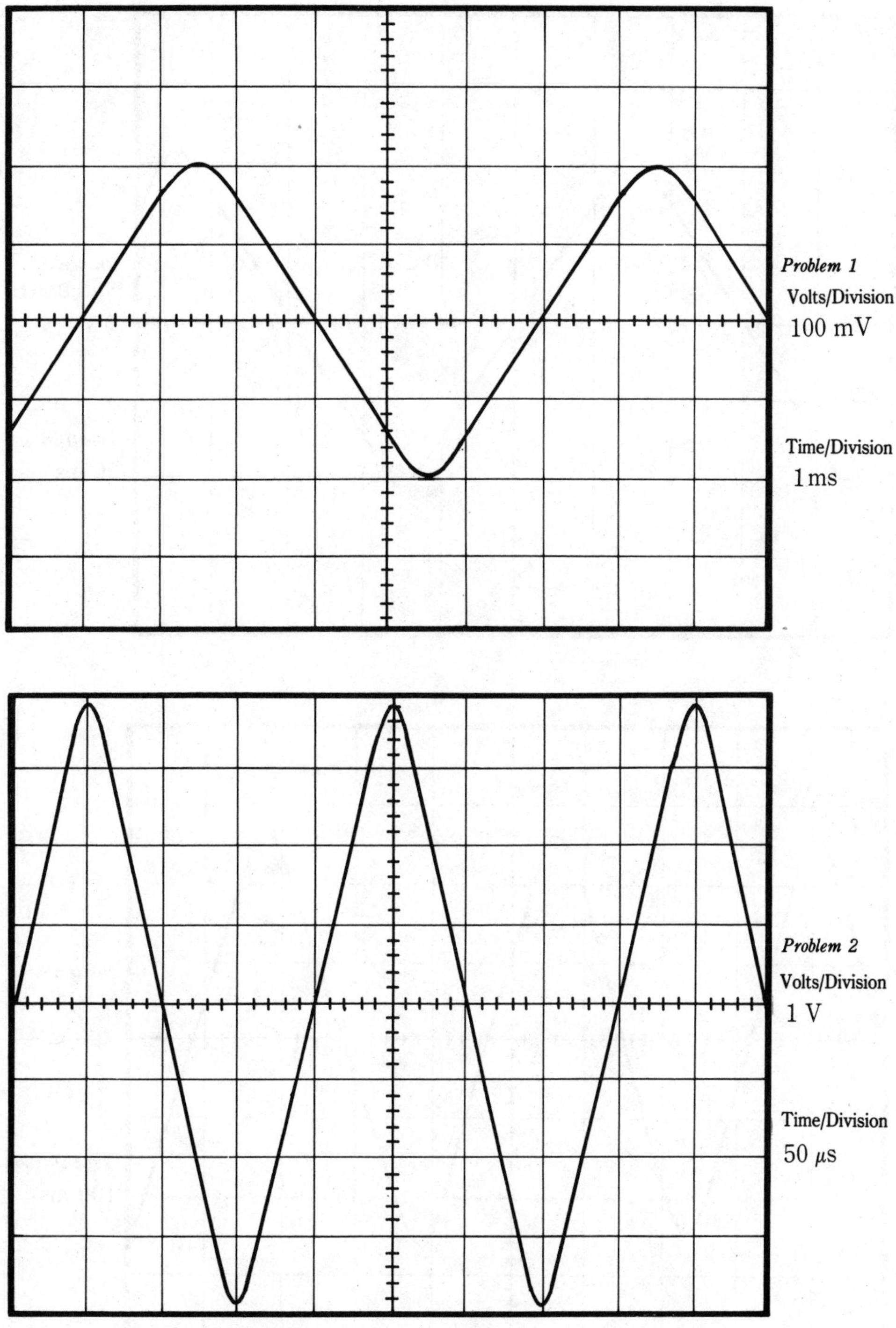
Problem 1
Volts/Division
100 mV
Time/Division
1 ms
Problem 2
Volts/Division
1 V
Time/Division
50 μs

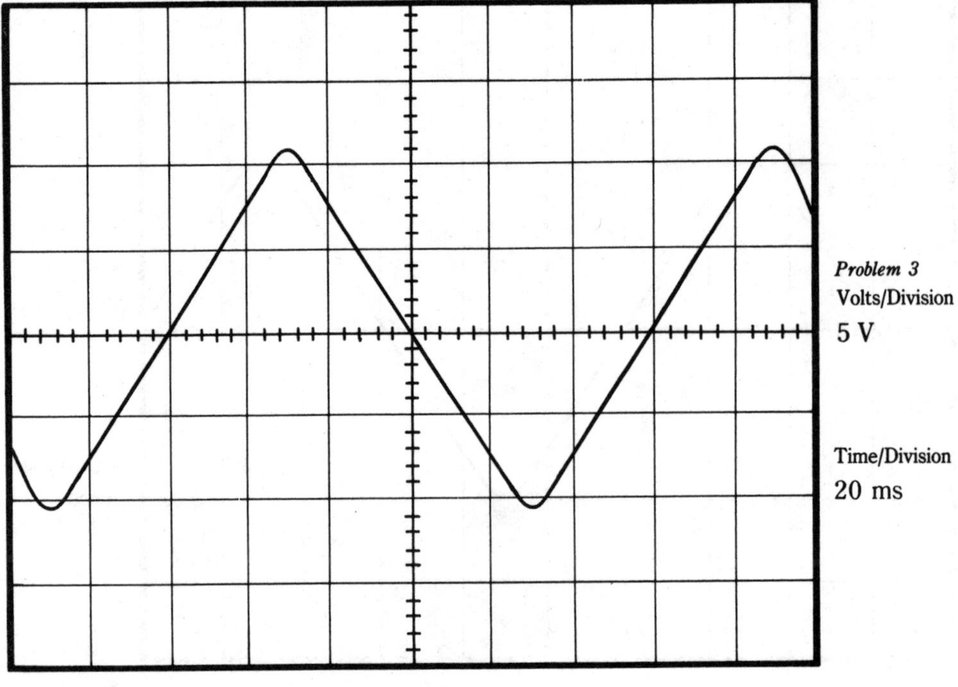

Problem 3
Volts/Division
5 V
Time/Division
20 ms

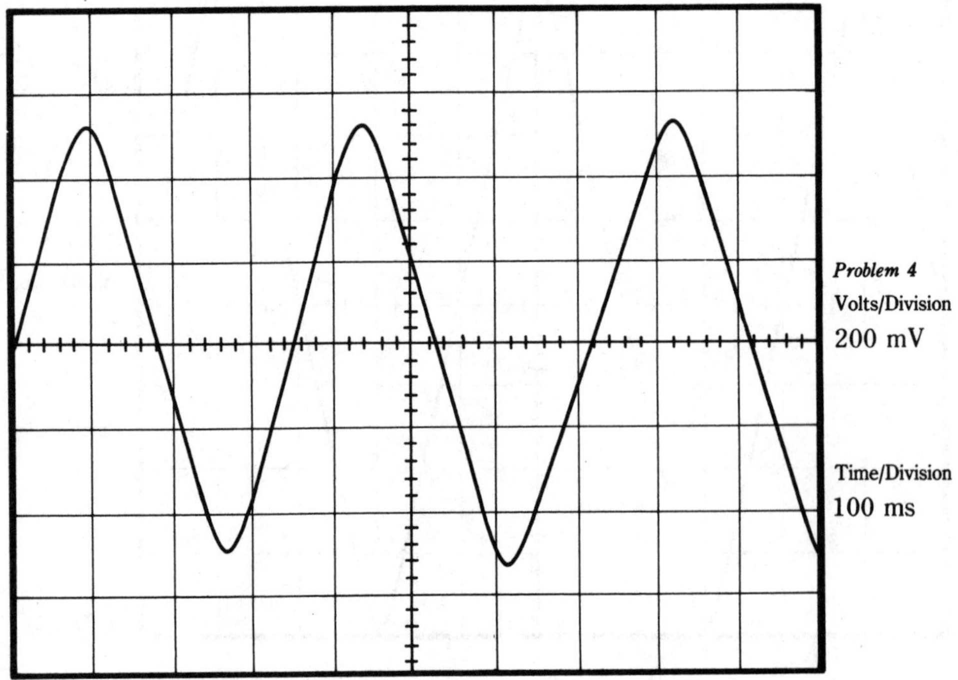

Problem 4
Volts/Division
200 mV
Time/Division
100 ms

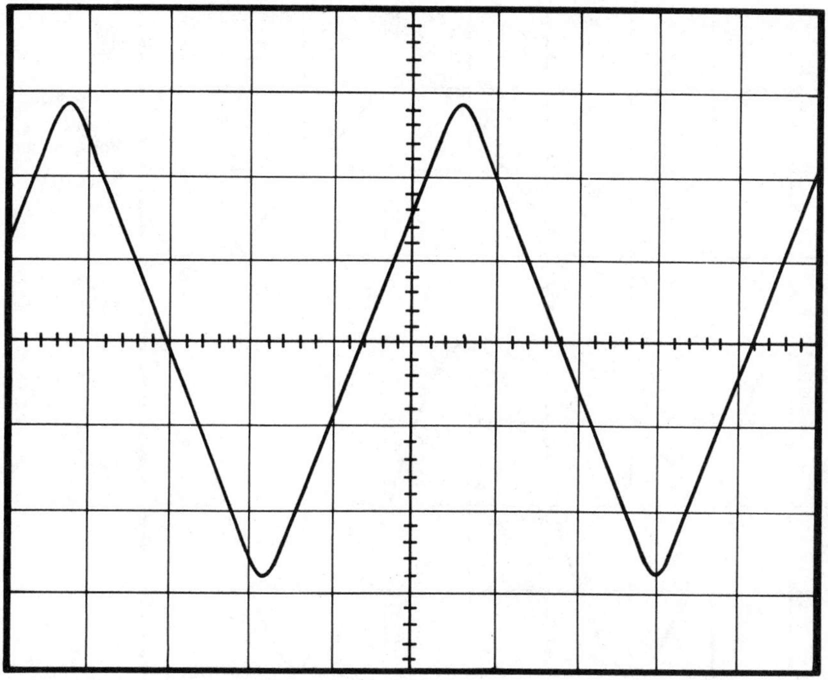

Problem 5
Volts/Division
2 V

Time/Division
500 μs

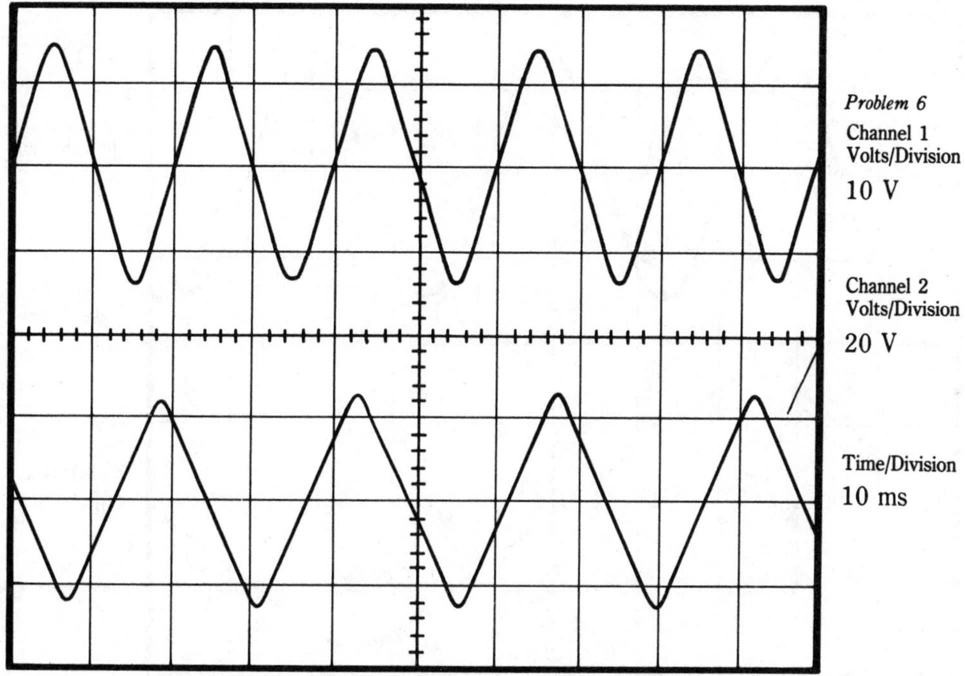

Problem 6
Channel 1
Volts/Division
10 V

Channel 2
Volts/Division
20 V

Time/Division
10 ms

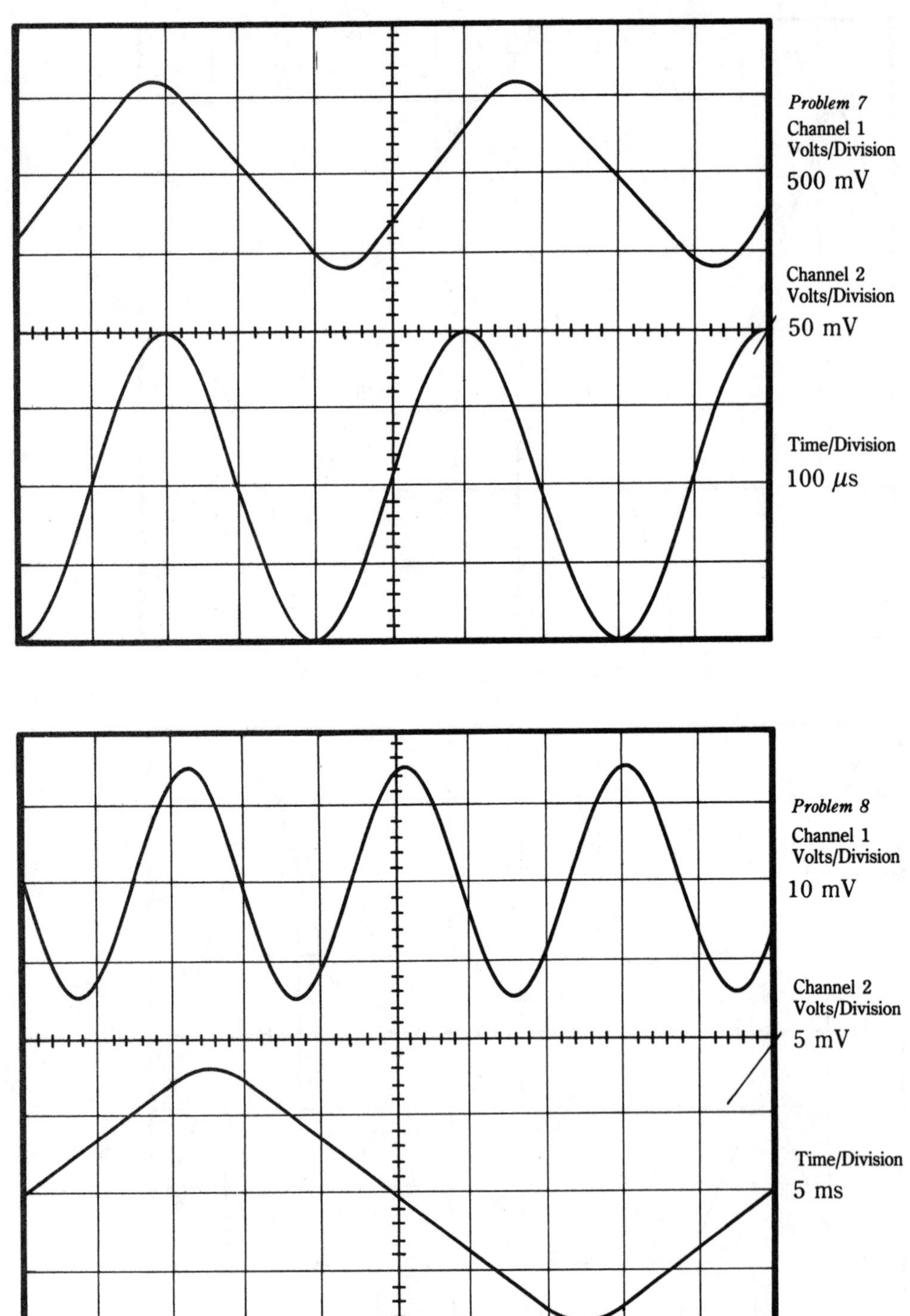
Problem 7
Channel 1
Volts/Division
500 mV
Channel 2
Volts/Division
50 mV
Time/Division
100 μs
Problem 8
Channel 1
Volts/Division
10 mV
Channel 2
Volts/Division
5 mV
Time/Division
5 ms

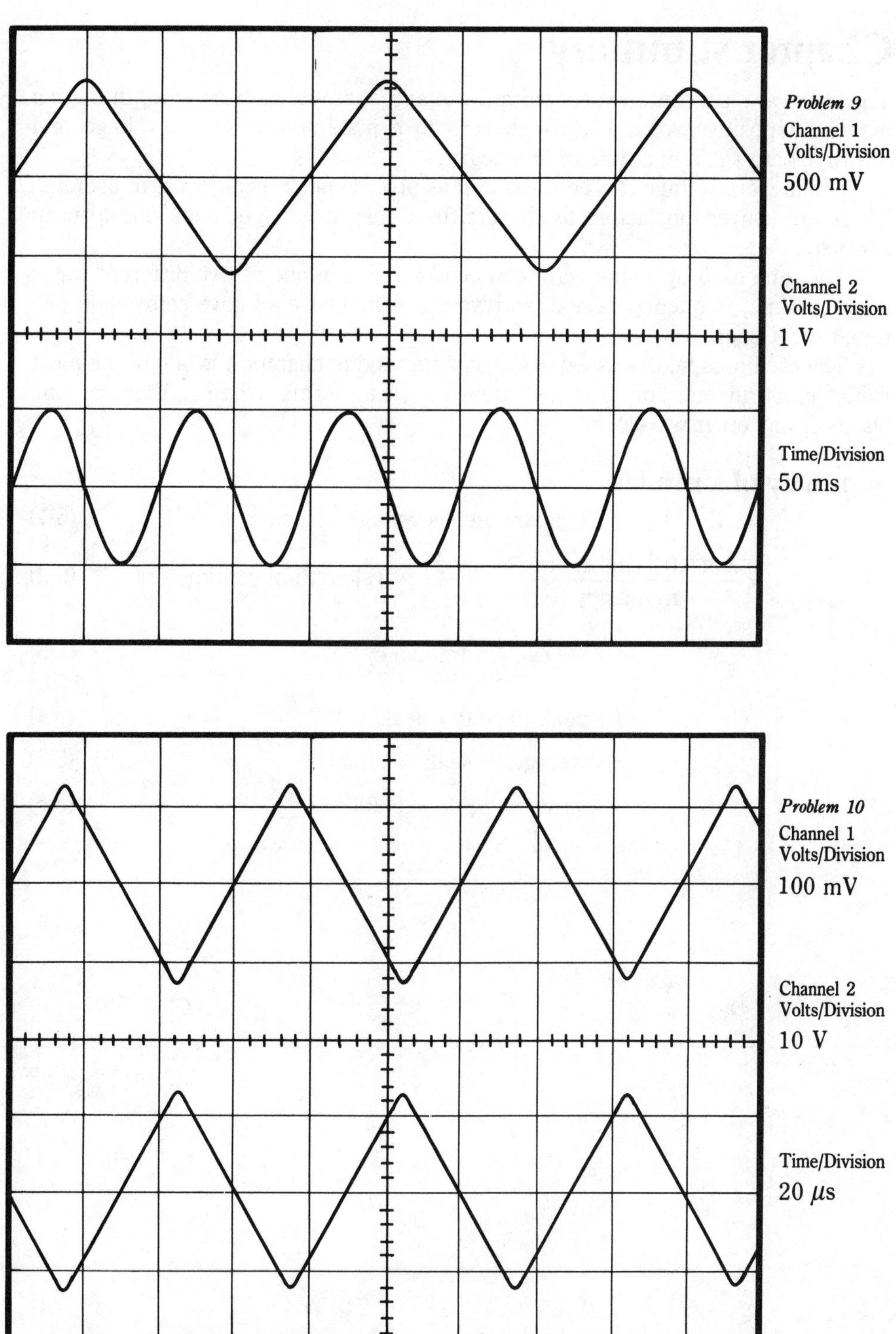
Problem 9
Channel 1
Volts/Division
500 mV
Channel 2
Volts/Division
1 V
Time/Division
50 ms
Problem 10
Channel 1
Volts/Division
100 mV
Channel 2
Volts/Division
10 V
Time/Division
20 μs

Chapter summary

The sine wave is a continuously varying voltage, that will go from zero, through a maximum positive voltage, through zero to a maximum negative voltage and return to zero again in a time of 360 degrees.

A sine wave voltage can be measured as peak to peak, peak, rms, or average. There are conversion factors to convert the different voltages from one form to another.

The time used by a sine wave can be expressed in one of five different ways: degrees, time, frequency, period, and wavelength. These all have conversion factors, except degrees.

The oscilloscope, discussed in this chapter and in chapter 4 is one of the most valuable instruments for use when measuring waveforms, whether they are sine waves or any other waveform.

Summary of formulas

$$V = V_m \sin \Theta \text{ instantaneous voltage of a sine wave} \qquad (5\text{-}1)$$

$$\lambda = \frac{3 \times 10^{10} \text{ centimeters/second}}{\text{frequency (Hz)}} \text{ wavelength in centimeters} \qquad (5\text{-}2)$$

$$f = \frac{1}{T} \text{ frequency} \qquad (5\text{-}3)$$

$$\text{peak to peak} = \text{peak} \times 2 \qquad (5\text{-}4)$$

$$\text{average} = \text{peak} \times 0.636 \qquad (5\text{-}5)$$

$$\text{rms} = \text{peak} \times 0.707 \qquad (5\text{-}6)$$

6
Transformers

TRANSFORMERS ARE FOUND IN ALMOST EVERY PIECE OF ELECTRONIC equipment. A transformer is usually used in the power supply section of a circuit, therefore, the study of transformers is fairly significant.

A transformer is basically two coils of wire wound together in such a way that there is a near perfect magnetic coupling as possible between the two coils of wire.

The mathematics of transformers deals with the turns ratio, voltage ratio, power relationships, and efficiency.

Turns ratio, voltage ratio, current ratio

The *turns ratio* is the mathematical ratio that describes the relationship of the number of turns in the primary of the transformer (input) to the number of turns in the secondary (output).

The voltage ratio is based on the turns ratio and is used to calculate the amount of voltage found at the secondary when a certain voltage is applied to the primary.

The current ratio has an inverse relationship to the turns ratio and is used to calculate the current of the secondary when current is applied to the primary.

Turns ratio

The turns ratio is expressed as the number of turns in the primary to the number of turns in the secondary. The capital letter N is used to represent the number of turns.

$$\frac{N_p}{N_s} \quad \text{transformer turns ratio} \qquad (6\text{-}1)$$

For example, if a particular transformer has 600 turns in the primary and 60 turns in the secondary, the ratio is:

$$\frac{600}{60} = \frac{10}{1} \text{ or 10:1 primary to secondary}$$

Because it is a ratio, it has no units and does not indicate the actual number of turns, only the ratio of primary to secondary.

Voltage ratio

The voltage ratio is the most common way of stating the capability of a transformer. The voltage ratio is equal to, and directly proportional to, the turns ratio. The reason voltage is often stated as the rating of a transformer is that it can be related to how the transformer will be used.

$$\frac{N_p}{N_s} = \frac{V_p}{V_s} \quad \text{voltage ratio} \qquad (6\text{-}2)$$

Example: a particular transformer has a turns ratio of 10:1 and 120 V rms is applied to the primary. What is the secondary voltage?

$$\frac{N_p}{N_s} = \frac{V_p}{V_s} \qquad (6\text{-}2)$$

$$\frac{10}{1} = \frac{120}{V_s}$$

$$V_s = \frac{120 \times 1}{10}$$

$$= 12 \text{ V rms}$$

Current ratio

Current ratio is helpful when determining the size of fuse to use to protect a circuit. The current ratio is the inverse of the turns ratio or voltage ratio. This means a current in the secondary of a *step up* (*step up* or *step down* always refers to the voltage) transformer will be higher than the current of the primary.

$$\frac{N_p}{N_s} = \frac{I_s}{I_p} \quad \text{current ratio} \qquad (6\text{-}3)$$

Example: a step-down transformer has a turns ratio of 6:1 and 120 V is applied to the primary. If 2 A flows in the secondary, what is the primary current?

$$\frac{N_p}{N_s} = \frac{I_s}{I_p} \quad \text{voltage given will not be used}$$

$$\frac{6}{1} = \frac{2}{I_p}$$

$$I_p = \frac{2 \times 1}{6}$$

$$= 0.333 \text{ A}$$

Example: a transformer has a turns ratio of 5:2 with a primary voltage of 120 V. Find the secondary voltage.

$$\frac{N_p}{N_s} = \frac{V_p}{V_s} \qquad (6\text{-}2)$$

$$\frac{5}{2} = \frac{120}{V_s}$$

$$V_s = \frac{120 \times 2}{5}$$

$$= 48$$

Voltage and current ratios calculated by the methods shown here assume a perfect transformer. Note that actual transformers have losses. The voltage ratio is affected by the *coefficient of coupling*. That means the ability of the transformer winding to couple the magnetic field from the primary to the secondary. The current used by the secondary in relation to the current developed in the primary also is affected by the various losses in a transformer. Any calculations made in this section assume perfect transformers. Calculating transformer efficiency is discussed later.

Practice problems

Complete the table.

	Turns ratio (pri : sec)	Primary volts (rms)	Secondary volts (rms)	Primary current (A)	Secondary current (A)
1	10:1	120	---	0.1	---
2	6:1	---	20	---	0.6
3	1:5	30	---	---	0.2
4	---	24	96	---	1
5	---	---	27	0.2	0.6
6	1:7	17	---	---	.03
7	2:5	10	---	0.3	---
8	3:4	---	160	---	1.3
9	4.5:2	---	53.3	0.5	---
10	7:6	28	---	---	0.21

Power relationships in a transformer

If a transformer is considered to have no losses (losses do exist in an actual transformer) then, there is a 1:1 relationship of power in the primary to power in the secondary. That is, power in the primary is equal to power in the secondary. If the primary is considered as the input and the secondary is considered as the output, $P_{in} = P_{out}$. As an example, some of the practice problems from a previous section are shown below.

Primary volts = 120 V, Secondary volts = 12 V
Primary current = 0.1 A, Secondary current = 1? A

$$P = I \times E \qquad (2\text{-}2)$$

$$P_{pri} = 0.1 \times 120$$

$$P_{pri} = 12 \text{ W}$$

$$P_{sec} = 1 \times 12$$

$$P_{sec} = 12 \text{ W}$$

As this example demonstrates, the power in the primary equals the power in the secondary when there are no transformer losses. In fact, it is from this power relationship that the inverse current relationship can be demonstrated.

$$P_{pri} = P_{sec}$$

$$I_p E_p = I_s E_s$$

$$\frac{E_p}{E_s} = \frac{I_s}{I_p}$$

Transformer efficiency

Because no device can operate without losses, nothing is 100 percent efficient. Transformer windings cannot have perfect coupling, the core material cannot be a perfect conductor of magnetic lines of force, and the wire cannot be a perfect conductor of electricity. In other electrical subject areas, the losses are usually small enough to be ignored. However, in transformers, if there is significant power transfer, there can be significant losses.

Transformer losses usually result in heat, which is a direct loss of power. Efficiency is stated in terms of percent. The closer the efficiency is to 100 percent, the better the system or the smallest amount of losses.

Efficiency of any system is calculated by:

$$\text{Eff} = \frac{P_{out}}{P_{in}} \times 100\% \qquad (6\text{-}4)$$

The formula states that power out is always less than or equal to power in, never larger.

Example

$$P_{out} = 300 \text{ W}, P_{in} = 400 \text{ W}$$

$$\text{Eff} = \frac{P_{out}}{P_{in}} \times 100\%$$

$$= \frac{300}{400} \times 100\%$$

$$= 75\%$$

Along with transformer efficiency is the application of voltage and current in a normal, working transformer. When a transformer has no load connected to the secondary, no current is being drawn from either the secondary or the primary. With no load, the transformer output voltage is at its maximum value. This is the equivalent of 100 percent efficiency. Because there is no current being drawn, there is no power.

As the transformer load current increases, the current flowing in the windings will cause I^2R drops in the wire to increase, resulting in a power loss. Also, current increase in the secondary causes a current increase in the primary, which results in I^2R losses in the primary, reducing its power. Current flowing in the windings also increase the core losses, further decreasing the efficiency.

When a transformer is in actual operation, the main concern is of the output voltage. This discussion on transformer efficiency is given to prove that a transformer output voltage will decrease as more current is drawn from the secondary. The following key point.

- The output voltage of a transformer decreases as the load current is increased.

Testing a transformer with an ohmmeter

A transformer can be thought of as two coils of wire; the primary and the secondary. Some transformers have a multi-tap secondary, which means there is more than one secondary winding.

Consider the primary first. Remember, an ohmmeter must be used in a circuit with no voltage applied. Some transformers will have a primary with a tap for connecting to a higher voltage. Consider using the ohmmeter on the primary without a tap first.

The ohmmeter reads the dc resistance of the winding, which is nothing more than a continuous piece of wire. If the coil is normal, it will read a dc resistance of between 1 and 200 Ω, depending on the coil. The most likely trouble that will happen to a transformer winding is an open coil. With a condition of an open, the ohmmeter will read infinite resistance. If the primary winding has a tap, it will have three wires coming out. The tap is usually at a point of approximately one-half the total resistance of the full coil.

The secondary is usually made with taps from the same coil. Often, the secondary will have several completely different coils. The individual coils will have to be tested with the ohmmeter on an individual basis. If the secondary is different coils, one faulty coil does not affect any other coil.

If a primary coil is open, there will be no voltage induced in the secondary.

Chapter summary

Transformers have a magnetic coupling between the primary winding and the secondary winding. An ac voltage applied to the primary winding will induce a voltage in the secondary winding, resulting in a transfer of power from the primary to the secondary.

An actual transformer will have an efficiency of less than 100 percent because of the losses in the transformer construction.

Note the following key point.

- The output voltage of a transformer decreases as the load increases.

Summary of formulas

$$\frac{N_p}{N_s} \quad \text{transformer turns ratio} \tag{6-1}$$

$$\frac{N_p}{N_s} = \frac{V_p}{V_s} \quad \text{voltage ratio} \tag{6-2}$$

$$\frac{N_p}{N_s} = \frac{I_s}{I_p} \quad \text{current ratio} \tag{6-3}$$

$$\text{Eff} = \frac{P_{\text{out}}}{P_{\text{in}}} \times 100\% \quad \text{efficiency in a transformer} \tag{6-4}$$

7
Inductors and capacitors

ALONG WITH RESISTORS, INDUCTORS AND CAPACITORS ARE THE THREE most basic components in the study of electricity and electronics. Inductors and capacitors are important because of their capability of storing electrical energy, called a *charge*.

These components have very useful properties in both dc and ac circuits. Inductance is the principle by which a transformer works. The coil in the ignition system of an automobile is an inductor that stores energy to fire the spark plug. Capacitors make filters for use with a power supply, fire the electronic flash unit on a camera, and are used to form timing circuits.

Properties of inductance

Whenever electrical current is flowing in a wire, there will be a magnetic field developed around the wire. *Inductance* is the ability of a conductor to use the magnetic field developed to induce a voltage in itself. The only time a coil can develop voltage within itself is when there is a changing current flow in the conductor. This changing current does not need to be an ac current flow; dc current will have the same effect if the current flow is changing. The voltage that is developed in an inductor is opposing the applied voltage and is called *back emf* (electromotive force).

- Back emf is developed in a coil of wire only if there is changing current in the coil.

When current first starts to flow, it must build the magnetic field in the inductor. This process of trying to build the magnetic field causes the back emf to oppose this changing current. Then, when the current is decreasing, the magnetic field that has been charged will discharge in the same direction as the current flow, adding to the current, which has the effect of opposing the decrease of current.

- Inductance is the characteristic that opposes any change in current.

Like resistors, it is possible to connect inductors in series or parallel in order to obtain a resultant, total inductance, similar to the idea of a total resistance.

Although inductors could be used for various applications in dc circuits, inductors serve an important function when used with ac circuits, especially when used with a sine wave. Inductors can be thought of when used as a transformer, but a single coil of wire can be used to form an induction electric motor. Also, the property of a magnetic field being developed around any coil of wire can also be used to make an electromagnet.

To demonstrate the strength an electromagnet can have when a large enough coil is used with a large enough current (remember it is current flow that develops the magnetic field), consider the following example. Large electromagnets are connected to the end of cable and are used to lift very large pieces of metal, such as cars. A dc voltage is applied to the coil, and the coil would be rather large. As long as the current is flowing, the magnetic field will hold solid; the moment power to the coil is removed, the magnetic field will collapse.

The example shown above is on a large scale, but it does serve to demonstrate the ability of the magnetic field that can be developed in a coil of wire. In electronics however, the size of the inductance is usually quite small.

The unit for inductance is the *henry* (H). A medium-size transformer or a filter choke, which is a coil used in power supply filters, will have inductance in the range of 1 or 2 H. An inductor used as part of a low-frequency resonant circuit will be about 100 mH. A high-frequency coil would be in the range of 1 or 2 μH. A high-frequency coil will have a small number of henries.

Inductors in series

More than one coil can be connected in series (Fig. 7-1) to produce different values of inductance. It is possible, however, to have the two coils located near each other to cause the magnetic field of one coil to interact with the field of the other coil. When this happens, it produces some transformer action, which is called *mutual inductance*. Unless the mutual inductance is in a transformer, mutual inductance is undesirable.

The equation for calculating two or more inductors in series assumes no mutual inductance.

$$L_T = L_1 + L_2 + L_3 \ldots \text{etc.} \tag{7-1}$$

Mutual inductance

As with transformers, mutual inductance is helpful in producing an end result that is not available from any single coil. This is found in the secondary of the transformer when it is intended for coils to be connected with either series-aiding or

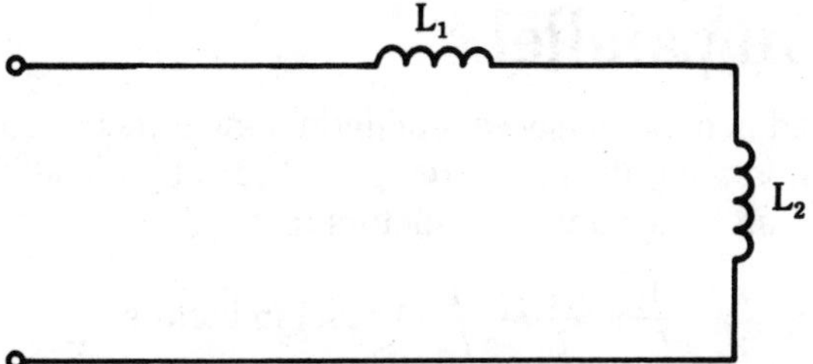

7-1 Two inductors in series. No mutual inductance.

series-opposing field to have a *phasing dot* on the schematic diagram. Figure 7-2 shows coils wound on a core material (this is not a schematic diagram) showing the direction of the windings. The large dots just above and to one side of the coil are the phasing dots. Another way of thinking about the application of the phasing dots is to consider the sine wave with its positive half cycle and negative half cycle. When one phasing dot is positive, the phasing dot on the neighboring coil will also be positive. After drawing in the instantaneous polarity of the coils, think of them in terms of series-aiding or series-opposing batteries.

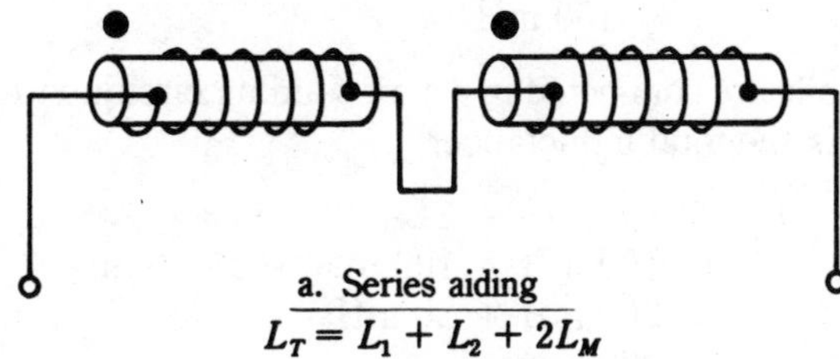

a. Series aiding

$L_T = L_1 + L_2 + 2L_M$

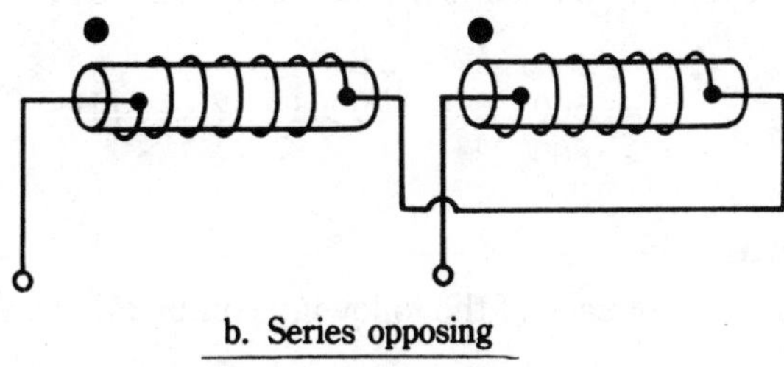

b. Series opposing

$L_T = L_1 + L_2 - 2L_M$

7-2 Inductors connected in series, using phasing dots to indicate polarity.

The formula for calculating total inductance when the coils are in series with mutual inductance is:

$$L_T = L_1 + L_2 + 2L_M \qquad \text{series-aiding mutual inductance} \qquad (7\text{-}2)$$

$$L_T = L_1 + L_2 - 2L_M \qquad \text{series-opposing mutual inductance} \qquad (7\text{-}3)$$

Inductors in parallel

Inductors connected in parallel have a total inductance calculated in the same manner as resistors in parallel, using the reciprocal formula. The short-cut formulas used in resistance can be used with inductors in parallel.

$$\frac{1}{L_T} = \frac{1}{L_1} + \frac{1}{L_2} \text{ parallel inductors} \qquad (7\text{-}4)$$

It is possible to have inductors connected in parallel with mutual inductance. However, because of the inverse relationships, the calculations become quite complicated, and the predictions of the end result make a parallel connection almost never used in actual practice.

Example problems

What is the equivalent inductance of three inductors connected in series with no mutual inductance? The values of the inductors are; 25 mH, 50 mH, 75 mH.

$$\begin{aligned} L_T &= L_1 + L_2 + L_3 \qquad (7\text{-}1) \\ &= 25\text{ mH} + 50\text{ mH} + 75\text{ mH} \\ &= 150\text{ mH} \end{aligned}$$

Two 100 mH coils are connected in a series-aiding circuit with a 10 mH mutual inductance. What is the total inductance?

$$\begin{aligned} L_T &= L_1 + L_2 + 2L_M \qquad (7\text{-}2) \\ &= 100\text{ mH} + 100\text{ mH} + 2 \times 10\text{ mH} \\ &= 200\text{ mH} + 20\text{ mH} \\ &= 220\text{ mH} \end{aligned}$$

A 500 μH inductor is connected in a series-opposing circuit with a 400 μH inductor with 50 μH mutual inductance. What is the effective inductance?

$$\begin{aligned} L_T &= L_1 + L_2 - 2L_M \\ &= 500\ \mu\text{H} + 400\ \mu\text{H} - 2 \times 10\ \mu\text{H} \\ &= 800\ \mu\text{H} \end{aligned}$$

Practice problems

Find the total inductance of each of the following combinations. Assume $L_M = 0$ if no value is given.

1. Series circuit; 500 μH, 700 μH
2. Series circuit; 10 mH, 50 mH, 30 mH
3. Series circuit; 100 mH, 0.2 H, 0.15 H, L_M = 10 mH, aiding
4. Series circuit; 1 H, 1.5 H, 1200 mH, L_M = 25 mH, opposing
5. Parallel circuit; 100 mH, 200 mH

6. $L_M = 0$

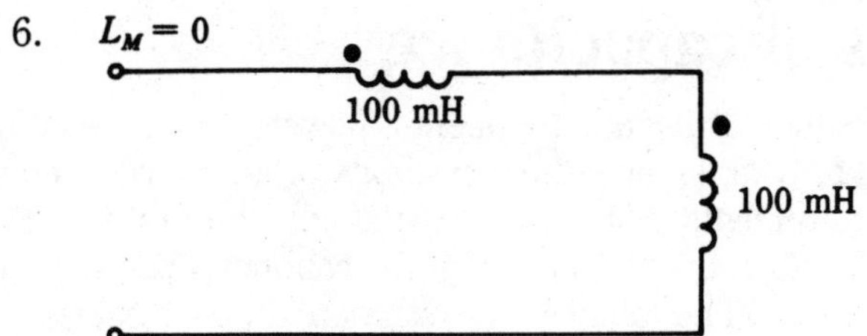

7. $L_M = 10$ mH

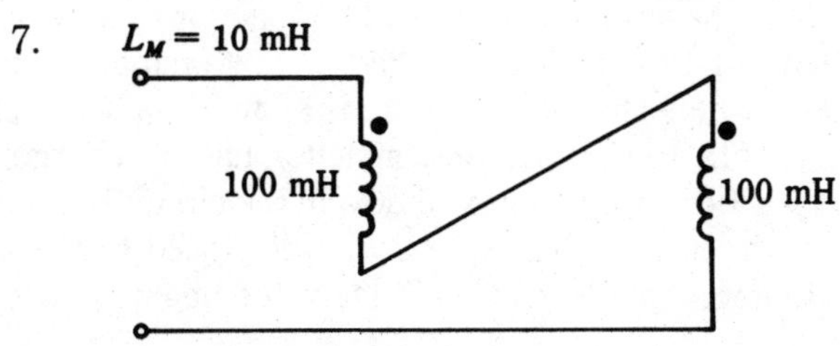

8. $L_M = 10$ mH

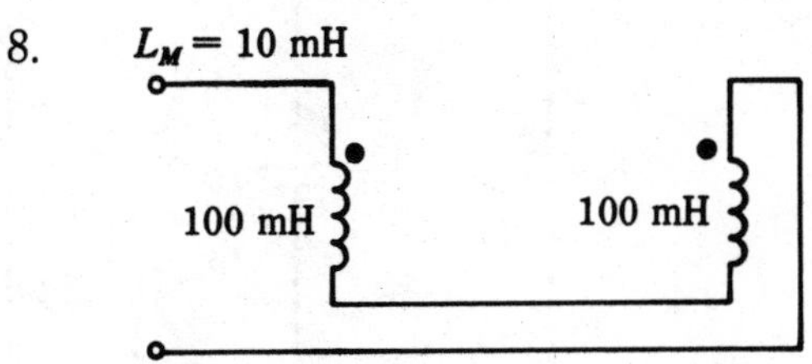

9. $L_M = 10$ mH

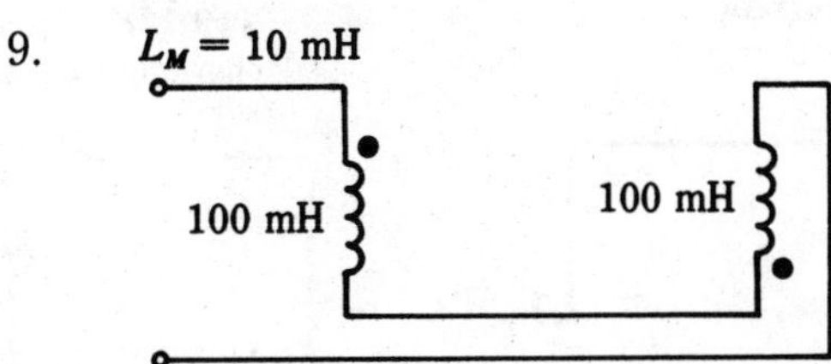

10. $L_M = 10$ mH

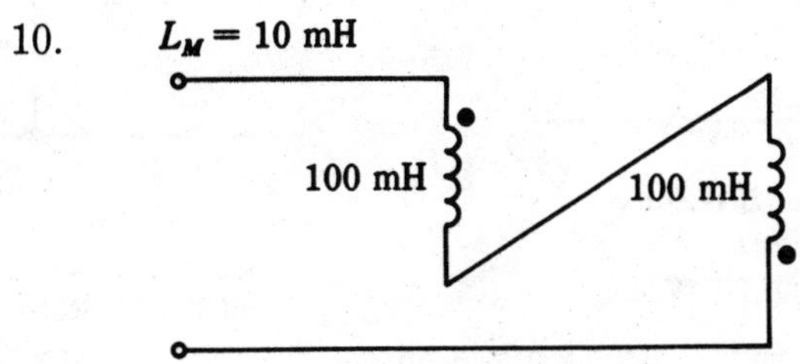

Properties of capacitance

Capacitors are frequently used components; however, many electronics students find it difficult to fully grasp how the capacitor works. In order to form a capacitor there are two requirements that must be satisfied. The first is that there must be two separate conductors, the second is that the conductors be separated by an insulator, called a *dielectric*. The following description of how a charge is stored is simplified, but it should help you better understand the capacitor.

When a dc voltage is applied to a capacitor, the capacitor will take on a charge with a voltage equal to the applied voltage. The capacitor requires a period of time to charge. (The charging time is discussed in the next chapter on time constants.) For this discussion, assume that enough time is allowed for a full charge.

Refer to Fig. 7-3. In Fig. 7-3A, the capacitor has no charge across it. The switch is in the off position, no current will flow in the circuit because there is not a complete current path. In Fig. 7-3B, the switch is closed to the a position. The capacitor will start charging. Electrons will leave the negative side of the battery,

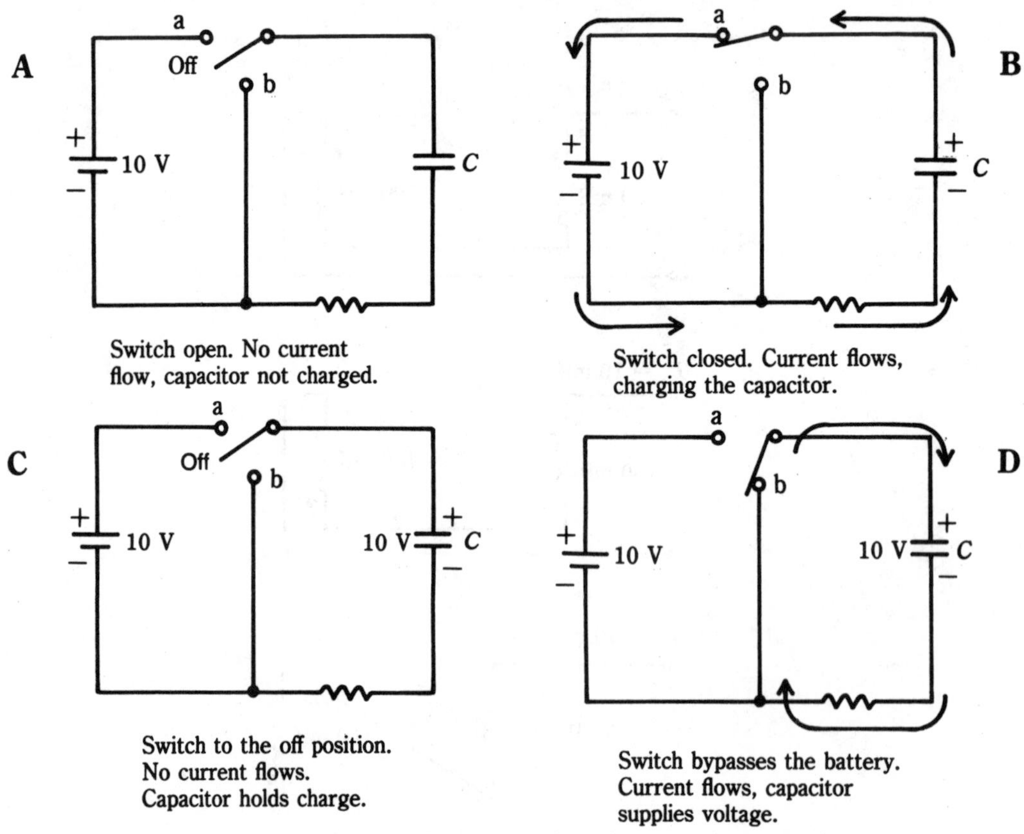

7-3 A dc circuit showing how a capacitor charges and stores voltage.

travel to the capacitor to the plate marked negative. This plate will allow the electrons to gather. The opposite plate, marked positive, will allow electrons to leave the plate and travel to the battery. For every electron that leaves the negative side of the battery and stores on the negative plate of the capacitor, there will be an equal number that will vacate the positive side of the capacitor and return to the battery. The battery considers this situation to be a closed circuit. The current will continue to flow until the capacitor has stored enough electrons on the negative terminal and removed enough electrons from the positive to have a difference in potential between the two plates equal to the battery—in this circuit, it is 10 V. When the capacitor has stored the 10 V, current will stop flowing in the circuit, even though the switch remains closed.

In Fig. 7-3C, the switch is returned to the off position. The capacitor had a full charge, and this charge will remain across the capacitor because there is no path for current to flow to allow the electrons to balance the difference in potential between the two plates. In actual practice, the dielectric, the insulator between the plates, is not a perfect insulator, no matter what it is made of. Because it is not a perfect insulator, some electrons can travel through the insulator and eventually discharge the capacitor. This is called *leakage*, and is generally so low that it does not cause problems.

In Fig. 7-3D, the switch is moved to the b position. In this position, the capacitor has a current path. Note the battery is removed from the circuit, and current through the resistor is in the opposite direction that it was in Fig. 7-3B. Current is now supplied by the capacitor. Its starting voltage was 10 V; however, as electrons leave the negative terminal and return to the positive plate, the potential difference across the plates is lowered, neutralized. Current will continue to flow, although the force drops off, until the capacitor is fully discharged.

Because a capacitor stores a voltage, it is the property of a capacitor to try to maintain a constant voltage in the circuit. A circuit voltage cannot change quickly because it must wait for the capacitor to charge. Notice also that current can flow in a dc circuit only while the capacitor is charging (or discharging).

- The property of a capacitor is to oppose any change in voltage.
- When a capacitor is in a dc circuit, current flows only when the capacitor is charging (or discharging).

The unit of capacitance is the *farad*. The farad is a measure of the ability of a capacitor to store a charge. The two factors that affect the ability of a capacitor to store a charge are the plate area and the dielectric thickness. The larger the plate area, the more charge can be stored. The closer the plates (thinner dielectric) the stronger is the electric field developed resulting in a larger charge.

The unit farad is much too large of a unit to use in an actual circuit. The most common unit size is the microfarad (μF), and the next most common is the picofarad. The microfarad is 10^{-6} farads, and the picofarad is 10^{-12} farads. Capacitors are almost always measured in one of these two unit sizes.

Capacitors in parallel

When capacitors are connected in parallel, the resultant, effective capacitance is the addition of the individual capacitors.

$$C_T = C_1 + C_2 + C_3 \quad \text{parallel capacitance} \qquad (7\text{-}5)$$

In Fig. 7-4, two capacitors are shown in parallel. By connecting the capacitors in parallel, the effective plate area is increased, thus increasing the capacitance value. Because it is a parallel connection, voltage is the same across both capacitors; therefore, they will charge to the same voltage. If the capacitance values were different, the stored voltage would be the same, but the stored number of electrons would be different. The smaller capacitor would store a smaller number of electrons, which simply means the voltage that is stored will discharge quicker.

When capacitors are connected in parallel (or series) to make a new equivalent capacitance, the capacitors should be of the same type. For example: two electrolytic, two disc, or two mica. The reason you should try to match the types is current-storage capacity. A comparison to this might be connecting a large automotive battery in series with a small flashlight battery. The small battery could be destroyed by the current capacity of the larger battery.

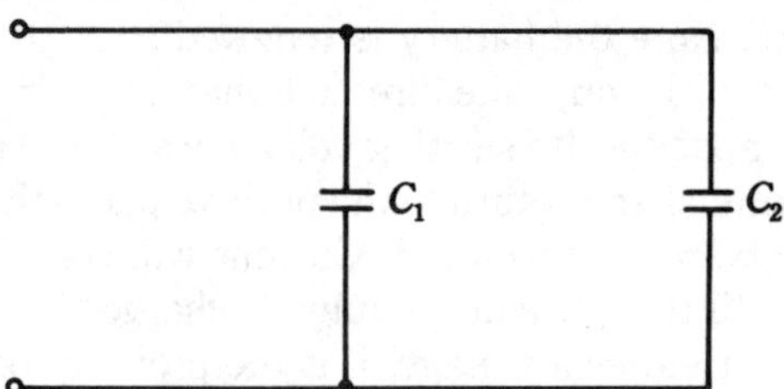

7-4 Two capacitors connected in parallel.

Capacitors in series

When capacitors are connected in series, the effect is to increase the thickness of the dielectric, the insulator between the plates, resulting in a lowered value of capacitance. The reciprocal formula is used to calculate the effective capacitance.

$$\frac{1}{C_T} = \frac{1}{C_1} + \frac{1}{C_2} \ldots \text{etc.} \quad \text{series capacitance} \qquad (7\text{-}6)$$

Any of the short-cut formulas apply to the reciprocal formula, regardless whether it is used for parallel resistors, parallel inductors, or series capacitors.

Often, it is difficult to visualize the current path in a circuit with series capacitors. For every electron that leaves the negative terminal of the battery and travels to the negative plate of the first capacitor, an equal number of electrons will depart the positive plate of the positive terminal of the first capacitor are collected on the negative terminal of the series capacitor. Then an equal number of electrons will leave the positive terminal of that capacitor and return to the power supply. Because an equal number of electrons that left the battery from the negative terminal return to the positive terminal, there is a complete current path. Even though it appears that the wire between the two capacitors receives no current, it is important to note that in a series circuit, current is the same throughout the series circuit. Figure 7-5 shows two capacitors in series.

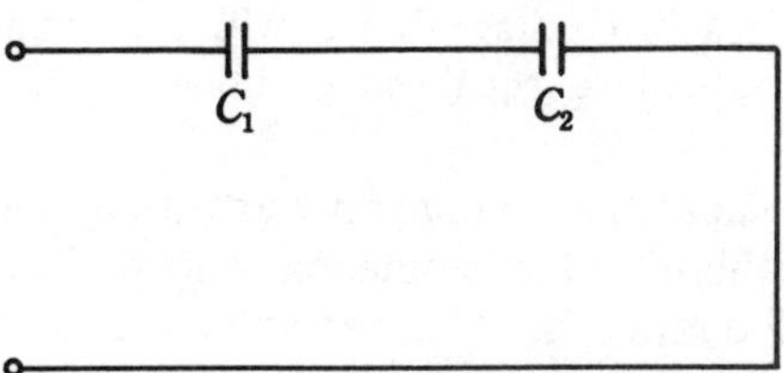

7-5 Two capacitors connected in series.

Capacitive voltage dividers

One very useful function of capacitors in series is the capacitive voltage divider. Voltage drops in any voltage divider is determined by the ratio of the components, not the applied voltage or the current flow.

$$V_{C1} = \frac{C_2}{C_1 + C_2} \times V \qquad \text{voltage divider formula for finding the voltage across } C_1 \tag{7-7}$$

$$V_{C2} = \frac{C_1}{C_1 + C_2} \times V \qquad \text{voltage divider formula for finding the voltage across } C_2 \tag{7-8}$$

The two formulas shown above are simply a modification of the resistive voltage divider formula. As with any voltage divider, it is important to notice the ratio is based on the ratio of the individual component values to the total value, found by adding the components values. This ratio is then multiplied by the applied voltage.

Figure 7-6 shows a capacitive voltage divider with two equal components, capacitive values. The voltage drops across each should be the same.

Substitute values and prove.

$$V_{C1} = \frac{C_2}{C_1 + C_2} \times V \tag{7-7}$$

$$= \frac{1\ \mu\text{F}}{1\ \mu\text{F} + 1\ \mu\text{F}} \times 100\ \text{V}$$

$$= \frac{1}{2} \times 100$$

$$= 50\ \text{V}$$

So if $V_{C1} = 50$ V (voltage across C_1), then $V_{C2} = 50$ V.

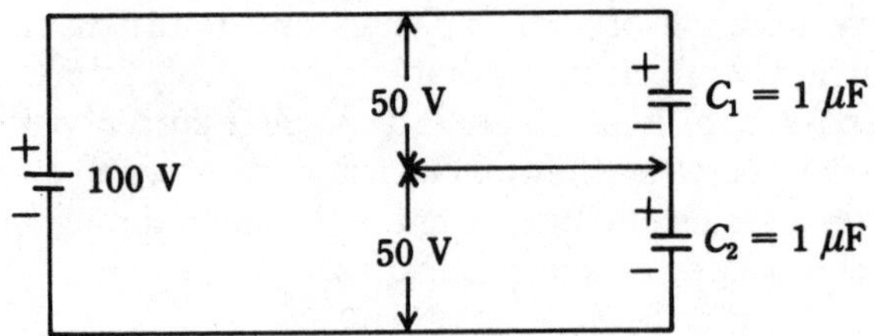

7-6 Capacitive voltage divider with equal values of capacitance, equal voltage drops.

Figure 7-7 shows a capacitive voltage divider with unequal capacitance values. It stands to reason that there should be unequal voltage drops, but the big question when trying to set up the ratios is which capacitor will get the highest voltage.

- In a capacitive voltage divider, the greater value capacitor gets the least voltage.
- In a capacitive voltage divider, the least value capacitor gets the greatest voltage.

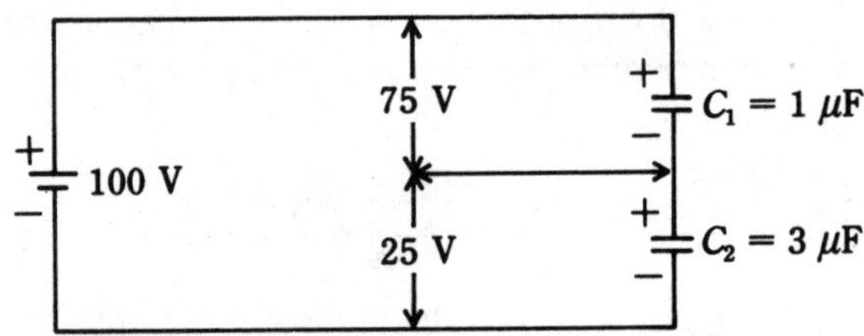

7-7 Capacitive voltage divider with unequal values of capacitance. Larger capacitor, smaller voltage.

Find V_{C1} for Fig. 7-7.

$$
\begin{aligned}
V_{C1} &= \frac{C_2}{C_1 + C_2} \times V \\
&= \frac{3\ \mu F}{1\ \mu F + 3\ \mu F} \times 100 \\
&= \frac{3}{4} \times 100 \\
&= 75\ V
\end{aligned}
$$

Find V_{C2} for Fig. 7-7.

$$
\begin{aligned}
V_{C2} &= \frac{C_1}{C_1 + C_2} \times V \\
&= \frac{1\ \mu F}{1\ \mu F + 3\ \mu F} \times 100
\end{aligned}
$$

$$= \frac{1}{4} \times 100$$

$$= 25 \text{ V}$$

Practice problems

In questions 1 through 5, find the total capacitance.

1. Parallel circuit: 2 μF, 3 μF
2. Parallel circuit: 5 pF. 0.001 μF, 10 pF
3. Parallel circuit: 0.01 μF, 0.001 μF, 0.22 μF
4. Parallel circuit: 47 μF, 47 μF, 47 μF
5. Parallel circuit: 100 μF, 100 μF, 100 μF, 100 μF, 100 μF

Questions 6 through 10 are all series circuit voltage dividers. Given are the capacitor values and the supply voltage. Find the total capacitance and voltage drops across each capacitor.

6. $C_1 = 1\ \mu\text{F}, C_2 = 2\ \mu\text{F}, V = 10\ \text{V}$
7. $C_1 = 5\ \text{pF}, C_2 = -10\ \text{pF}, V = 10\ \text{V}$
8. $C_1 = 0.01\ \mu\text{F}, C_2 = 0.1\ \mu\text{F}, V = 5\text{V}$
9. $C_1 = 10\ \mu, C_2 = 10\ \mu\text{F}, C_3 = 10\ \mu\text{F}, V = 15\ \text{V}$
10. $C_1 = 0.1\ \mu\text{F}, C_2 = 0.2\ \mu\text{F}, V = 20\ \text{V}$

Chapter summary

Inductors and capacitors have properties that make them valuable in certain applications. This chapter was intended to discuss the hows and whys of these properties, not to discuss the applications. In later chapters, some of the applications are discussed.

Although inductors and capacitors are often thought of as being useful only to ac circuits, dc circuits provide an understandable discussion on how they work.

- Back emf is developed in a coil of wire if there is changing current in the coil.
- Inductance is the characteristic that opposes any change in current.
- The property of a capacitor is to oppose any change in voltage.
- When a capacitor is in a dc circuit, current flows only when the capacitor is charging (or discharging).
- In a capacitive voltage divider, the higher-value capacitor gets the smallest voltage drop.
- In a capacitive voltage divider, the lower-value capacitor gets the largest voltage drop.

Summary of formulas

$$L_T = L_1 + L_2 + L_3 \ldots \text{etc.} \quad \text{series inductors} \tag{7-1}$$

$$L_T = L_1 + L_2 + 2L_M \quad \text{series-aiding mutual inductance} \tag{7-2}$$

$$L_T = L_1 + L_2 - 2L_M \quad \text{series-opposing mutual inductance} \tag{7-3}$$

$$L_T = \frac{1}{L_1} + \frac{1}{L_2} \quad \text{parallel inductors} \tag{7-4}$$

$$C_T = C_1 + C_2 + C_3 \quad \text{parallel capacitors} \tag{7-5}$$

$$\frac{1}{C_T} = \frac{1}{C_1} + \frac{1}{C_2} \ldots \text{etc.} \quad \text{series capacitors} \tag{7-6}$$

$$V_{C1} = \frac{C_2}{C_1 + C_2} \times V \quad \text{voltage-divider formula for the the voltage across } C_1 \tag{7-7}$$

$$V_{C2} = \frac{C_1}{C_1 + C_2} \times V \quad \text{voltage-divider formula for the voltage across } C_2 \tag{7-8}$$

8
Time constants

WHEN A VOLTAGE IS APPLIED TO AN INDUCTOR OR A CAPACITOR, A PERIOD of charge is required. In fact, during the first few moments that the voltage is applied, or removed, the building, or collapsing of the charge will always follow a curve with the same shape. This is called the *universal time constant curve.*

The time involved in the charging or discharging of an inductor or capacitor is called a time constant. The time constant finds a great many applications in electronic circuitry.

Universal time constant curve

The *time constant* is a mathematical formula used to calculate the time required to reach a percentage of full charge or discharge. Time constant formulas will be shown later.

- A time constant is defined as the length of time, required to reach 63 percent of full charge or discharge.
- Five time constants is defined as the time required to reach full charge or discharge.

The universal time constant curve is actually made up of two curves (refer to Fig. 8-1). The bottom axis of the curve is the number of time constants, with 5 time constants indicating a full charge or discharge. The vertical axis on the left side is labeled in percentage of full charge. On the discharge curve, the percentage of full charge actually indicates the charge still remaining after a period of discharge.

The purpose of the universal time constant curve is to predict the voltage of either charge or discharge for any time constant between 0 and 5 time constants, even for decimal time constants. The curve provides a solution by graphical analysis,

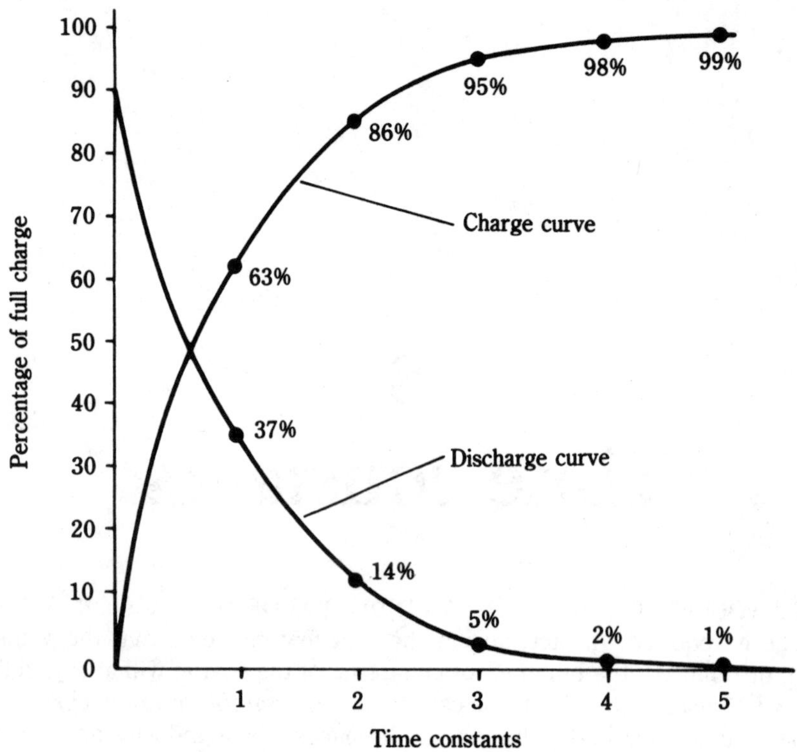

8-1 Universal time constant curve.

which although it is somewhat inaccurate, is often accurate enough and actually shows the charging/discharging action.

Plotting the charge curve

There are two ways to plot the curves. The first method is to use the 63 percent as the definition of a time constant. This is an approximation. The second method of plotting the curve is through the use of the time constant equation.

The following is the calculations to plot the charge curve using 63 percent as one time constant. The results are shown in Table 8-1.

Starting from a time of 0, the charging starts toward 100 percent of the full charge.

Step 1 At the first time constant, 63 percent of the full charge has been completed, leaving 37 percent.

Step 2 After the second time constant, 63 percent of the 37 percent remaining from step 1 will be completed ($0.63 \times 0.37 = 0.23$). This means 23 percent of the full charge is completed during this step. Adding to what has been done so far, 63 percent + 23 percent = 86 percent completed. Four percent remains of the full charge.

Step 3 The third time constant will charge 63 percent of the remaining portion ($0.63 \times 0.14 = 0.09$). This means that 9 percent of the full charge is completed during this step. Adding to what has been done so far, 63 percent + 23 percent + 9 percent = 95 percent completed; 5 percent remains.

Step 4 The fourth time constant will charge 63 percent of the remaining portion. ($0.63 \times 0.05 = 0.03$). This step provides 3 percent of the full charge. Adding to what has been done so far; 63 percent + 23 percent + 9 percent + 3 percent = 98 percent completed; 2 percent remains.

Step 5 The fifth time constant will charge 63 percent of the remaining portion. ($0.63 \times 0.02 = 0.01$). This step provides 1 percent of the full charge. Adding to what has been done so far; 63 percent + 23 percent + 9 percent + 3 percent + 1 percent = 99 percent completed; 1 percent (actually less because of rounding) remains to be fully charged.

Note that if each time constant only does 63 percent of what remains, the unit will never reach a full charge. This makes sense since an actual component, regardless of how good it is will have some losses. Therefore, 5 time constants is considered to be as close to a full charge as is possible. Any further attempt at charging will only overcome the losses.

The next set of calculations is based on the formula for calculating the instantaneous charge.

$$\% \text{ charge} = 1 - e^{-T} \times 100\% \quad \text{instantaneous charge} \tag{8-1}$$

e is the natural log
T is the time constant

Formula 8-1 is called the instantaneous charge formula because it gives the exact value anywhere along the charge curve. It does not have to be used at the whole number time constants, but can be used at any point, for example; 2.25 T. The capital letter T is used to indicate the time constant. The formula uses a $-T$. When the calculator is used to calculate the e^{-T} a decimal value will result which will be the equivalent of the percentage of the charge remaining. By subtracting this from 1, the value will then be the decimal of the percentage charged during that time. Multiplying by 100 converts the decimal to a percent.

Table 8-1. Plotting the universal time constant curve.

Time constant	Charging 63%	Charging % = $1-e^{-T}$	Discharging 63%	Discharging % = e^{-T}
0	0	0	100%	100%
1	63	63.2	37	36.8
2	86	86.5	14	13.5
3	95	95.0	5	5.0
4	98	98.2	2	1.8
5	99	99.3	1	0.67

The results of the following calculations are in Table 8-1. The calculations shown are for the instantaneous charge, calculated at the whole number time constants.

Note e^{-T} is shown on the calculator as e^x; simply replace the e^x with the proper value of $-T$.

Calculate the first time constant.

$$\begin{aligned}\% &= 1 - e^{-1} \times 100\% \\ &= 1 - 0.368 \times 100\% \\ &= 63.2\% \text{ of full charge at the first time constant}\end{aligned}$$

Calculate the second time constant.

$$\begin{aligned}\% &= 1 - e^{-2} \times 100\% \\ &= 1 - 0.135 \times 100\% \\ &= 86.5\% \text{ of full charge at the second time constant}\end{aligned}$$

Calculate the third time constant.

$$\begin{aligned}\% &= 1 - e^{-3} \times 100\% \\ &= 1 - 0.0498 \times 100\% \\ &= 95\% \text{ of full charge at the third time constant}\end{aligned}$$

Calculate the fourth time constant.

$$\begin{aligned}\% &= 1 - e^{-4} \times 100\% \\ &= 1 - 0.0183 \times 100\% \\ &= 98.2\% \text{ of full charge at the fourth time constant}\end{aligned}$$

Calculate the fifth time constant.

$$\begin{aligned}\% &= 1 - e^{-5} \times 100\% \\ &= 1 - 0.0067 \times 100\% \\ &= 99.3\% \text{ of full charge at the fifth time constant}\end{aligned}$$

Comparing the results of the calculations in Table 8-1, recognize that even though the equation using the natural log, e^{-T}, is a more accurate calculation, the approximation method of 63 percent is close. Therefore, the definition of one time constant being 63 percent is a good approximation.

Plotting the discharge curve

In plotting the discharge curve portion of the universal time constant curve, there are two ways to find the location of the whole-number time constants on the curve. The first method is to use the time constant equals 63 percent. The second method is to use the following formula.

$$\%\text{ of full charge} = e^{-T} \times 100\% \qquad \text{instantaneous discharge} \tag{8-2}$$

e is the natural log
T is the time constant

Formula 8-2 will give the exact instantaneous value of the discharge for any point along the curve. Notice it is stated in terms of the percent of full charge.

Calculate using the method of the 63 percent time constant.

Time zero of the discharge curve is located at the maximum charge point, 100 percent.

Step 1 Starting from 100 percent, discharge the first time constant brings the curve to 100 − 63 = 37 percent of the full charge.

Step 2 Starting the second time constant from the point left off after the first time constant, discharge 63 percent of what is remaining. (0.63 × 0.37 − 0.23). The second time constant discharges 23 percent, which means the point of the discharge curve is now 100 − 63 = 37. 37 − 23 = 14 percent of full charge remaining.

Step 3 Time constant 3 will discharge 63 percent of the 14 percent remaining. (0.63 × 0.14 = 0.09). The third time constant discharges 9 percent. 14 − 9 = 5 percent of full charge remains.

Step 4 The fourth time constant will discharge the remaining portion 63 percent. (0.63 × 0.05 = 0.03). The fourth time constant discharges 3 percent of the total. 5 − 3 = 2 percent remaining of the original full charge.

Step 5 The fifth time constant will discharge the remaining portion 63 percent. (0.63 × 0.02 − 0.01). The fifth time constant discharges 1 percent of the total. 2 − 1 = 1 percent remaining of the full charge after 5 time constants.

Formula 8-2 is used to calculate the plotting of the discharge curve in order to determine the exact values of each point, because the 63 percent method is only an approximation. The results of these calculations are found in Table 8-1.

Calculate the first time constant.

$$\begin{aligned}\text{\% of full charge} &= e^{-1} \times 100\% \\ &= 0.368 \times 100\% \\ &= 36.8\% \text{ remains of full charge} \\ &\qquad \text{after one time constant}\end{aligned}$$

Calculate the second time constant.

$$\begin{aligned}\% &= e^{-2} \times 100\% \\ &= 1.35 \times 100\% \\ &= 13.5\% \text{ remains after two time constants}\end{aligned}$$

Calculate the third time constant.

$$\begin{aligned}\% &= e^{-3} \times 100\% \\ &= 0.0498 \times 100\% \\ &= 5.0\% \text{ remains after three time constants}\end{aligned}$$

Calculate the fourth time constant.

$$\begin{aligned} \% &= e^{-4} \times 100\% \\ &= 0.0183 \times 100\% \\ &= 1.8\% \text{ remains after four time constants} \end{aligned}$$

Calculate the fifth time constant.

$$\begin{aligned} \% &= e^{-5} \times 100\% \\ &= 0.0067 \times 100\% \\ &= 0.67\% \text{ remains of full charge} \\ &\qquad \text{after five time constants} \end{aligned}$$

The calculations shown above are all used to plot the universal time constant curve. It is a valid curve for any circuit, regardless of the component values. However, the standard means of calculating time constants involves the application of actual component values for a particular circuit.

L/R time constant

Whenever an inductor has electricity applied to it, there will be a magnetic field developed around the coil. The magnetic field does not charge instantly, but rather takes a period of time based on the value of the inductor and the amount of resistance that is in the circuit. The same also applies when the electricity is removed, the inductor will discharge through the path provided for discharging. The length of time involved will depend on the value of the inductor and the resistance in the circuit.

Even though the actual length of time is different for each different circuit, one time constant is still 63 percent of full charge and it takes five time constants to either charge or discharge.

The formula for calculating the time constant of an inductive circuit is:

$$T = \frac{L}{R} \quad \text{inductive time constant} \tag{8-3}$$

Formula 8-3 gives the time constant for an inductor with series resistance. Time is in seconds if resistance is in ohms and inductance is in henries.

Refer to Fig. 8-2. Whenever analyzing a time constant circuit, trace the current path for charge and discharge.

The charge path of the inductor is from the battery negative terminal, through the inductor, through the resistor, and returning to the positive side of the battery through the switch in the top position. When charged, the inductor will have an effective polarity as shown with the plus at the top.

The discharge path is always opposite the charge path, as far as the inductor is concerned. The discharge path starts at the inductor, on the bottom side, through the center path to the bottom side of the switch, through the resistor and returning to the inductor.

The calculations for the charge and discharge time constants for this circuit are exactly the same because the value of inductance and resistance are the same for both charge and discharge.

$$T = \frac{L}{R} \tag{8-3}$$

$$= \frac{0.2\ \text{H}}{1000\ \Omega}$$

$$= 0.0002 \text{ seconds}$$

$$= 0.2 \text{ ms—the value of one time constant}$$

One time constant equals 0.2 ms; therefore full charge is reached 5 × 0.2 ms (5 time constants) = 1 ms for full charge. The discharge time constant is exactly the same for this circuit.

The analysis of Fig. 8-2 is not complete with just calculating the value of time constant. The property of an inductor is to build a magnetic field by having current flow through the windings. The inductor opposes any change in current, and this accounts for the time constant. Because the magnetic field cannot build or collapse instantly if the power is instantly removed from the inductor, the magnetic field will try to maintain the same current through its discharge path.

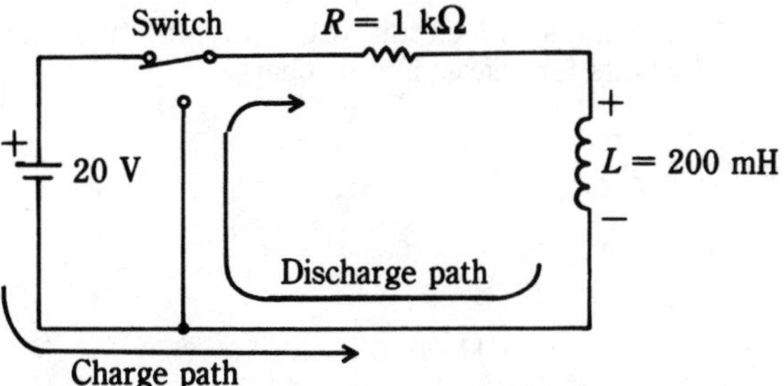

8-2 A circuit to demonstrate the *L*/*R* time constant.

In Fig. 8-2, the inductor is represented as having zero ohms coil resistance, with the resistor being a separate component. Current in the charge path, therefore, is calculated by using Ohm's Law:

$$I = \frac{E}{R} \tag{2-1A}$$

$$= \frac{20\ \text{V}}{1000\ \Omega}$$

$$= 20 \text{ mA—current after full charge}$$

The current calculated here is after full charge is reached because prior to that the inductor is opposing. A 20 mA current is flowing after full charge is reached, as long as the battery is connected. When the switch is changed to the discharge position, the inductor will try to maintain the same current. The voltage across the inductor will switch polarity, allowing the current to discharge the inductor in the reverse direction. If the resistor is considered the load, the current through the resistor will develop a voltage drop which can be calculated.

The voltage developed across the resistor at the very start of discharge will be equal to the battery voltage because the current is the same and the same resistor is used for charge and discharge. As the magnetic field decays, as calculated with the time constants, the current will also decrease at the same rate. Because the current decreases, the voltage developed across the resistor will also decrease.

Figure 8-3 is an inductive circuit that has different resistances in the charge and discharge paths. This circuit will have two different time constants and the voltage developed across the discharge resistor should be higher. Calculate time constants first.

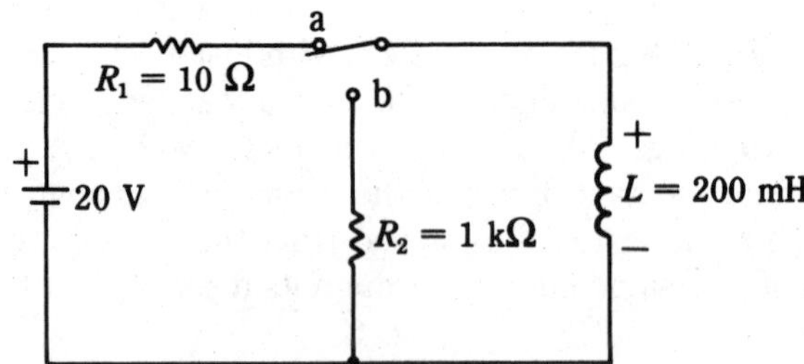

8-3 An inductive circuit with different time constants for charge and discharge.

Charge time constant

$$T = \frac{L}{R} \tag{8-3}$$

$$= \frac{0.2\ \text{H}}{10\ \Omega}$$

$$= 0.02 \text{ seconds}$$

$$= 20 \text{ ms—time constant for charge}$$

$$5 \times T = 100 \text{ ms (for full charge)}$$

Discharge time constant

$$T = \frac{0.2\ \text{H}}{1000\ \Omega}$$

$$= 0.0002 \text{ seconds}$$

$$= 0.2 \text{ ms—time constant for discharge}$$

$$5 \times T = 1 \text{ ms (full discharge)}$$

For the calculations involved here, assume the switch is in the proper position for a period of time much greater than is required for five time constants. In other words, assume the inductor will reach full charge and discharge. Next, calculate the charge current.

$$I = \frac{E}{R} \qquad (2\text{-}1A)$$

$$= \frac{20\ V}{10\ \Omega}$$

$$= 2\ A$$

When the switch is moved to position b, the battery will be disconnected and the inductor will be allowed to discharge through resistor R_2. Not only does the discharge path have a shorter time constant, but notice what happens with the voltage developed across the resistor at the instant the switch changes positions. The current flowing at the moment the switch changes will continue to flow because of the back emf in the inductor caused by the magnetic field.

$$E = I \times R$$
$$= 2\ A \times 1000\ \Omega$$
$$= 2{,}000\ V \text{ (voltage developed across the resistor at the moment of switch change)}$$

The 2,000 V developed across the resistor is an instantaneous value at the very start of discharge. If it is desired to determine the developed voltage at any other point in time, use the universal time constant curve to determine the instantaneous current during discharge and the instantaneous voltage can then be calculated.

High voltage produced by opening an RL circuit

Refer to Fig. 8-4. This circuit represents the ignition system of an automobile. The switch is the points that open and close to allow the coil to charge and then discharge through the spark plug. R_1 represents the dc resistance of the coil windings. R_2 represents the resistance of the spark plug gap. This resistance value is purely fictitious since the spark plug is actually an open circuit. Some value is needed to be given in the discharge path, therefore, this example and 250 kΩ to represent an open circuit.

The important calculation here is the voltage developed across the spark plug when the switch allows the inductor to discharge.

Current at full charge must first be calculated and then, using the charge current, calculate the voltage across the spark plug at the first moment of discharge.

Charge current.

$$I = \frac{E}{R} \qquad (2\text{-}1A)$$

$$= \frac{120\ V}{100\ \Omega}$$

$$= 0.12\ A \text{ full charge current}$$

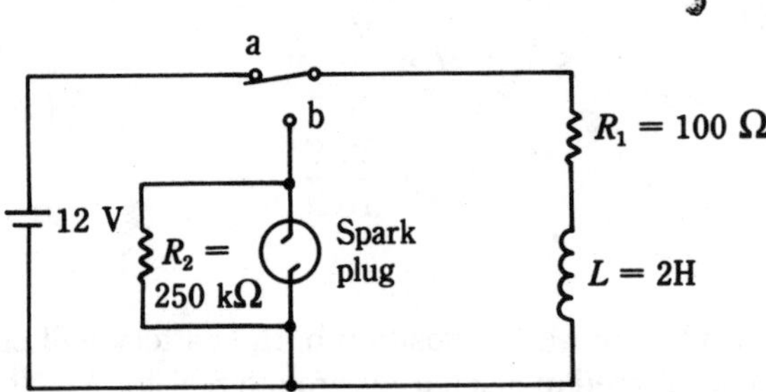

8-4 High voltage developed by opening an *LR* circuit.

Discharge voltage

$$E = I \times R \text{ (formula for voltage during discharge)} \quad (2\text{-}1)$$
$$= 0.12\text{ A} \times 250{,}000\ \Omega$$
$$= 30{,}000\text{ V (voltage developed at the instant of the start of discharge)}$$

The calculation of the voltage developed across a spark plug are somewhat simplified here, but the principle of operation is accurate. It takes an extremely large voltage to cause the spark to jump the gap of a spark plug. The modern electronic ignition systems work much the same with the major difference being the points (the switch) has been replaced with an electronic switch that does not wear out.

When large motors are used in a circuit, it is necessary to provide protection for the switch. The electric motor is a very large inductor. When the circuit is turned off, there must be a discharge path provided, or the switch will become the discharge path and a spark will result across the contacts of the switch. This spark usually causes considerable damage to the switch.

RC time constant

Whenever the term *time constant* is mentioned, the first thing that should come to mind is the capacitive circuit. The reason for this is that there are many more applications for the capacitive time constant circuit than there are for the inductive circuit. This does not in any way diminish the practical inductive circuits; it simply means capacitors are more common.

The property of a capacitor is to store a charge that can be said to be the same as storing a voltage. Because a capacitor stores a charge of voltage it will oppose any change in voltage. What this means is that the capacitor must try to store a charge equal to the change in voltage, and this cannot be done instantly. The length of time involved in the storage of the charge is called the time constant of the RC circuit.

The formula for calculating the time constant of a capacitive circuit is:

$$T = R \times C \quad \text{capacitive time constant} \quad (8\text{-}4)$$

Formula 8-4 gives the time constant for a capacitor with its series resistance. Time is in seconds, resistance is in ohms, and capacitance is in farads.

Remember, when using any formula, be sure to keep in mind that the formula will have the correct final units if you use the correct units in the calculations. Notice, in the RC time constant formula, the capacitance is in farads. Actual capacitor values are seldom, if ever, in farads but rather in microfarads or picofarads.

Figure 8-5 shows a circuit that demonstrates the charge and discharge paths for a sample RC time constant. The charge path is from the negative side of the battery, to the bottom, negative plate of the capacitor. For every electron that leaves the negative terminal of the battery and collects on the negative terminal of the capacitor, an equal number of electrons leave the positive (top) plate of the capacitor, travel through the 100 Ω resistor and return to the positive terminal of the battery. This causes the capacitor to charge by having a storage of electrons on the negative plate and a shortage of electrons on the positive plate. The storage of electrons creates an electric field and is a difference in potential, which is the definition of voltage. Therefore, the capacitor is storing voltage.

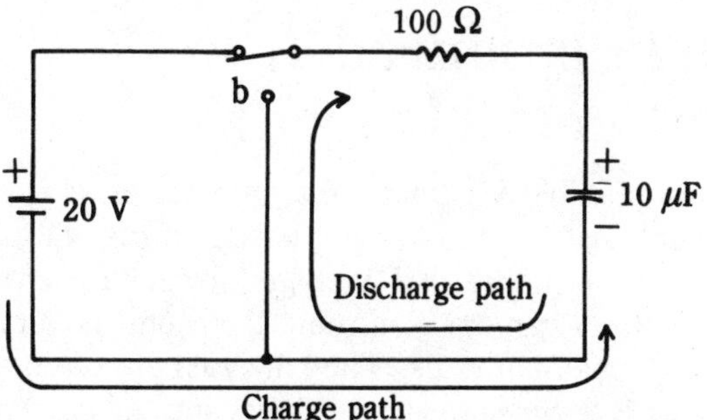

8-5 A circuit to demonstrate the *RC* time constant.

The discharge path is when the switch is placed in the b position. The battery is then disconnected and the capacitor is allowed to neutralize the difference in potential of stored electrons. The discharge path is from the negative plate of the capacitor, through the center path, to the b contact of the switch, through the resistor, and replacing the missing electrons on the positive plate.

When calculating the time constant for the RC circuit, both the charge path and discharge path have the same values of capacitance and resistance. Therefore, the time constant for both charge and discharge are the same.

$$\begin{aligned} T &= R \times C \qquad (8\text{-}4) \\ &= 100\ \Omega \times 10\ \mu\text{F} \\ &= 1000\ \mu\text{s} \\ &= 1\ \text{ms} \\ 5 \times 1\ \text{ms} &= 5\ \text{ms for a full charge} \end{aligned}$$

When the capacitor reaches full charge after five time constants, the voltage will be 20 V across the capacitor. Once the capacitor is charged to full value, current can no longer flow in the circuit because the battery and the capacitor will have no difference in potential. If the battery remains connected after full charge is reached, even though it is said to have zero current flow, an actual capacitor will have a very slight leakage through the dielectric. The battery will have a very slight leakage current to replace any lost charge. Another point to make at full charge is that because no current flows (leakage current is so small, it is usually considered zero) then there cannot be any voltage developed across the resistor.

When the switch is placed in position b the capacitor will be allowed to discharge. At the first moment of discharge, the voltage developed across the resistor will be equal to the capacitor voltage, because it is a simple series circuit. At the first moment of discharge, the voltage is at its highest value and the current through the resistor, known as the instantaneous discharge current, can be calculated using Ohm's Law.

When examining the inductor and L/R time constants, a different discharge was used than the charge path to generate a very high voltage. The capacitor circuit and RC time constants can be used to also have a different discharge path. The difference will be that the capacitor will produce a high current.

High current produced at first instant of discharge

Figure 8-6 shows a possible schematic diagram for an electronic flash unit in a camera. Light bulbs have a light intensity based on the wattage. There are three different ways to develop the needed wattage (power); raise the voltage, raise the current, or raise both. A camera electronic flash unit is very light and portable because it uses very small batteries. The batteries are often the AAA type. This type of battery has 1.5 V per battery and a very limited current capacity. It would be impossible to use a battery of this type to ignite a very high intensity flash bulb very many times without the capacitor.

Figure 8-6 shows the charge path for the capacitor through a 100 Ω resistors, to charge the 250 μF capacitor. This gives a time constant for charge of 25 ms. The reason the resistor is needed is to limit the charge current. When the switch is moved to position b the capacitor will charge current. When the switch is moved to position b, the capacitor will discharge through the light bulb, shown as having 1 Ω resistance. This gives a discharge time constant of 250 μs. Keep in mind the flash on a camera is very fast.

When the capacitor is fully charged, it will have a voltage of 3 V across it. At the time of discharge, the 3 V stored in the capacitor will be seen across the 1 Ω flash bulb. Instantaneous discharge current will then be 3 A. Flash bulbs are made in such a way as to produce a high intensity light with this current.

Chapter summary

The universal time constant curve is a very useful tool to graphically show the charge or discharge of an inductive or capacitive circuit. Note that 63 percent of full charge or discharge represents 1 time constant. It takes five time constants to reach full charge or full discharge.

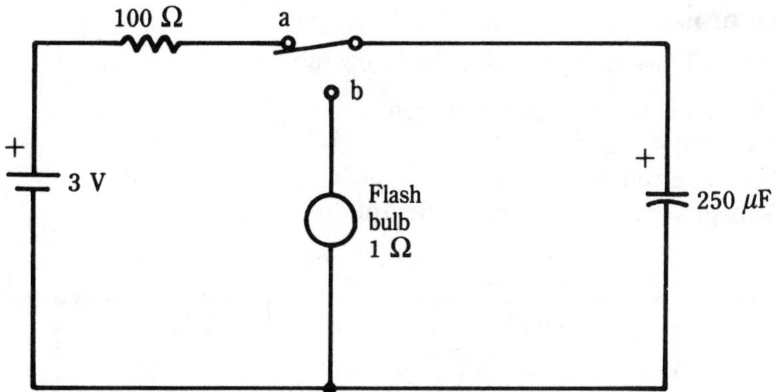

8-6 Schematic for an electronic flash unit in a camera.

When an inductor is placed in series with a resistor, there will be a length of time for the inductor to reach full charge or discharge; this is the time constant. An inductor needs to build the magnetic field by current flowing in the turns. An inductor opposes any change in current. At full charge, the inductor will have current limited by the circuit resistance.

At the first moment the inductor is caused to discharge, the current will be equal to the charge current. This current will flow through whatever discharge path is provided. If the discharge path has a very high resistance, there will be a very high voltage developed at the first moment of discharge.

When a capacitor is placed in a series with a resistor, there will be a length of time required for the capacitor to charge or discharge, this is the time constant. A capacitor needs to store electrons, and therefore a voltage. At full charge, the dc current will stop flowing, except for a very small leakage current. A capacitor opposes any change in voltage. At full charge the capacitor will store the applied voltage.

At the first moment the capacitor is discharged, the voltage applied to the discharge path will equal the last voltage stored in the capacitor. If the discharge path has a very low resistance, there will be a very high current during discharge through the resistor.

Summary of formulas

$$\text{\% of full charge} = 1 - e^{-T} \times 100\% \quad \text{instantaneous charge} \tag{8-1}$$

$$\text{\% of full charge remaining} = e^{-T} \times 100\% \quad \text{instantaneous discharge} \tag{8-2}$$

$$T = \frac{L}{R} \quad \text{time constant for an inductor in series with a resistor} \tag{8-3}$$

$$T = R \times C \quad \text{time constant for a capacitor in series with a resistor} \tag{8-4}$$

Practice problems

In circuits 1–5, use the schematic diagrams to find:

a. One time constant during charge
b. Full charge current
c. One time constant during discharge
d. Voltage developed at first instant of discharge

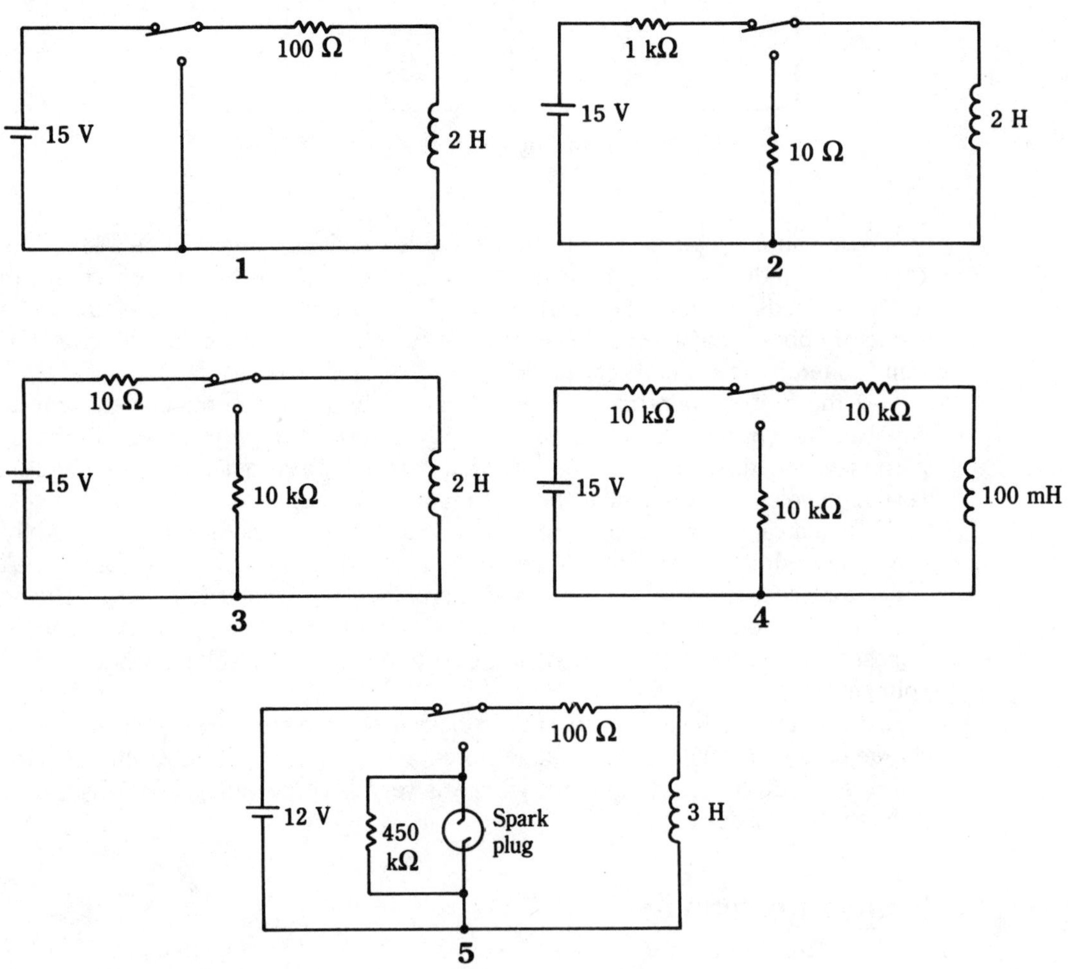

In circuits 6–10, use the schematic diagrams to find:

a. One time constant during charge
b. Full charge voltage
c. One time constant during discharge
d. Current developed at first instant of discharge

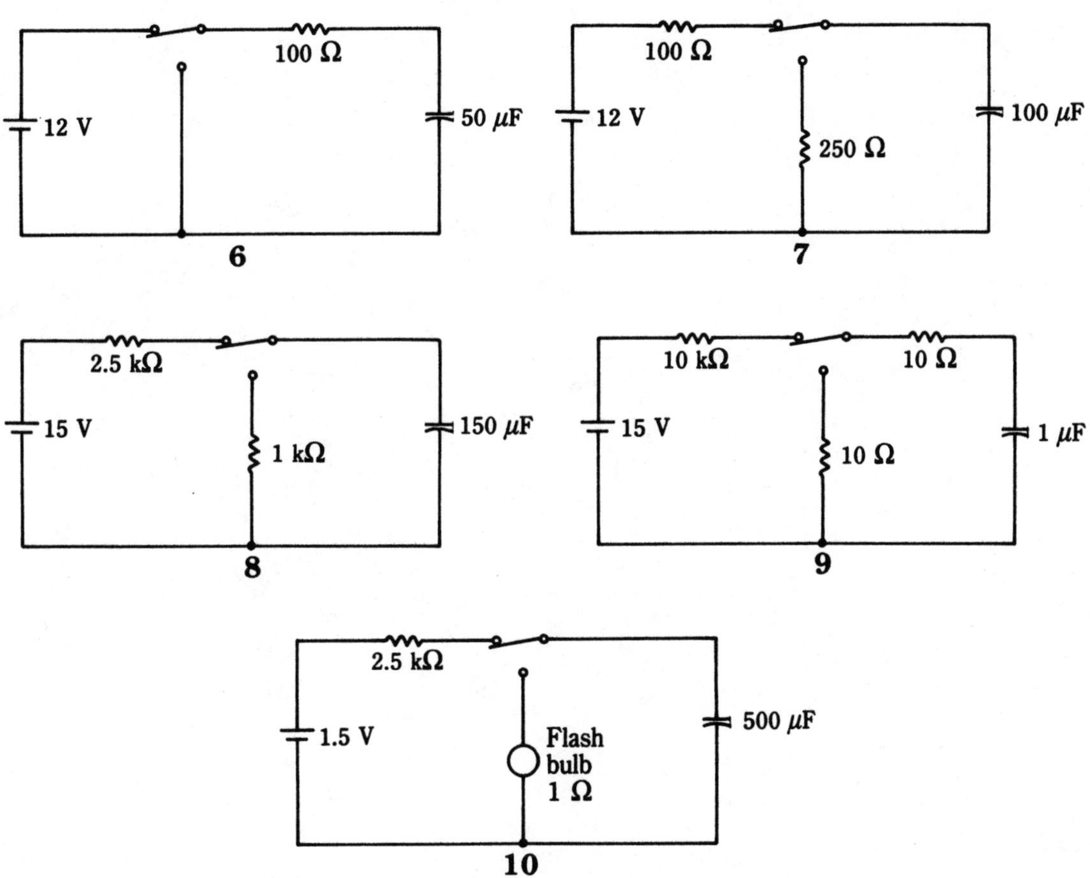
100 Ω
12 V
50 μF
6
100 Ω
12 V
250 Ω
100 μF
7
2.5 kΩ
15 V
1 kΩ
150 μF
8
10 kΩ
10 Ω
15 V
10 Ω
1 μF
9
2.5 kΩ
1.5 V
Flash
bulb
1 Ω
500 μF
10

9
RC waveshaping

WAVESHAPING CIRCUITS ARE USEFUL APPLICATIONS OF THE BASIC resistor-capacitor time constant. The greatest application is passing a square wave through the series circuit, with an output taken in parallel with the resistor. The result is a negative spike. The negative spike has direct applications in digital circuits that require a negative trigger.

Differentiation/integration

Differentiation and *integration* are two words used to describe the RC time constant circuit. This is the main topic of the chapter.

Compare the circuits shown in Figs. 9-1 and 9-2. Figure 9-1 is a differentiator circuit. It can be identified by where V_{out} is taken in the circuit. When the output is taken across, in parallel with the resistor, the circuit is called a differentiator. It is not necessary to use a square wave input, but it best demonstrates the function of the circuit.

The RC combination is usually selected for a short time constant when it is to be used as a differentiator. The primary advantage is through the use of the spike seen across the resistor. Because the output is taken directly across the resistor, the output voltage will always be exactly the same as the voltage across the resistor. A resistor allows the voltage to change instantaneously.

Figure 9-2 shows an integrator circuit. The output is taken directly across the capacitor and will, therefore, follow the charge and discharge of the capacitor. The integrator circuit will usually use a medium to long time constant to allow a more pronounced curve of the charge and discharge of the capacitor.

Figures 9-3, 9-4, and 9-5 show circuits with their output waveforms (V_c). Notice that the output can be taken either across the capacitor or across the resistor

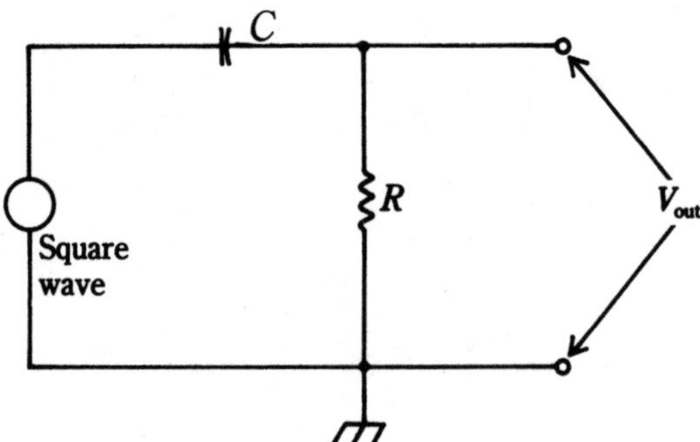

9-1 Differentiator. V_{out} across the resistor.

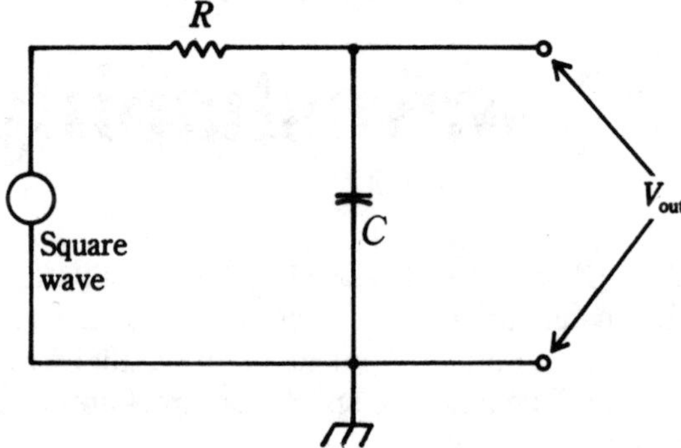

9-2 Integrator. V_{out} across the capacitor.

(V_R) and the resultant waveform is shown. The circuits are drawn this way to simplify the drawings and explanation. Keep in mind if a differentiator is desired, the output is across the resistor. If an integrator is desired, the output is across the capacitor.

A square wave is used in each of the four time constant circuits because it is a very effective way to use a battery that switches on and off at a precise frequency. The square wave used here is a 1 kHz wave. The square wave used is entirely positive. That means the waveform goes from zero to a peak positive voltage and returns to zero but does not go negative at all. The advantage of using this type of voltage is that it acts like a dc source that is switched on and off at a specified rate. During the time that the input is switched to 0 V, it is considered the off time. It will be treated as a short circuit to allow the capacitor a path to discharge.

Short time constant

Figure 9-3 shows a short time constant circuit. It is classified as a short time constant because the time for five time constants, full charge and discharge time, is shorter than the time allowed for charge and discharge. In other words, the on and off times of the square wave are each 0.5 ms long. One time constant for this circuit

$T = R \times C$

1 time constant = 0.02 ms
5 time constants = 0.1 ms

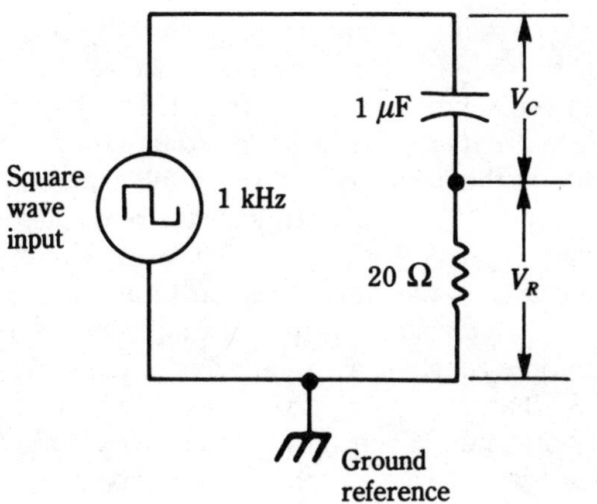

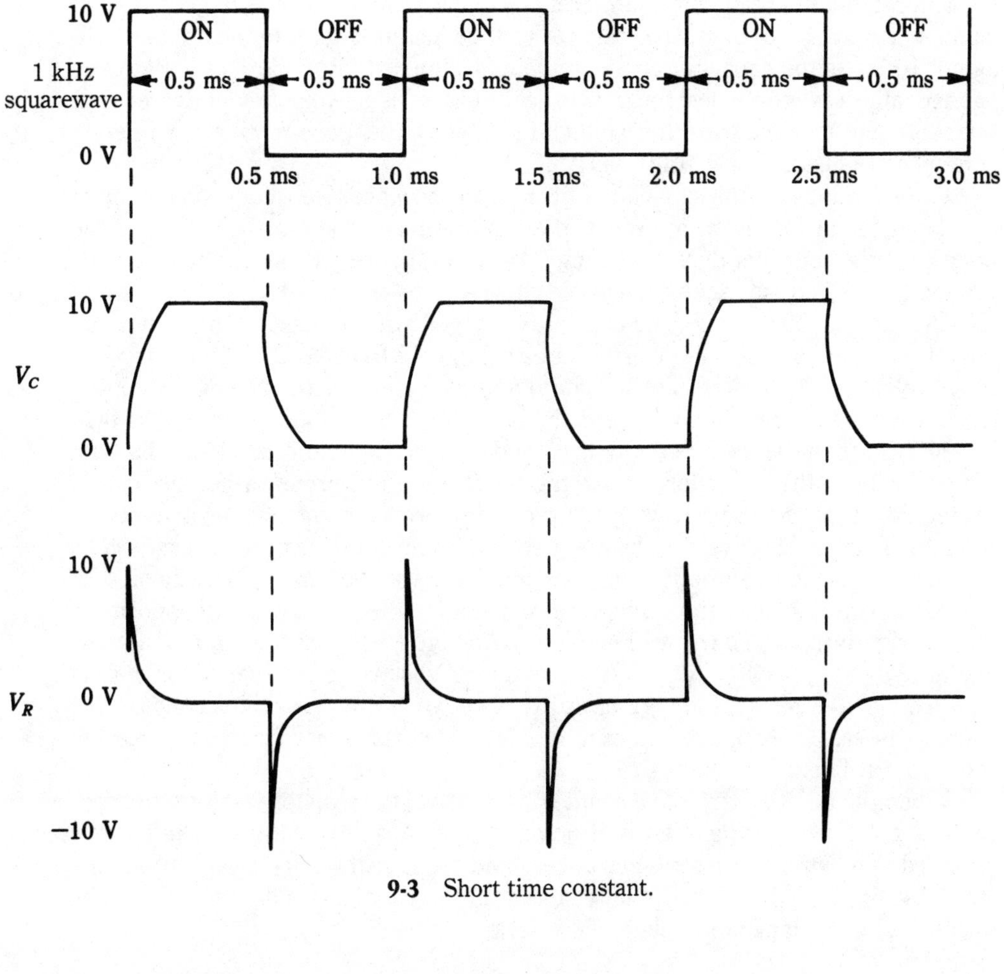

9-3 Short time constant.

is 0.02 ms which, makes five time constants 0.1 ms. The on and off times are five times longer than what is needed to reach full charge or discharge.

The curve of V_C follows the universal time constant curve used in chapter 8. At the start of the on time, the capacitor starts its charge cycle through the 20 Ω resistor. The time constant is so short that the capacitor will reach full charge in short time. Once it reaches full charge, it will stay at full charge for the remainder of the on position of the input square wave.

When the input square wave comes to the start of the off time, the capacitor will start its discharge, through the 20 Ω resistor. The power supply will have what appears to be a short circuit. The time constant with the value of resistance is so short, the capacitor reaches full discharge in a very short time. It will remain discharged for the remainder of the off time. When the input changes again to the positive voltage, the capacitor will again charge and the cycle is repeated.

The output taken across the resistor, V_R, provides a unique output waveform, especially with the short time constant circuit. Because the input is a square wave, the voltage across the resistor is always whatever voltage is not across the capacitor, when the input is during the on time. At the very start of the square wave, the capacitor is not charged and will therefore be 0 V. The resistor voltage is able to change instantly; therefore, the resistive voltage instantly goes to the applied voltage of 10 V. As the capacitor charges, the capacitor will drop more of the applied voltage, and less will be left for the resistor. The resistor voltage will decrease at the same rate that the capacitor voltage increases. The resistor voltage will go to zero when the capacitor is fully charged.

Keep in mind that the capacitor will charge with a positive voltage on the top of the capacitor in the drawing. When the input voltage goes to the off time, the polarity of the capacitor does not change. However, the relationship it had with the ground reference point does change.

During the charge time, the capacitor charged with a positive on the top and negative on the bottom. The resistor had a voltage with the positive on the top and the negative on the bottom. The ground reference point was then a negative polarity. Everything in the circuit, except the capacitor can change polarity instantly. Therefore, when the power supply is turned off, the capacitor acts like a battery. The polarity of the capacitor is with the positive on the ground reference point, during the discharge time. Having the ground reference point as a positive makes the polarity across the resistor instantly switch to the opposite of the charge cycle.

In reference to the ground reference point, the voltage across the resistor, during the off time, or discharge time of the capacitor, will have a polarity opposite to the polarity during the on time. The resistor voltage will instantly jump to the voltage across the capacitor. At the first moment of discharge, the capacitor is fully charged and the resistor voltage jumps to a negative 10 V. As the capacitor discharges, the resistor voltage will exactly follow the capacitor voltage and return to the zero volt line.

When the input square wave again goes positive, the resistor voltage will again jump to positive 10 V; then it will drop off as the capacitor charges. The cycle is repeated. The input wave switches to zero and the resistor voltage jumps to negative 10 V because this is the voltage stored on the capacitor. The resistor voltage decreases as the capacitor voltage decreases.

Medium time constant

Figure 9-4 shows a circuit with a medium time constant. It is classified as a medium time constant because five time constants of the circuit is equal to the on and off times of the input waveform.

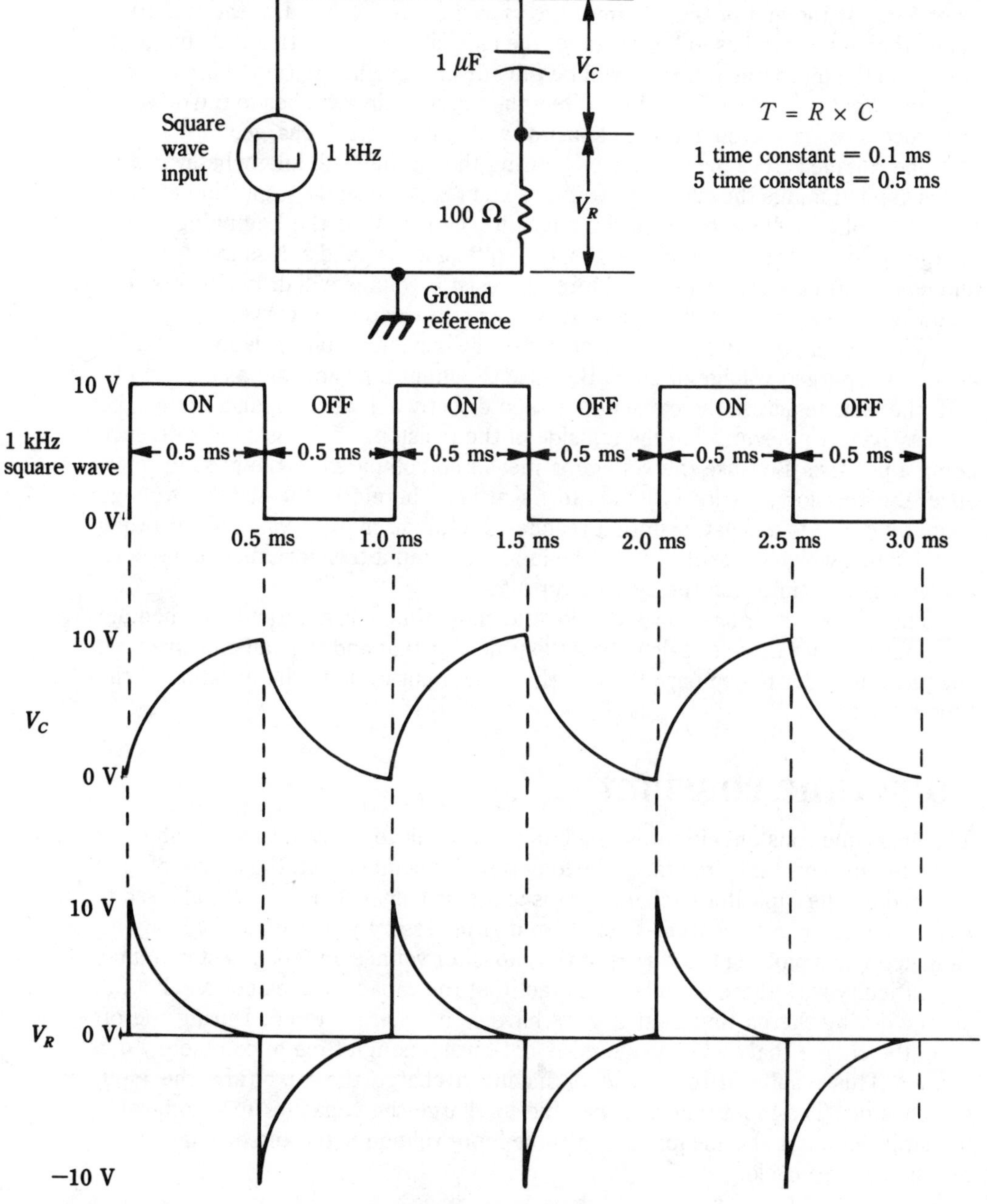

9-4 Medium time constant.

The medium time constant circuit has a waveform that displays a universal time constant curve. The waveform is useful when the positive or negative spikes need to be a longer duration than the short time constant would allow.

The voltage across the capacitor, V_C, starts at 0 V and when the input switches to the positive 10 V, the capacitor will begin the charging process. The input is on for 0.5 ms, which is the same time required for the capacitor to reach a full charge. Therefore, at the end of the on time, the capacitor has charged to the full 10 V. When the input switches to the off time, the capacitor is still with a full charge. It will begin the discharge time and will be fully discharged in a period of 0.5 ms, the time the input is in the off position. When the input again switches to the on time, the process repeats itself and the capacitor will again start to charge.

The voltage across the resistor, V_R, during the on time will take whatever voltage is applied minus the capacitor voltage. The resistor voltage jumps instantly to the full applied voltage because the capacitor is at 0 V at the beginning of the charge cycle. As the capacitor charges, the voltage is dropped across the capacitor and less is left for the resistor, therefore the resistor voltage will drop towards zero following the exponential curve, exactly opposite the capacitor curve.

When the input switches to the off time, the capacitor voltage is momentarily at the full charged voltage of 10 V. Because the input now appears as a short circuit, the voltage across the capacitor is seen directly across the resistor, with the polarity having a negative on the top side of the resistor and the ground reference being a positive. Because the voltage is measured compared to the ground reference, the resistor polarity switches to negative. Therefore, the resistor voltage instantly jumps to a negative until it reaches the full discharged value of zero volts after 0.5 ms, which is the off time of the input. The input then switches to the positive voltage, on time, and the cycle is repeated.

The only calculations necessary for the short time constant and the medium time constant are those of calculating the time constant and the value of five time constants in order to compare the length of the input cycle to the duration of the time constants.

Long time constant

The long time constant circuit is much more difficult to analyze than the short or medium time constant circuits. In the long time constant circuit, the length of time required for the capacitor to charge or discharge is longer than the time allowed to either charge or to discharge. Therefore, it is necessary to calculate each significant step when plotting the curves of the capacitor voltage and the resistor voltage the difficulty with these circuits is the fact that the capacitor does not reach a full charge during the on time, but it does have some charge stored. During the off time, the charge that has been stored is not given enough time to completely discharge. This results in the capacitor having a charge the next time the input switches on. The input will then be able to charge the capacitor to a somewhat higher level, but it also means there will be more voltage to try and discharge during the next off cycle.

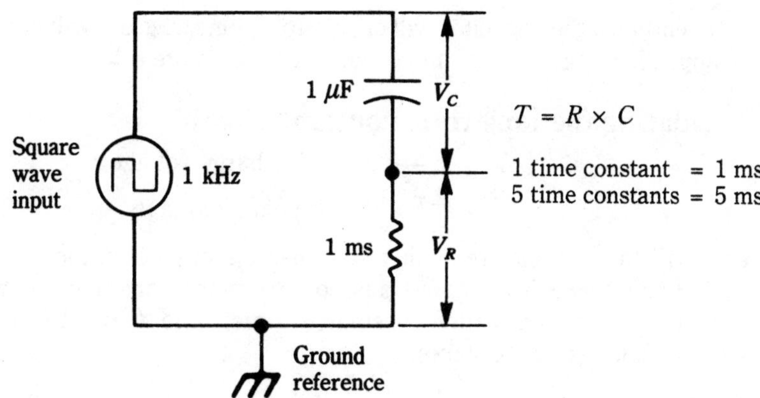

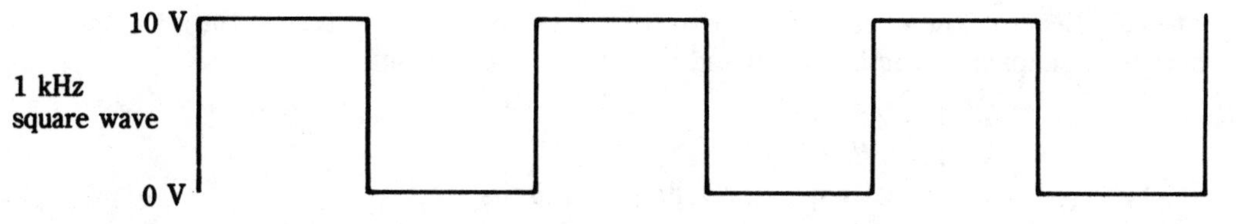

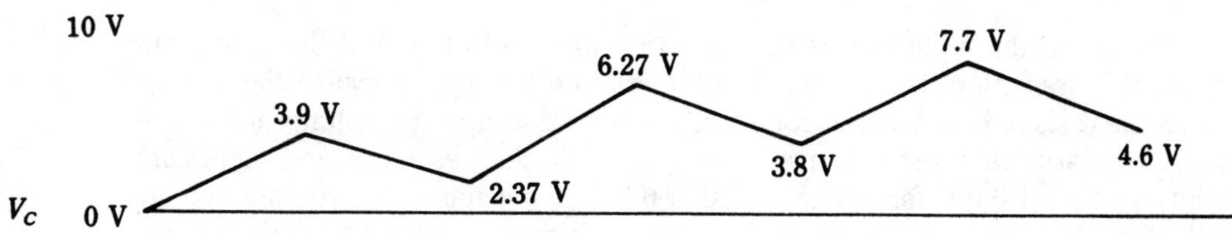

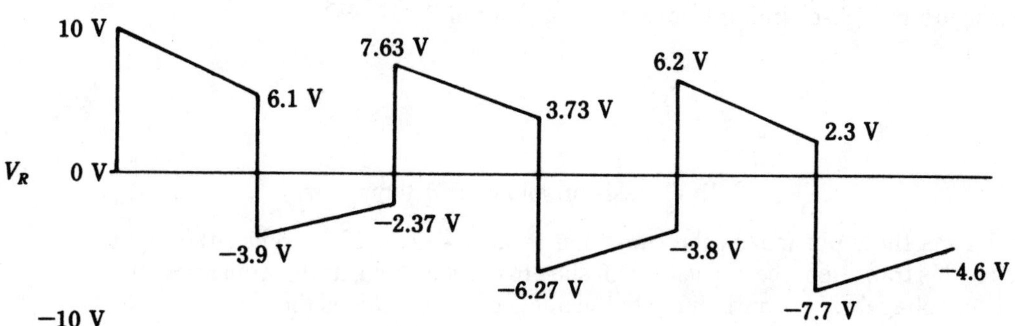

9-5 Long time constant.

Eventually, the capacitor will charge to a point where it will charge to not quite the applied voltage of 10 V and discharge to not quite 6 V.

Calculating the long time constant circuit

$$V = 1 - e^{-T} \times V \quad \text{charge voltage} \tag{9-1}$$

$$V = e^{-T} \times V \quad \text{discharge voltage} \tag{9-2}$$

Note All calculations are made at a transition (charge) in the input square wave. Each 0.5 ms is an input transition. An input transition of 0.5 ms equals 0.5 time constants (1 time constant = 1 ms). The T in the formula is replaced with 0.5 time constants.

At 0 ms, the input switches to +10 V. The capacitor starts to charge, starting from 0 V. The resistor will instantly go to the full applied voltage since the capacitor has 0 V across it.

At 0.5 ms, the capacitor has had 0.5 time constants to charge, and the input switches to 0 V. Because the capacitor has had 0.5 ms charge time constants, the voltage across the capacitor can be calculated with the charge formula.

$$\begin{aligned} V &= 1 - e^{-T} \times V && (9\text{-}1) \\ &= 1 - e^{-5} \times 10\text{ V} \\ &= 3.9\text{ V} \quad \text{capacitor voltage at 0.5 ms} \\ V_R &= V - V_C \\ &= 10 - 3.9 \\ &= 6.1\text{ V} \quad \text{resistor voltage at 0.5 ms} \end{aligned}$$

At 0.5 ms, the input makes the transition from +10 to 0 V. When the input makes this transition, the capacitor will start to discharge. Because the voltage across the resistor can switch instantly, it will have the capacitor voltage across it—with the polarity reversed. Then, as the capacitor discharges to the next significant point, which is 1.0 ms, for a time period of 0.5 time constants, the resistor voltage will follow the capacitor voltage because the capacitor is now supplying the voltage to the circuit.

At 1.0 ms, the voltage across the capacitor can be calculated using the discharge formula, discharging from a starting point of 3.9 V.

$$\begin{aligned} V &= e^{-T} \times V && (9\text{-}2) \\ &= e^{-0.5} \times 3.9 \\ &= 2.37\text{ V} \\ V_R &= -V_C \\ &= -2.37\text{ V} \quad \text{resistor voltage at 1.0 ms} \end{aligned}$$

At 1.0 ms the input makes the transition from 0 V to +10 V. When the input makes this transition, the capacitor will start to charge. Also, at this transition, the resistor voltage will instantly jump to the applied voltage minus the capacitor voltage.

The capacitor will again be allowed 0.5 time constants of charge time between 1.0 ms and the next transition at 1.5 ms.

The charge voltage that was calculated for all 0.5 ms steps will always be the same amount of charge voltage, with this charge cycle with a voltage of 2.37 already present. Therefore, the capacitor will charge the 3.9 V starting from the 2.37 V.

$$
\begin{aligned}
V_C &= 3.9 + \text{existing voltage} \\
&= 3.9 + 2.37 \\
&= 6.27 \text{ V} \quad \text{capacitor voltage at 1.5 ms} \\
V_R &= V - V_C \quad \text{applied voltage-capacitor voltage} \\
&= 10 - 6.27 \text{ V} \\
&= 3.73 \text{ V} \quad \text{resistor voltage at 1.5 ms}
\end{aligned}
$$

At 1.5 ms, the input switches from +10 V to 0 V. The capacitor will again start to discharge and the resistor will jump to the capacitor voltage in the negative polarity.

Notice the trend of the capacitor is toward a full charge. Eventually, the capacitor will reach a full charge during the charge portion and then during the discharge to approximately 6 V. It will then continue to go between 10 V and 6 V as long as the input is allowed to switch at the present rate. It never does charge to the full applied voltage of 10 V. However, it will get quite close.

The resistor voltage will become centered around the zero center line. The effect of this circuit is to pass the square wave, although its shape has been altered, and eliminate the center line being at a dc level.

Chapter summary

Following the universal time constant curve, there are some applications for the RC time constant circuits.

In this chapter, the voltage developed across the resistor is the main subject. It is explored by examining the negative voltage that is displayed when the capacitor holds the voltage and the input changes from a positive voltage to zero.

The short time constant circuit developed the sharpest negative (and positive spike). The spike can be made broader by making a longer time constant, up to the point of being classified as a medium time constant circuit.

The medium time constant circuit is an extension of the short time constant. The medium is classified right at, or very close to, the time constant being equal to the time of the applied voltage. Both the medium and short time constant circuits are used to generate the spikes.

The long time constant circuit is used to pass the square wave through to the output, with some change in the shape of the original waveform. The capacitor has the effect of stopping the dc voltage and passing only the voltage that is changing.

Differentiation is when the output is taken across the resistor. Integration is when the output is taken across the capacitor.

10 Inductive reactance

WHEN DISCUSSING REACTANCE, WHETHER IT BE INDUCTIVE OR CAPACITIVE, reactance is considered to be the ac resistance of the particular component. That is, will offer a different resistance to the flow of ac than it will to the flow of dc.

In addition to the fact that a reactive component will offer resistance to the ac signal, it will also cause a phase shift between the applied voltage and the voltage drop across the reactive components.

Inductive reactance, X_L

$$X_L = 2\pi fL \tag{10-1}$$

The factor 2π is a constant in this formula.
f is frequency, measured in hertz.
L is inductance, measured in henries.
X_L is inductive reactance, measured in ohms.

The formula states that there is a direct relationship between the inductive reactance, frequency, and inductance. Keep in mind that a direct relationship means that when one of the quantities (either frequency or inductance) is increased, the quantity it is compared to (inductive reactance) will also increase.

- Inductive reactance symbol, X_L is directly related to frequency and inductance.

The inductive reactance symbol, X_L, is read as "X = sub = C," which stands for reactance, has the subscript L, which stands for inductance. When the section on capacitive reactance is discussed, it will be X_C, read as "X sub C." These subscripts of X make it possible to use the same letter X to represent any reactance and to distinguish between inductive and capacitive through the use of subscripts.

Linear relationship of X_L versus frequency and inductance

Refer to Fig. 10-1. The figure shows two curves. The first shows X_L, plotted on the vertical axis with inductance plotted on the horizontal axis. The frequency for plotting this curve is 10 kHz. The second curve is X_L plotted on the vertical axis and frequency plotted on the horizontal axis, with the inductance for this curve being 16 mH.

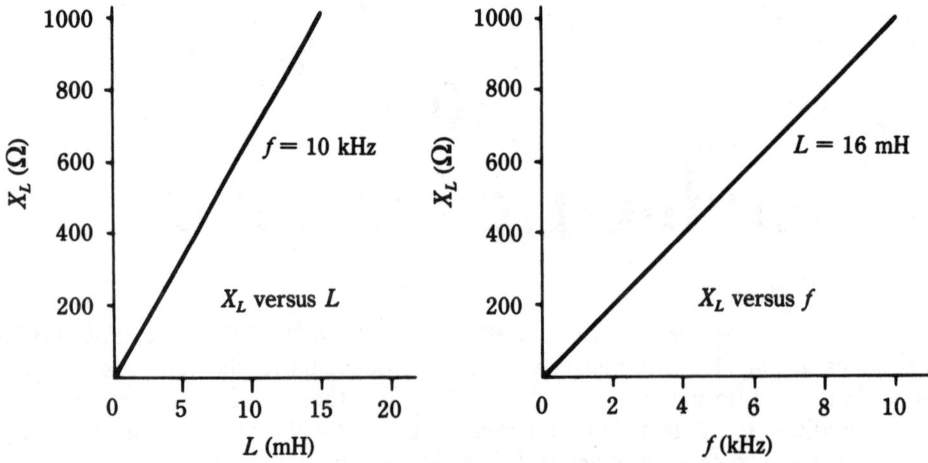

10-1 The linear relationship of X_L versus frequency and inductance.

The following is a sample calculation for the points on the curve.

For the curve of X_L versus L with $f = 10$ kHz, the X_L, was selected for the points where the line would cross the major divisions and the equation was solved for the value of L at that point.

$$X_L = 2\pi f L \tag{10-1}$$

$$L = \frac{X_L}{2\pi f}$$

$$= \frac{600}{2\pi \text{ kHz}}$$

$$= 9.5 \text{ mH}$$ inductance value when the frequency is 10 kHz and the X_L is 600 Ω

Note For calculations such as this one, or any that involve the constant π it is best to use the value of π in the calculator. In the event that the calculator being used does not have π, use 3.14 as the value of π.

For the curve of X_L versus f with $L = 16$ mH the X_L was again selected so the points would be where the line crosses the major divisions. The equation is solved for f with L being equal to 16 mH.

$$X_L = 2\pi f L \tag{10-1}$$

$$f = \frac{X_L}{2\pi L}$$

$$= \frac{800}{2\ \pi 16\ \text{mH}}$$

$$= 7960\ \text{Hz}$$ frequency when the value of inductance is 16 mH and the reactance is 800 Ω

The practice problems include some of the points for the two curves.

Practice problems

Find the value of the unknown quantity using the information given.

1. $f = 1000$ Hz, $L = 100$ mH; find X_L
2. $X_L = 1000\ \Omega, f = 10$ kHz; find L
3. $X_L = 200\ \Omega, f = 10$ kHz; find L
4. $X_L = 200\ \Omega, L = 16$ mH; find f
5. $X_L = 600\ \Omega, L = 16$ mH; find f
6. $f = 5000$ Hz, $L = 2$ *H;* find X_L
7. $X_L = 400\ \Omega, f = 10$ kHz; find L
8. $X_L = 1000\ \Omega, L = 16$ mH; find f
9. $L = 50\ \mu$H, $f = 3$ MHz; find X_L
10. $L = 50$ mH, $f = 0$ Hz (dc); find X_L

Series or parallel inductive reactances

Because reactance is measured in ohms, the same as resistance, it stands to reason that whenever a reactance, either inductive or capacitive, is connected with another one of the same type, it will follow the rules for resistors when they are in series or parallel.

$$X_L \quad X_{Ls} = X_{L1} + X_{L2} + X_{L3} + \ldots \text{ series} \tag{10-2}$$

$$X_L \quad \frac{1}{X_{Lp}} = \frac{1}{X_{L1}} + \frac{1}{X_{L2}} + \frac{1}{X_{L3}} + \ldots \text{ parallel} \tag{10-3}$$

When it is necessary to calculate a voltage drop or the current through a reactance, use the same method as for resistors, following Ohm's Law. But use this method only if there is only reactance in the circuit. Note that resistance and reactances must all be the same type: either inductive or capacitive.

All the basic rules also apply. Current is the same throughout a series circuit. Current divides in a parallel circuit. Voltage drops in a series circuit are directly proportional to the size of the reactance. Voltage is the same throughout a parallel circuit.

Figure 10-2 shows a series circuit with two inductive reactances. To calculate the total inductive reactance, use the formula to add the series components, just like series resistors. Calculate total current using Ohm's Law with the applied voltage and total reactance. Voltage drops across each of the reactance is also found by using Ohm's Law, using total current and the individual reactances.

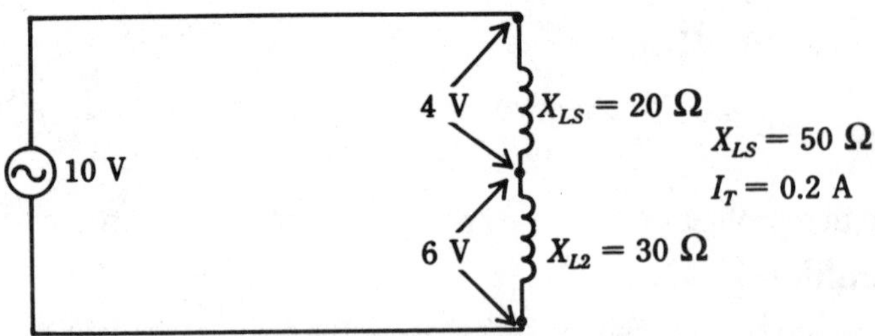

10-2 Series inductive reactance.

Figure 10-3 shows to inductive reactances in parallel. Find total reactance using the reciprocal formula. Find total current using Ohm's Law with the total reactance and the applied voltage. Voltage for each branch is the same because they are parallel branches. Calculate branch current using Ohm's Law with the branch voltages and the reactance of the individual branches. In this case branch voltage is the same as the applied voltage. Branch currents should add up to the total current.

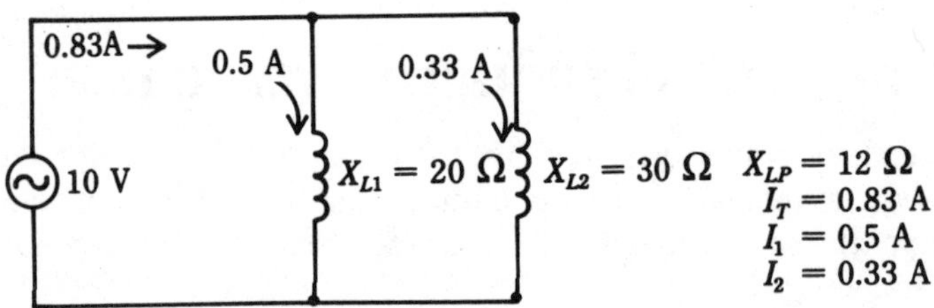

10-3 Parallel inductive reactance.

Series X_L and R

In all series circuits, the current is the same throughout the circuit. When an ac current is flowing in a circuit containing inductance, the inductor magnetic field will expand and collapse following the applied sine wave. Because it is current that causes the magnetic field to develop and this requires a period of time, the voltage developed across the inductor will appear before the current can cause the magnetic field. The current flow, then, is delayed by the inductor voltage drop. This delay is measured in degrees of the sine wave. Inductive voltage is always 90 degrees ahead of the inductive current. This can also be said that the current lags the voltage by 90 degrees. Because the current in the series circuit is the same throughout the circuit, the current in the resistor must be the same as the current in the inductor. The current through a resistor causes a voltage drop that will have no phase shift, in comparison to the current. If the current through the resistor is the same as the voltage, the voltage drop across the resistor must be 90 degrees out of phase with the voltage developed across the inductor. The inductor voltage will be 90 degrees ahead or the voltage across the resistor lags by 90 degrees.

- In a series circuit, I_L lags V_L, V_R lags V_L always by 90 degrees.

Figure 10-4 shows a sine wave analysis of the current through the series circuit, the voltage drops across the inductor and resistor, and the applied voltage. Notice how the current sine wave and the resistor sine wave, V_R, are exactly in phase, with only the amplitudes being different.

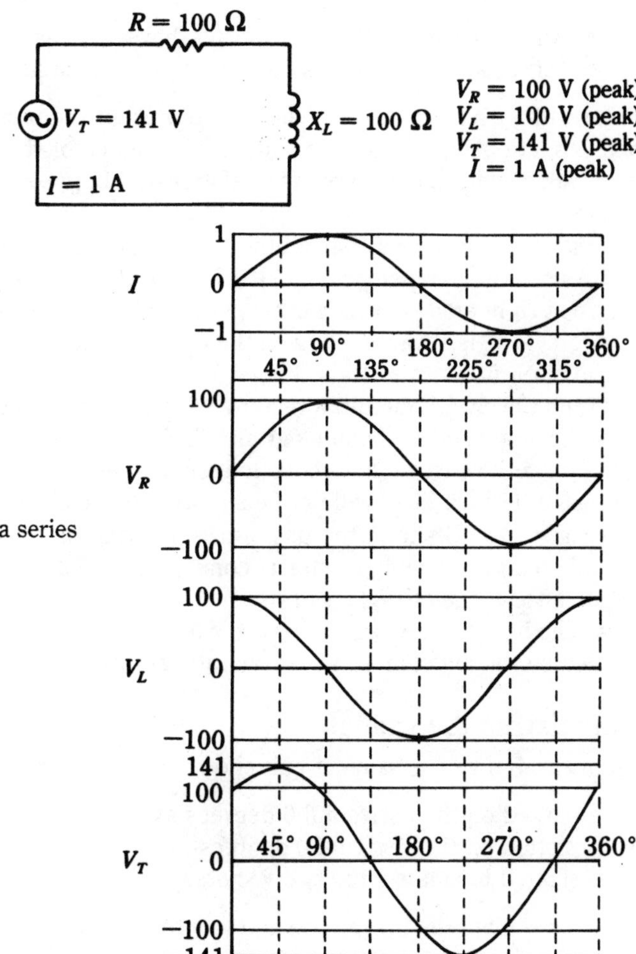

10-4 Sine wave analysis of a series X_L and R circuit.

The phase relationship of two sine waves can be determined by the maximum and minimum values and when they occur in terms on the number of degrees of the sine wave.

Because of the current and the resistive voltage are exactly in phase, they can be thought of as the reference. Usually the voltage across the resistor is chosen because it can be observed directly on an oscilloscope.

Notice in Fig. 10-4 how the sine wave representing the voltage across the inductor has its peak value occur at a point in time equal to 90 degrees earlier than the resistor voltage. Another way to say this is to say that the resistor voltage, V_R, has its peak value *lag* the inductor voltage, V_L, by 90 degrees.

The fourth sine wave in Fig. 10-4 is the representation of the total voltage. This sine wave is actually the summation of each point along the curves of V_R and V_L. Notice the applied voltage has its peak value occur at a point between the tor voltage and the resistor voltage. This will always be the case. The applied voltage in this particular circuit has a phase shift of 45 degrees.

- An inductive circuit will always have a phase shift value of between 0 degrees and 90 degrees, depending on the circuit values.

Figure 10-5 uses vectors or phasors to represent the relative magnitude (size) and phase relationship of the sine waves. The values of the sine waves are shown as peak values with the vectors. The relationship between the two vectors shows the phase.

Figure 10-5 shows two methods of drawing the phasor diagrams. Figure 10-5A uses the parallelogram method. Here the two basic vectors, V_R and V_L, are drawn at right angles to each other. V_L is plotted up to show that it leads the V_R by 90 degrees. V_R is always plotted on the horizontal to indicate a 0 degree phase shift, or the reference vector. Dotted lines are connected to the vectors to form a parallelogram (for an RL circuit, a rectangle). Then a new vector is drawn between the opposite angles of the parallelogram. The new vector represents the applied voltage. Notice how it will always be longer than either of the other two vectors, and the magnitude, or size, will not be simply the two sides added together.

Figure 10-5B shows the use of a right triangle to connect the vectors of the circuit voltages. The V_L vector is connected to the end of the V_R vector and the hypotenuse of the triangle is drawn by connecting the ends of the two voltage vectors. The hypotenuse forms the vector to represent the applied voltage. The angle formed by the hypotenuse is the operating voltage, called theta, Θ.

Calculating the triangle

Figure 10-5 uses the same circuit shown in the schematic of Fig. 10-4.

V_R is plotted on the horizontal 0 degrees axis.
V_L is plotted on the vertical 90 degrees axis.
V_T is plotted to connect the two vectors.

To calculate the magnitude and length of V_T use:

$$V_T^2 = V_R^2 + V_L^2 \text{ (Pythagorean theorem)}$$

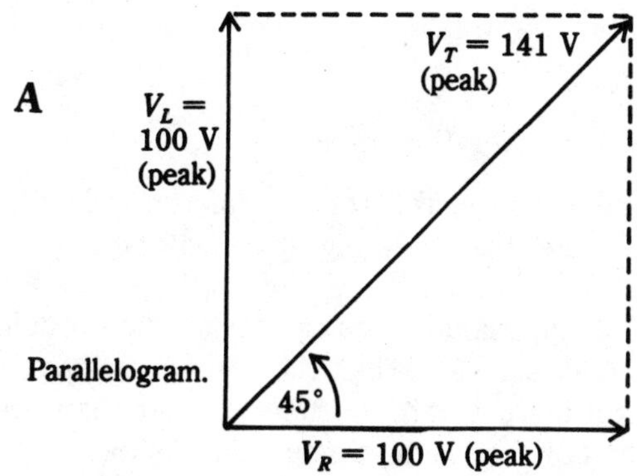

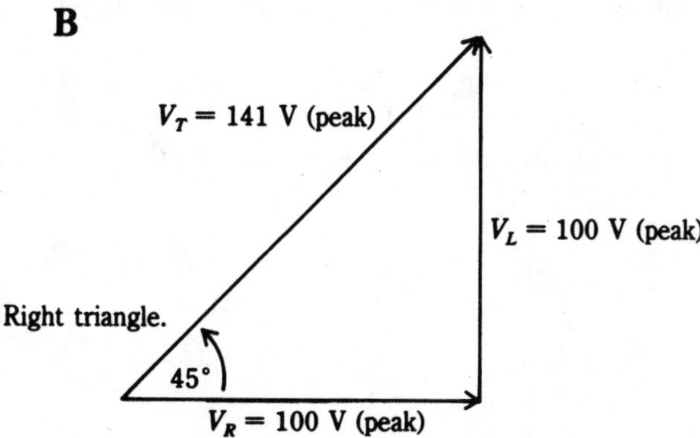

10-5 Vector analysis of a series X_L and R circuit.

The total voltage in an inductive circuit is:

$$V_T = \sqrt{V_R^2 + V_L^2} \qquad (10\text{-}4)$$

$$= \sqrt{100^2 + 100^2}$$

$$= \sqrt{10{,}000 + 10{,}000}$$

$$= 141 \text{ V}$$

To calculate the operating angle, the angle formed by the hypotenuse, use:

$$\tan \Theta = \frac{V_L}{V_R} \quad \text{tangent of the angle equals the opposite over the adjacent} \qquad (10\text{-}5)$$

$$\Theta = \tan^{-1} \frac{V_L}{V_R} \quad \text{operating angle formula arranged for use directly with a calculator}$$

$$= \tan^{-1} \frac{100}{100} \quad \text{substitute values}$$

$$= 45 \div \text{ operating angle}$$

Notice in the example above, the voltages of the inductor and resistor are equal. Whenever this occurs, the operating angle will be 45 degrees and the two voltages will each be 70.7 percent of the applied voltage.

By observing the voltage triangle, you can make some predictions. If the voltage of the inductor increases greater than the resistor voltage, the phasor representing the inductor will increase and become greater than the resistor phasor. The height of the triangle will become greater than its base. The operating angle will increase toward 90 degrees.

If the voltage of the inductor is made small compared to the resistor voltage, the base of triangle is greater than the height. The operating angle will decrease towards 0 degrees.

Figure 10-6A shows a triangle where the V_L is large compared to the V_R. Figure 10-6B shows a triangle where the V_L is small compared to the V_R. Calculations are performed using the same methods as demonstrated with Fig. 10-5.

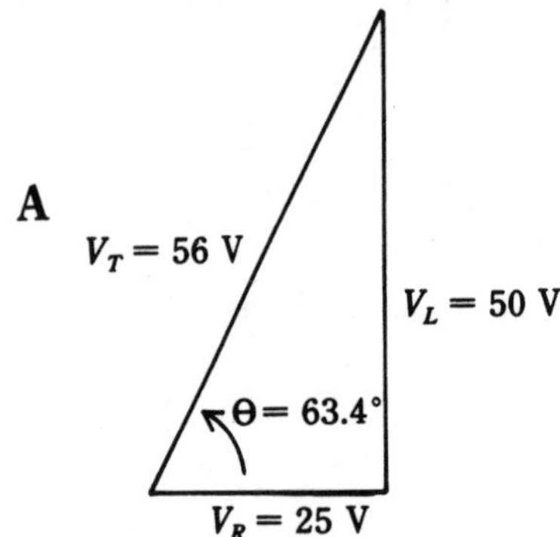

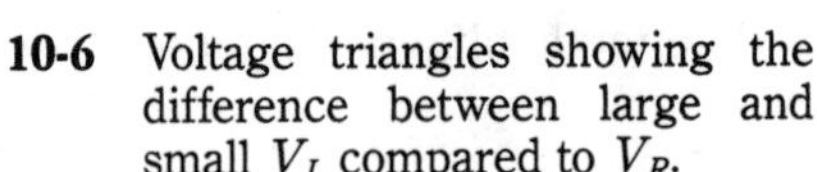
10-6 Voltage triangles showing the difference between large and small V_L compared to V_R.

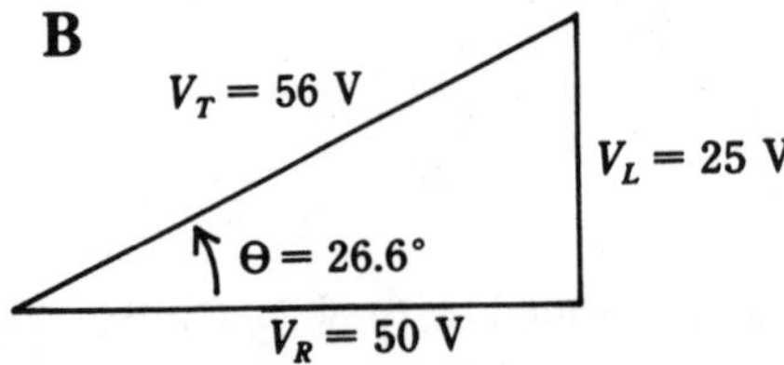

Calculating the impedance in a series circuit

In order to complete the calculations of the series circuit with inductance and resistance, it is necessary to determine the total impedance of the circuit.

The word *impedance* is used rather than resistance or reactance because neither of these terms would properly describe the total effect of the circuit.

- Impedance (Z) is the ac resistance of a circuit containing both resistance and reactance. It has an angle equal to the circuit phase angle.

It is necessary to determine the impedance of a series circuit through the use of vectors. The vector for impedance will look similar to the voltage triangle. Resistance is always plotted on the horizontal and reactance is always plotted on the vertical. Inductive reactance is plotted up. Impedance is the resultant by hypotenuse, calculated by use of Pythagorean theorem:

$$Z = \sqrt{R^2 X L^2} \quad \text{impedance of a series circuit} \qquad (10\text{-}6)$$

The operating angle can be found using the impedance triangle. It will be exactly the same as the voltage triangle.

$$\Theta = \tan^{-1} \frac{X_L}{R} \quad \text{operating angle of a series circuit} \qquad (10\text{-}7)$$

Calculating a complete series circuit

The given values of the circuit in Fig. 10-7 are: $V = 90$ V, $f = 60$ Hz, $L = 400$ mH, $R = 100\ \Omega$. Find X_L, Z, Ω, I_T, V_R, V_L. Calculate the X_L.

$$\begin{aligned} X_L &= 2\pi f L \\ &= 2\pi\,(60)\ 400\ \text{mH} \\ &= 250\ \Omega \end{aligned} \qquad (10\text{-}1)$$

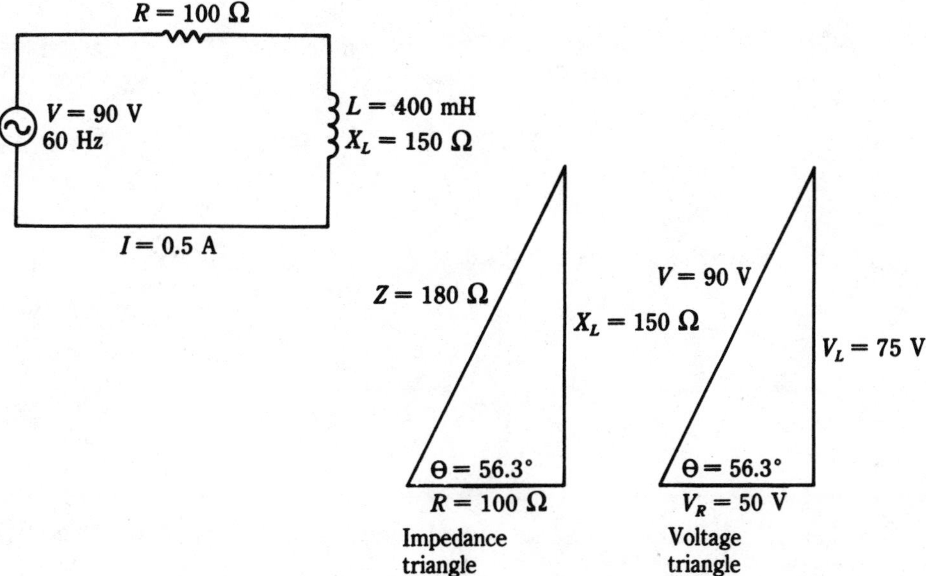

10-7 Solving a series *RL* circuit using phasor triangles.

Calculate impedance (Z) using the impedance triangle. Plot X_L up and R horizontal.

$$Z = \sqrt{R^2 + X_L^2} \tag{10-6}$$

$$= \sqrt{100^2 + 150^2}$$

$$= \sqrt{10,000 + 22,500}$$

$$= \sqrt{32,500}$$

$$= 180\ \Omega$$

Calculate the operating angle, Θ.

$$\Theta = \tan^{-1} \frac{X_L}{R} \tag{10-7}$$

$$= \tan^{-1} \frac{150}{100}$$

$$= \tan^{-1}\ (1.5)$$

$$= 56.3$$

With the total impedance calculated and the supply voltage given, calculate the current.

$$I = \frac{V}{Z} \quad \text{Ohm's Law modified}$$

$$= \frac{90\ \text{V}}{180\ \Omega}$$

$$= 0.5\ \text{A}$$

Use the current to find the voltage drops.

$$\begin{aligned} V_R &= I \times R \\ &= 0.5\ \text{A} \times 100\ \Omega \\ &= 50\ \text{V} \end{aligned} \tag{2-1}$$

$$\begin{aligned} V_L &= I \times X_L \\ &= 0.5\ \text{A} \times 150\ \Omega \\ &= 75\ \text{V} \end{aligned} \tag{2-1}$$

- The voltage drops cannot be simply added. They must be added through vector addition.

Use vector addition to compare the calculated voltage drops with the given applied voltage.

$$V = \sqrt{V_R^2 + V_L^2} \tag{10-4}$$

$$= \sqrt{50^2 + 75^2}$$

$$= \sqrt{2500 + 5625}$$

$$= 90 \text{ V}$$

Numbers are rounded for easy use. Without rounding, any calculations should be very close to what is expected.

Summary for series circuit:

- Calculate the inductive reactance, X_L, based on the frequency and inductance values, unless X_L is given.
- Calculate the impedance, Z, using the impedance triangle. Vector addition requires the use of the Pythagorean theorem:

$$Z = \sqrt{R^2 + X_L^2} \tag{10-6}$$

- Calculate the operating angle, Θ, based on the impedance triangle and the tangent function:

$$\Theta = \tan^{-1} \frac{X_L}{R} \tag{10-5}$$

- Calculate the series current using the applied voltage and the impedance. If the applied voltage is not given, enough other information will be given to use Ohm's Law to calculate voltage.
- Use the current to calculate the voltage drops across each component. If applied voltage is not given for calculating current, enough other information should be given.
- Use vector addition to calculate the applied voltage, based on the voltage drops.

$$V = \sqrt{V_R^2 + V_L^2} \tag{10-4}$$

- Θ can also be found using the voltage triangle rather than the impedance triangle.

Practice problems

Use the two schematics shown to calculate the answers to the problems.

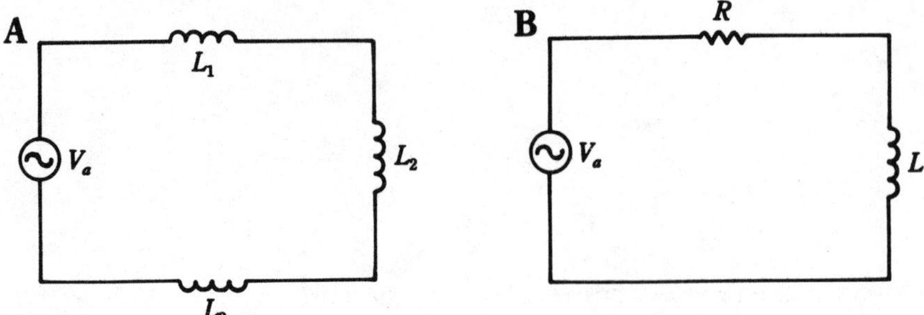

Use schematic *A* to answer problems 1 and 2. $f = 60$ Hz. Find: L_T, X_{L1}, X_{L2}, X_{L3}, and X_{LT}.

1. $L_1 = 1$ H, $L_2 = 2$ H, $L_3 = 3$ H
2. $L_1 = 20$ mH, $L_2 = 40$ mH, $L_3 = 80$ mH

Use schematic *B* to answer problems 3 through 8. $V_a = 100$ V. Find: *Z, I,* V_R, Θ, draw impedance triangle

3. $R = 100\ \Omega$, $X_L = 100\ \Omega$
4. $R = 25\ \Omega$, $X_L = 50\ \Omega$
5. $R = 10\ \Omega$, $X_L = 100\ \Omega$
6. $R = 75\ \Omega$, $X_L = 5\ \Omega$
7. $R = 75\ \Omega$, $X_L = 25\ \Omega$
8. $R = 0\ \Omega$, $X_L = 50\ \Omega$

Use schematic *B* to answer problems 9 and 10. Find *R*, X_L, V_a, Θ. Draw voltage triangle.

9. $V_R = 20$ V, $V_L = 40$ V, $I = 0.25$ A
10. $V_R = 25$ V, $V_L = 15$ V, $I = 0.333$ A

Parallel X_L and *R*

The basic rules of parallel circuits shows that the voltage is the same throughout the parallel branches and the current divides inversely to the resistance of the individual branches. Because the voltage is the same across all the branches, there cannot appear to be any phase shift in the voltage of the components compared to the applied voltage.

The current of the inductor will lag the current of the resistor by 90 degrees because it is the current that causes the inductor to charge. Another way to think of this is to realize that the current of an inductor always lags the inductor voltage by 90 degrees. The voltage cannot have a phase shift, the current of the inductor must lag the current of the resistor, which is always the reference, by 90 degrees.

- In a parallel circuit, I_L lags I_R by 90 degrees.

Figure 10-8 is the sine wave analysis of a circuit with equal values of R and X_L. Equal values are chosen to simplify the drawings. Because the values are equal, the current through each branch should be equal, as calculated by Ohm's Law. I_T is the line current from the power source. I_T is the vector summation of I_R and I_L at every point along the sine wave. For ease of calculations, the peak value is chosen to perform any calculation.

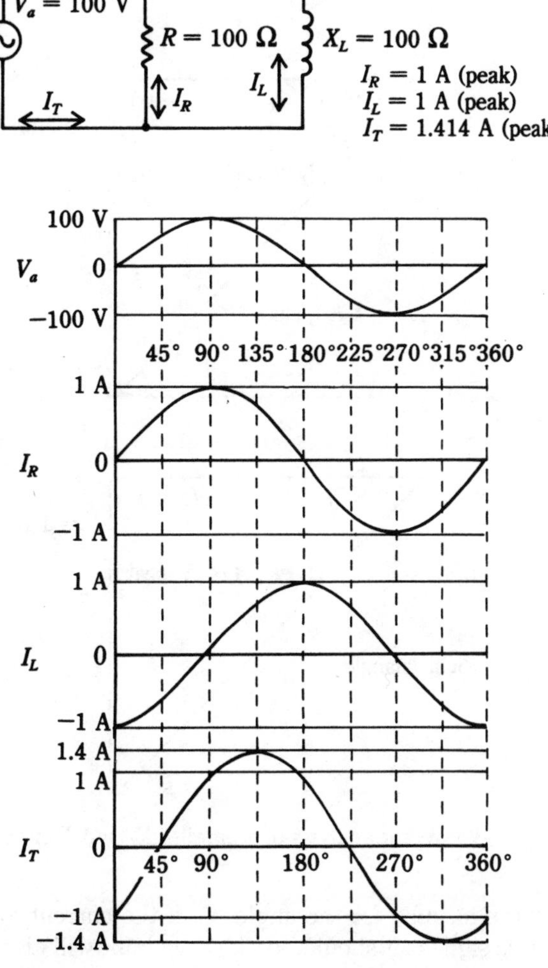

10-8 Sine wave analysis of a parallel X_L and R circuit.

Because I_L lags I_R by 90 degrees, the sine wave representing I_L starts 90 degrees later than the sine wave for I_R. The operating angle, Θ, is determined by the sine wave for I_T, which will always be between 0 and negative 90 degrees. The negative angle indicates a lagging condition, or a delayed condition.

Figure 10-9 shows the vector diagram of the current in this same circuit. The resistive current is again plotted on the horizontal, at the reference line. The vector, or phasor, for I_L is plotted at a 90 degree angle to I_R. It is plotted down to show it is lagging by 90 degrees. The hypotenuse of the triangle is completed and the result is I_T at the operating angle.

$$I_T = \sqrt{I_R^2 + I_L^2} \quad \text{total current in a parallel circuit} \qquad (10\text{-}8)$$

$$\Theta = \tan^{-1} \frac{-I_L}{I_R} \quad \text{operating angle} \qquad (10\text{-}9)$$

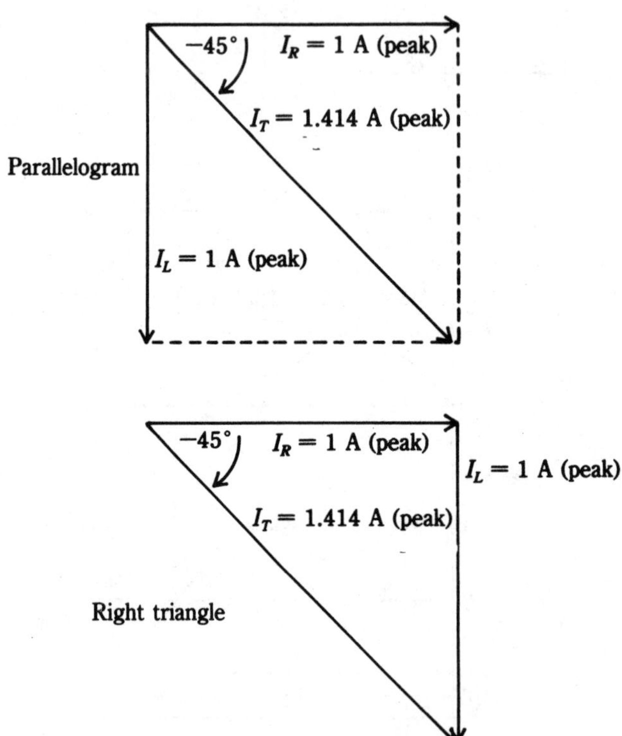

10-9 Vector analysis of a parallel X_L and R circuit.

When using the formula for the angle, Θ, if the current for the inductance is assigned a negative sign, as it should be, the calculator will give the angle as a negative angle. If the negative sign is neglected, the angle will be the same value, except it will not be negative from the calculator.

The only quantity remaining to be calculated in the parallel circuit is the value of total impedance. Because the resistor and reactance are in parallel, it becomes quite complicated to combine these through the reciprocal formula, considering they have to be combined with vectors. The easiest way to calculate the total Z is to use the total current and the applied voltage.

$$Z = \frac{V_a}{I_T} \quad \text{total impedance in a parallel circuit} \qquad (10\text{-}10)$$

In the section on series circuits, you analyzed what would happen if the X_L were made greater or less than the value of R. Refer to Fig. 10-6 as a reminder. In a series circuit, the voltage drop is increased across the larger resistor. In a parallel circuit, the current is decreased with the larger value of resistance (or reactance). Therefore, if the X_L in a parallel circuit were made larger than the resistor, the current through the inductor would decrease. The operating angle would also decrease, toward 0 degrees. If the value of X_L were made smaller than the value of R, the current through the inductor would be larger than the current through the resistor and the operating angle would increase toward negative 90 degrees.

Figure 10-10 shows two current triangles. One triangle, Fig. 10-10A, shows a small inductive current in comparison to the resistive current. Because the current of the inductor is small, that means the X_L is large, in comparison to the resistor. Notice that the operating angle, Θ, is smaller than 45 degrees.

Figure 10-10B shows a large inductive current compared to the resistive current. This means the inductive reactance, X_L, is smaller than the value of the resistor. Notice that the operating angle, Θ, is larger than 45 degrees.

A

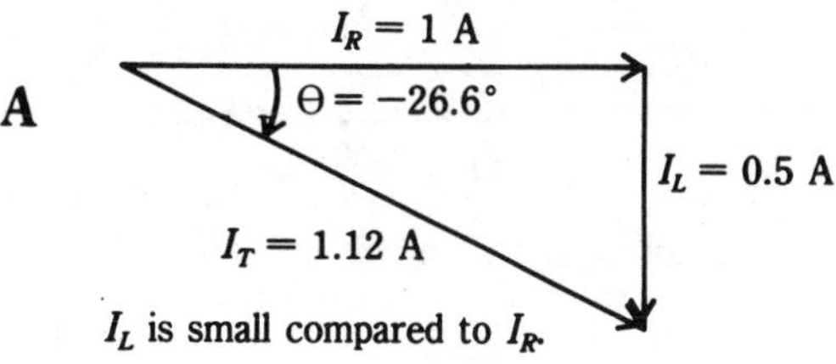

B

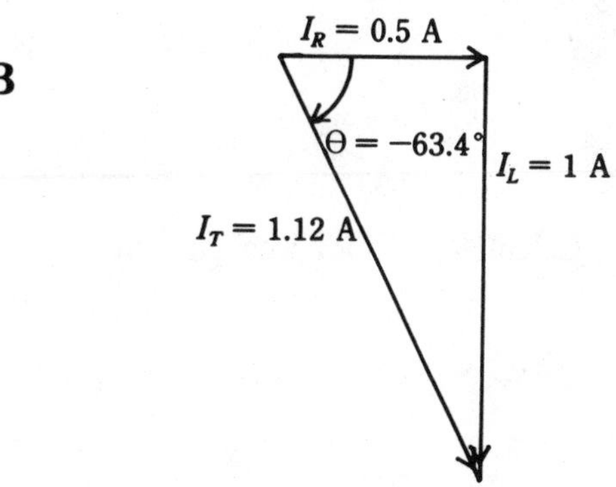

10-10 Current triangles showing the difference between large and small I_L compared to I_R.

Calculating a parallel circuit

Figure 10-11 shows a parallel circuit containing inductance and resistance. Included in the figure is the current triangle to determine the total amount of current using vector summation. When working with a parallel circuit, remember that the voltage across all branches is the same. Given values: V = 10 V, R, – 10 Ω, X_L = 7 Ω. Find: I_R, I_L, I_T, Θ, Z.

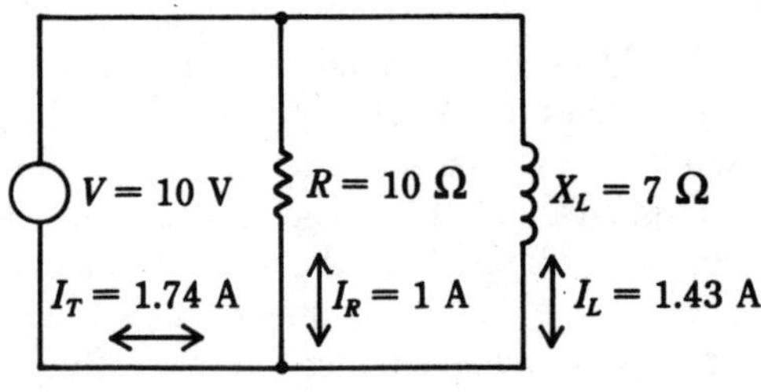

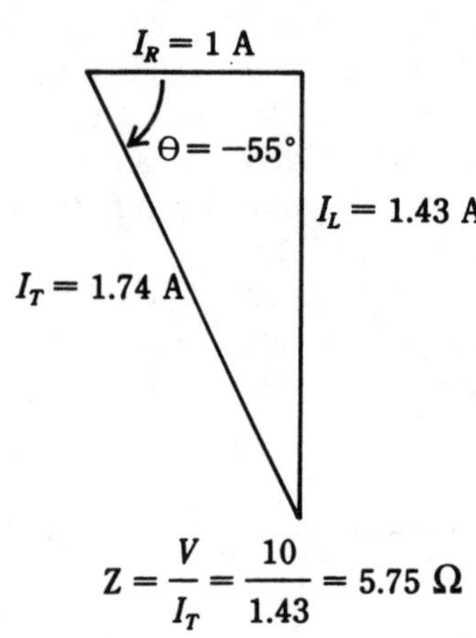

$$Z = \frac{V}{I_T} = \frac{10}{1.43} = 5.75\ \Omega$$

10-11 Using the current triangles to solve a parallel circuit with inductance and resistance.

Calculate the resistive current, I_R.

$$I_R = \frac{V}{R} \qquad \text{(2-1A)}$$

$$= \frac{10\text{ V}}{10\ \Omega}$$

$$= 1\text{ A plotted at 0 degrees}$$

Calculate the inductive current, I_L.

$$I_L = \frac{V}{X_L} \qquad \text{(2-1A)}$$

$$= \frac{10\text{ V}}{7\ \Omega}$$

$$= 1.43\text{ A plotted at 90 degrees down}$$

Calculate the total current using Pythagorean theorem.

$$I_T = \sqrt{I_R^2 + I_L^2} \tag{10-8}$$

$$= \sqrt{I^2 + 1.43^2}$$

$$= 1.74 \text{ A}$$

Calculate the operating angle, Θ.

$$\Theta = \tan^{-1} \frac{-I_L}{I_R} \tag{10-9}$$

$$= \tan^{-1} \frac{-1.43}{1}$$

$= -55$ degree — angle is negative because it is plotted down. Calculator can give a negative angle I_L is used as a negative

Calculate total impedance, Z. Because the total current has been calculated and the applied voltage is given, this is the easiest method to find Z.

$$Z = \frac{V}{I_T} \tag{10-10}$$

$$= \frac{10 \text{ V}}{1.74 \text{ A}}$$

$= 5.75\ \Omega$ — should be less than the smallest 5 Ω, which is the resistance of the parallel branches

Summary for parallel circuits

- Calculate the values of the resistive current and the inductive current. These are also called the branch currents.
- Calculate the total current using the branch currents to form a current triangle.
- Calculate the operating angle based on the current triangle. The angle in an inductive circuit will be negative.
- Calculate the total impedance using the total current and the applied voltage with Ohm's Law.

Practice problems

Use the two schematics shown to calculate the answers to the problems. Use schematic A to answer problems 1 and 2. Frequency is 50 Hz. Find L_T, X_{L1}, X_{L2}, X_{L3}, and X_{LT}.

1. $L_1 = 1$ H, $L_2 = 2$ H, $L_3 = 3$ H
2. $L_1 = 20$ mH, $L_2 = 40$ mH, $L_3 = 80$ mH

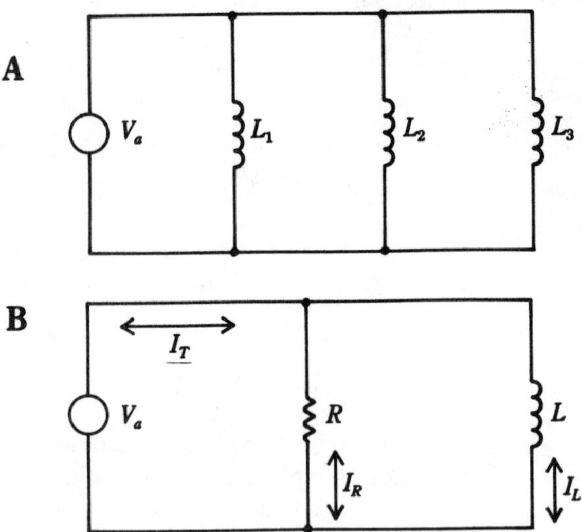

Use schematic B to answer problems 3 through 8. V_a = 100 V. Find I_R, I_L, I_T, Θ, Z, and draw the current triangle.

3. R = 100 Ω, X_L = 100 Ω
4. R = 25 Ω, X_L = 50 Ω
5. R = 10 Ω, X_L = 100 Ω
6. R = 75 Ω, X_L = 5 Ω
7. R = 75 Ω, X_L = 25 Ω
8. R = 5 Ω, X_L = 5 Ω

Use schematic B to answer questions 9 and 10. V_a = 100 V. Find: R, X_L, I_R, Θ, Z.

9. I_{R}, = 1 A, I_L = 3 A
10. I_{R}, = 25 mA, I_L = 10 mA

Power in a reactive circuit

When working with circuits that contain a reactance, your calculations of voltage, current and total resistance, or impedance, become complicated in comparison to dc circuits, or any circuit containing only pure resistance. The calculations for power in an ac circuit containing resistance and reactance are more complicated than when the circuit has only pure resistance.

In an ac circuit there are three ways of describing the power: real power, reactive power, and apparent power.

- Real power (also called true power) is the power dissipated in pure resistance, the unit is W (watts).

- Reactance power is the power dissipated in pure reactance; the unit is *VARS* (volt-ampere-reactive).
- Apparent power is the power of a circuit containing both resistance and reactance. It is calculated by: $I \times E$ (multiplying total current by the applied voltage), unit is *VA* (volt-ampere).
- Power factor is a ratio of the real power to the total power. It is a pure number, with no units and will always be between 0 and 1. It is calculated by taking the cosine of the operating angle; cos Θ.

When working with the power relationships in an ac circuit, follow the basic rules of power, in reference to series or parallel circuits, as you would for a circuit with only resistors. The big difference is that it is necessary to deal with the voltage or current triangles.

In a series circuit, the voltage triangle will give each of the three types of powers and the operating angle can be used to calculate the power factor.

In a parallel circuit, because voltage is the same throughout the parallel circuit, the current triangle is used to calculate the powers and the operating angle is again used to determine the power factor.

The following sample calculations are using the sample circuits in earlier sections of this chapter. (See Fig. 10-7.) Calculate real power using the resistor voltage.

$$\begin{aligned} P &= I \times E \\ &= 0.5 \text{ A} \times 50 \text{ V} \\ &= 25 \text{ W resistive power} \end{aligned} \tag{2-2}$$

Reactive power, using inductive voltage.

$$\begin{aligned} P &= I \times E \\ &= 0.5 \text{ A} \times 75 \text{ V} \\ &= 37.5 \text{ VARS inductive power} \end{aligned} \tag{2-2}$$

Apparent power, using applied voltage.

$$\begin{aligned} P &= I \times E \\ &= 0.5 \text{ A} \times 90 \text{ V} \\ &= 45 \text{ VA apparent power of total circuit} \end{aligned} \tag{2-2}$$

Power factor, using the operating angle.

$$\begin{aligned} PF &= \cos \Theta \\ &= \cos 56.3 \text{ degree} \\ &= 0.555 \text{ ratio of total power consumed by the resistor} \end{aligned} \tag{10-17}$$

Resistive power, using current in resistor.

$$\begin{aligned} P &= I \times E \\ &= 1 \text{ A} \times 10 \text{ V} \\ &= 10 \text{ W true power} \end{aligned} \tag{2-2}$$

Reactive power, using inductive current.

$$\begin{aligned} P &= I \times E \\ &= 1.43 \text{ A} \times 10 \text{ V} \\ &= 14.3 \text{ VARS reactive power} \end{aligned} \quad (2\text{-}2)$$

Apparent power, using total current.

$$\begin{aligned} P &= I \times E \\ &= 1.74 \text{ A} \times 10 \text{ V} \\ &= 17.4 \text{ VA total circuit power} \end{aligned} \quad (2\text{-}2)$$

Power factor, using the operating angle.

$$\begin{aligned} PF &= \cos \Theta \\ &= \cos -55 \text{ degree} \\ &= 0.574 \text{ ratio of resistive power to total power} \end{aligned} \quad (10\text{-}17)$$

The method shown above for both the series circuit and the parallel circuit is only one way of performing the calculations. Following is a summary of the power formulas and some alternative formulas.

$$\begin{matrix}\text{Real Power} \\ (W)\end{matrix} = I_R \times E_R \quad (10\text{-}11)$$

$$\begin{matrix}\text{Real Power} \\ (W)\end{matrix} = I^2 \times R \quad (10\text{-}12)$$

$$\begin{matrix}\text{Real Power} \\ (W)\end{matrix} = VI \cos \Theta \quad (10\text{-}13)$$

$$\begin{matrix}\text{Reactive Power} \\ \text{(VARS)}\end{matrix} = X_L \times E_L \quad (10\text{-}14)$$

$$\begin{matrix}\text{Reactive Power} \\ \text{(VARS)}\end{matrix} = VI \sin \Theta \quad (10\text{-}15)$$

$$\begin{matrix}\text{Apparent Power} \\ \text{(VA)}\end{matrix} = V \times I \quad (10\text{-}16)$$

$$\begin{matrix}\text{Power Factor} \\ \text{(No units)}\end{matrix} = \cos \Theta \quad (10\text{-}17)$$

$$\begin{matrix}\text{Power Factor} \\ \text{(No units)}\end{matrix} = \frac{R}{Z} \quad (10\text{-}18)$$

$$\begin{matrix}\text{Power Factor} \\ \text{(No units)}\end{matrix} = \frac{I_R}{I_T} \quad (10\text{-}19)$$

Practice problems

Use the two schematics shown to calculate the answers to the problems.
Use schematic A for problems 1 through 5. $V = 100$ V. Find: Z, I, V_R, V_L, Θ, P_R, P_A, PF.

1. $R = 400\ \Omega$, $X_L = 500\ \Omega$
2. $R = 100\ \Omega$, $X_L = 250\ \Omega$
3. $R = 1000\ \Omega$, $X_L = 750\ \Omega$
4. $R = 1000\ \Omega$, $X_L = 500\ \Omega$
5. $R = 10\ \Omega$, $X_L = 20\ \Omega$

Use schematic B for problems 6 through 10. $V = 100$ V. Find: I_R, I_L, I_T, Z, Θ, P_R, P_X, P_A, PF

6. $R = 500\ \Omega$, $X_L = 500\ \Omega$
7. $R = 100\ \Omega$, $X_L = 250\ \Omega$
8. $R = 1000\ \Omega$, $X_L = 750\ \Omega$
9. $R = 1000\ \Omega$, $X_L = 500\ \Omega$
10. $R = 10\ \Omega$, $X_L = 20\ \Omega$

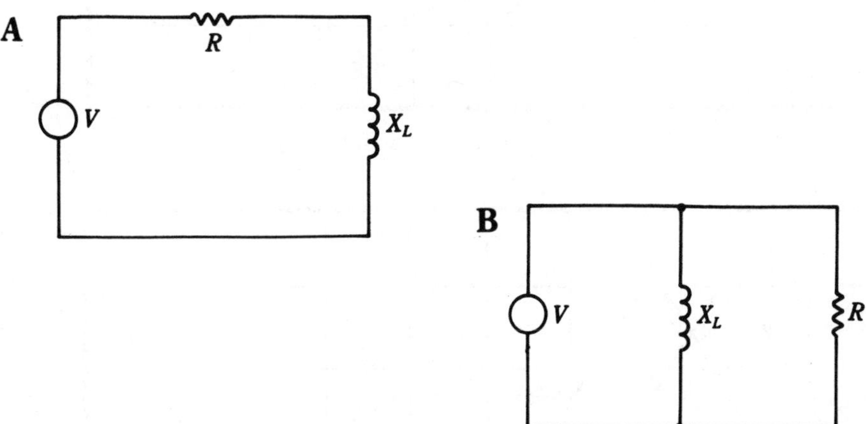

Measuring phase angle with an oscilloscope

The operating angle, or phase shift, of a circuit containing reactance is a measurement that is ideally suited for the dual trace oscilloscope. The drawings in the accompanying Figs. 10-12 through 10-17 help explain the calculation of the phase shift with a dual-trace oscilloscope. The circuit used is a simple series circuit with an inductor and resistor. Values of the individual components are not given because the calculations here are on the scope, not on the circuit. In the event that you wish to set up the circuit as an experiment and practice, you need to get the inductor. The resistor value can easily be calculated. Calculate the resistor value by working with the frequency to be used and determining X_L. From there, use the impedance triangle for a series circuit to obtain the value of resistance needed to complete the triangle at the desired phase angle.

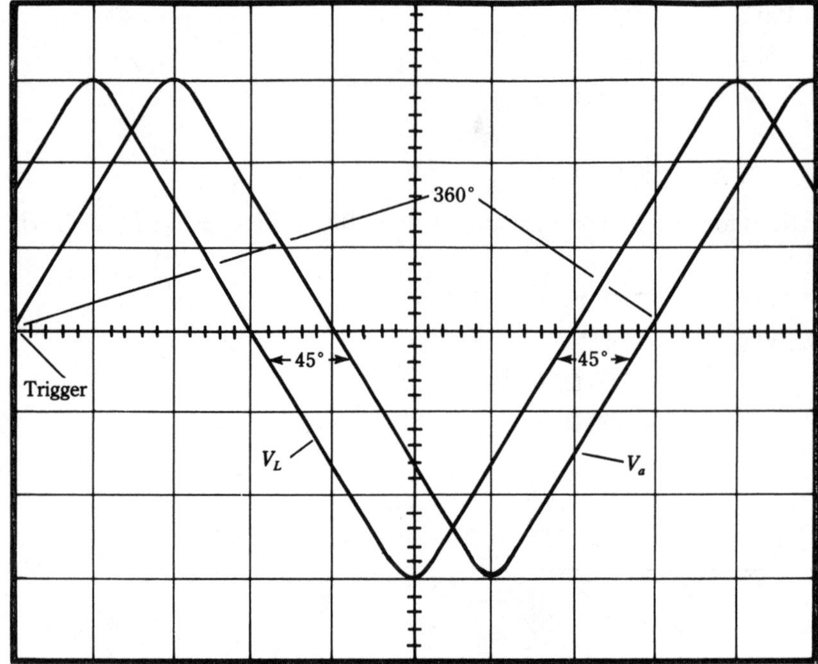

$R = X_L$
360° = 8 divisions
1 division = 45°
Difference = 1.0 division
Phase angle = 45°

10-12 Scope triggered on the input voltage.

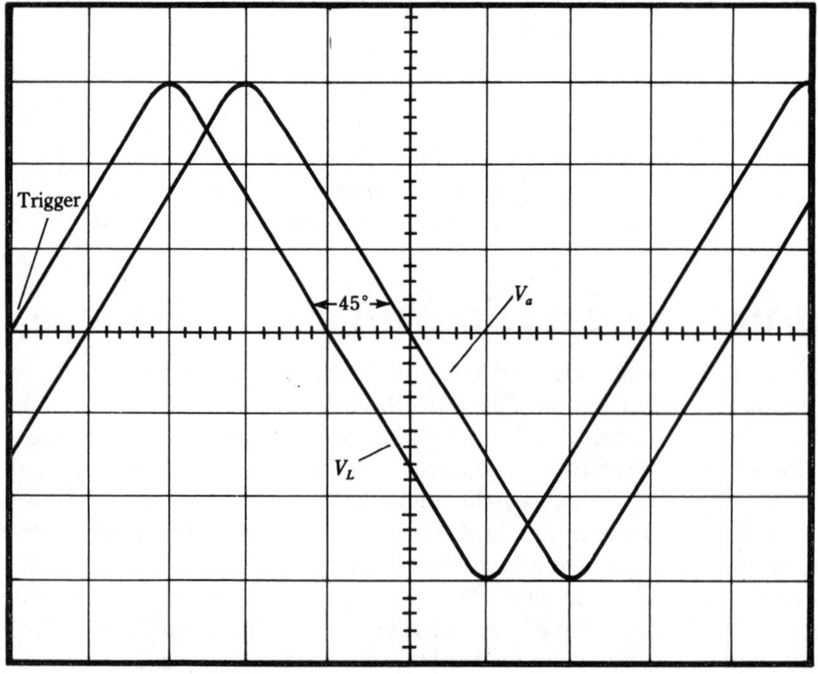

$R = X_L$
Phase angle = 45°

10-13 Scope triggered on the inductor voltage.

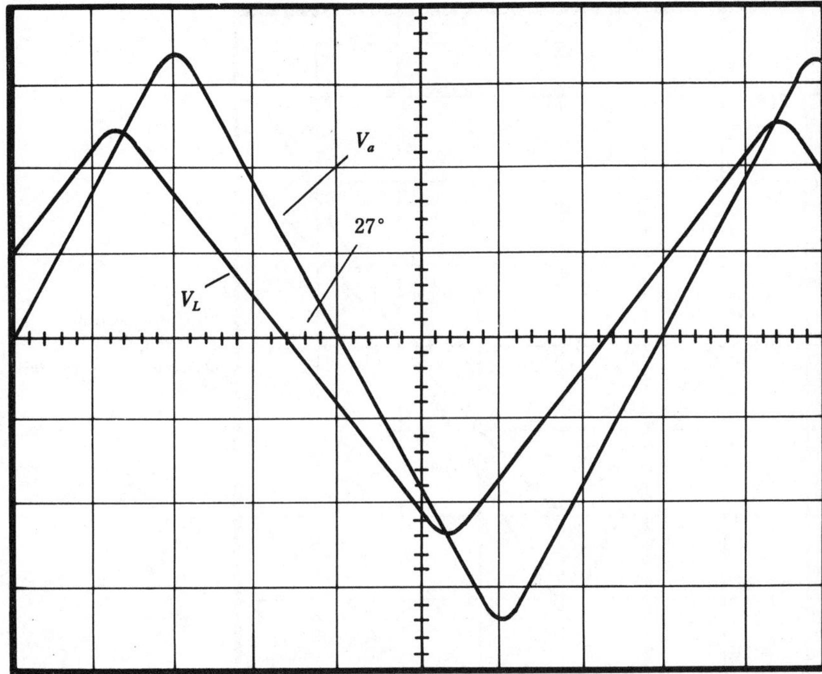

360° = 8 divisions
1 division = 45°
Difference = 0.6 division
Phase angle = 27°

10-14 *R* larger than X_L.

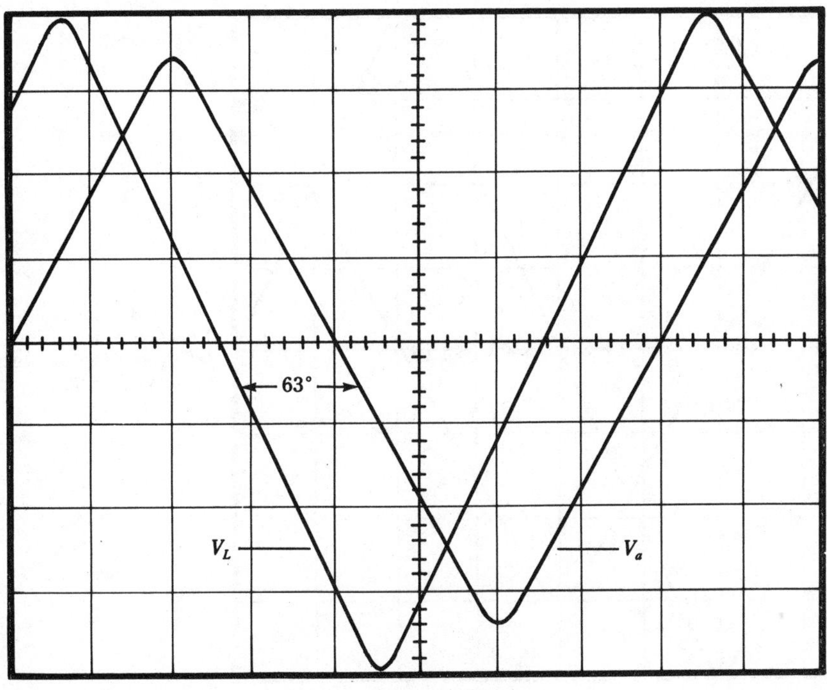

360° = 8 divisions
1 division = 45°
Difference = 1.4 divisions
Phase angle = 63°

10-15 *R* smaller than X_L.

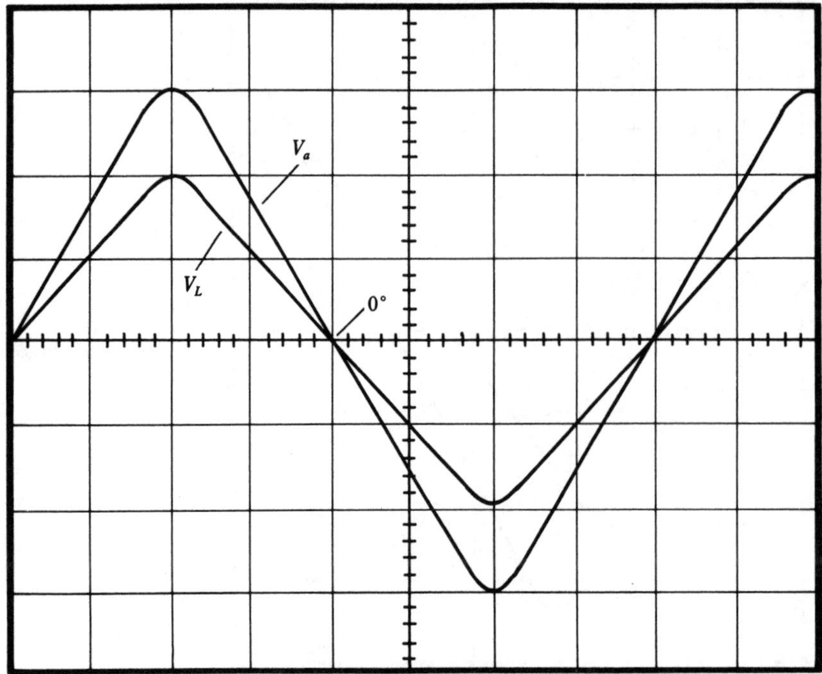

10-16 *R* much larger than X_L.

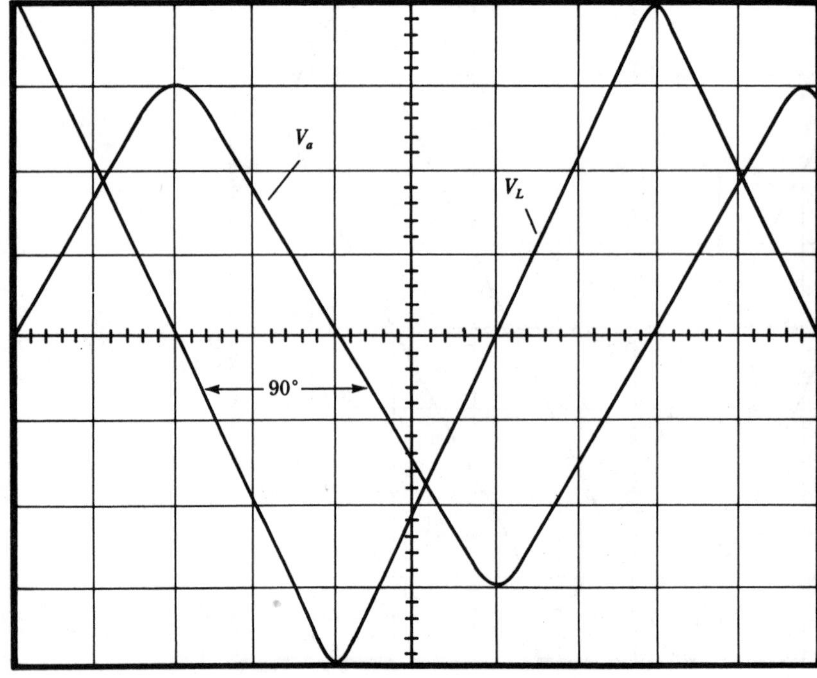

10-17 *R* much smaller than X_L.

Procedure for calculating the phase angle on an oscilloscope

This section is not intended to teach the use of the oscilloscope; therefore, it is assumed that you are familiar with the controls and basic operation.

Connect the dual trace scope—channel 1 across the entire circuit and channel 2 across the inductor. Connect the scope ground to the common ground point in the circuit.

The scope should display two sine waves. Channel 1 will display the applied voltage, V_a. Channel 2 will display the inductive voltage, V_L.

Because this is a series circuit with one voltage source, the frequency of both waveforms will be exactly the same. The amplitude of each waveform can be varied to make the measurements as easy as possible. The amplitude will not affect the phase angle to be measured. The waves should be as large as possible in order to have the best accuracy.

Adjust the scope to measure one full cycle, or a little more. Do not display too many cycles because this will make the readings difficult.

- Measure on the scope should be taken where the sine wave cross the zero reference line.

At the zero reference line, the sine wave will be the straightest of any point along the waveform. Peaks can be used but the accuracy is much less.

Follow the steps and examples to determine the phase angle in degrees:

Step 1 Count the number of divisions, accurate to tenths (0.1), in one complete cycle.

Step 2 Calculate the degrees per division by dividing 360 by the divisions in one cycle.

Step 3 At the zero reference line, measure the difference between the two waveforms, accurate to tenths of a division.

Step 4 Calculate the phase angle by multiplying the degrees per division by the difference.

Example, Fig. 10-12:

1. Divisions in one cycle = 8
2. $\dfrac{\text{degrees}}{\text{division}} = \dfrac{360 \text{ degrees}}{\text{number of divisions}}$

   $\dfrac{\text{degrees}}{\text{division}} = \dfrac{360}{8}$
3. Difference = 1.0 divisions
4. Phase angle = deg/div × diff

   = 45 deg/div × 1.0 div

   = 45 degrees

Note Both Figs. 10-12 and 10-13 have a 45-degree phase shift, but the two drawings look different. The reason is the way the scope is triggered. Figure 10-12 uses the V_a as a reference, and Fig. 10-13 uses V_L as the reference. The remainder of the sample drawings use V_a as the reference because this is the best way. Regardless of which method is used to trigger the scope, V_L will still be leading V_a by the phase shift.

In Figs. 10-12 and 10-13, the phase shift of 45 degrees indicates the values of X_L and R are equal.

Refer to Fig. 10-14. R is slightly larger than X_L. 8 blocks in 360 degrees. 45 degrees per division. V_L leads V_a by 0.6 divisions. $0.6 \times 45° = 27°$ phase shift.

Refer to Fig. 10-15. R is slightly smaller than X_L. 8 divisions = 360 degrees. V_L leads V_a by 1.4 divisions. $1.4 \times 360 = 63°$ phase shift.

Refer to Fig. 10-16. R is very large in comparison to X_L. Because R is so large, all the voltage would be dropped across the resistor and very little or none across the inductor. Resistors have no phase shift when compared to the applied voltage. Therefore, this drawing shows a 0-degree phase shift.

Refer to Fig. 10-17. The resistor is very small when it is compared to the X_L. Therefore, all of the voltage would be dropped across the inductor and none across the resistor. Since the inductor has a 90-degrees phase shift when compared to the applied voltage, the drawing shows a 90-degrees phase shift.

Chapter summary

The term *reactance* is used to describe the ac resistance of either a capacitor or an inductor. In this chapter, inductive reactance is investigated.

Inductive reactance varies with the frequency as well as the value of inductance. Inductors can be connected in series or parallel, just as resistors can, to arrive at different values of total reactance.

When an inductor is connected in series with a resistor, there is a phase shift between the voltage drops across the two components. The voltage across the inductor always leads the voltage across the resistor by 90 degrees. Another way to say this is to say the current lags the voltage by 90 degrees. A triangle can be formed by the voltages to give the total voltage of the series circuit. Another triangle can be formed to give the total impedance of the series circuit. Voltage or resistance and reactance cannot be simply added as in a dc circuit with only resistance. The adding must be performed using phasor triangles.

In a parallel circuit consisting of an inductor and a resistor, the current splits to the different parallel branches and the current of the inductor will lag the current of the resistor by 90 degrees. This develops a current triangle to solve for the total current.

Both series and parallel circuits have an operating angle or a phase shift that is the angle Θ, of the triangles used to solve the circuit.

When a circuit contains a reactive component, the power of the circuit is found in three terms; watts, *VARS, VA*. The resistance of the circuit is the only true power dissipated, with the unit watts. The reactive component has reactive power, with a unit of *VARS*. The total power of the circuit is the apparent power, with a unit of *VA*.

- Inductive reactance, X_L, is directly related to frequency and inductance.
- In a series circuit, I_L lags V_L, V_R lags V_L always by 90 degrees.
- An inductive circuit will always have a phase shift of between 0 degrees and 90 degrees, depending on circuit values.
- Impedance (Z) is the ac resistance of a circuit containing both resistance and reactance. It has an angle equal to the circuit phase angle.
- The voltage drops of a circuit containing resistance and reactance cannot simply be added. They must be added through vector addition.
- In a parallel circuit, I_L lags I_R by 90 degrees.
- Real power (also called true power) is the power dissipated in a pure resistance, unit is *W* (watts).
- Reactive power is the power dissipated in pure reactance, unit is VARS (volt-ampere-reactive).
- Apparent power is the power of a circuit containing both resistance and reactance, unit is *VA* (volt-ampere).
- Power factor is a ratio of the real power to the total power. It is a pure number with no units.
- Measurements made on the scope should be taken where the sine wave crosses the zero reference line.

Summary of formulas

$$X_L = 2\pi fL \quad \text{inductive reactance} \tag{10-1}$$

$$X_{LS} = X_{L1} + X_{L2} + X_{L3} + \ldots \quad \text{series reactances} \tag{10-2}$$

$$\frac{1}{X_{LP}} = \frac{1}{X_{L1}} + \frac{1}{X_{L2}} + \frac{1}{X_{L3}} + \ldots \quad \text{parallel reactances} \tag{10-3}$$

$$V_T \sqrt{V_R^2 + V_L^2} \quad \text{real voltage in a series circuit} \tag{10-4}$$

$$\Theta = \tan^{-1} \frac{V_L}{V_R} \quad \text{phase angle of a series circuit, using voltages} \tag{10-5}$$

$$Z = \sqrt{R^2 + X_L^2} \quad \text{total impedance of a series circuit} \tag{10-6}$$

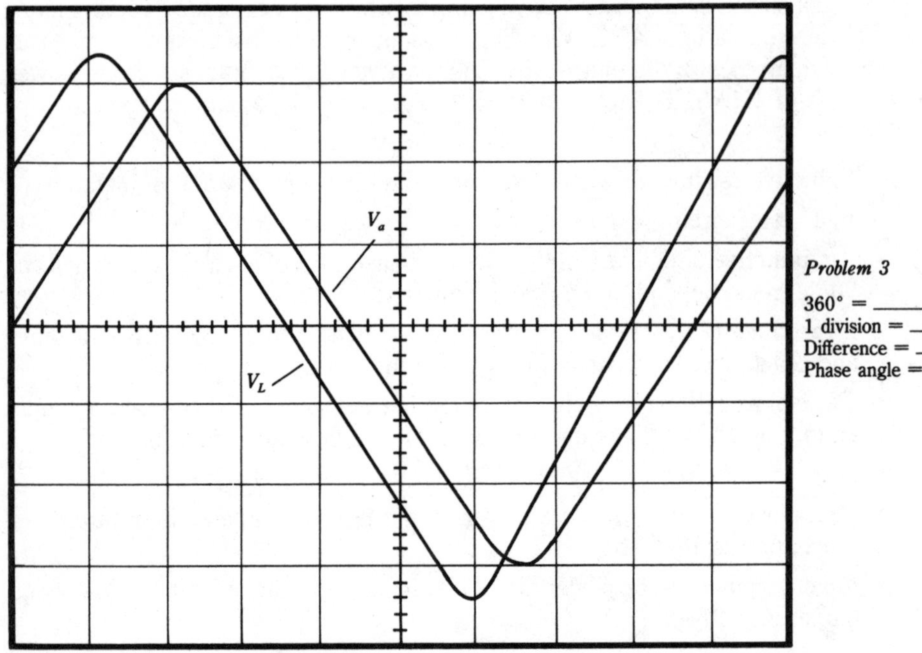

Problem 3

360° = _______ divisions
1 division = _______°
Difference = _______ divisions
Phase angle = _______°

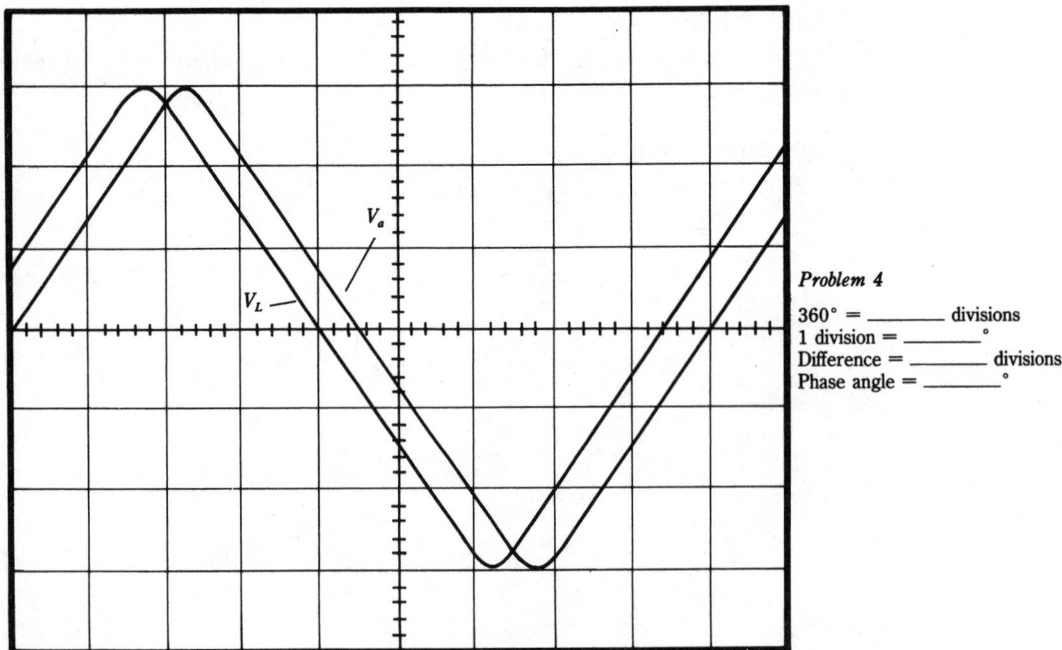

Problem 4

360° = _______ divisions
1 division = _______°
Difference = _______ divisions
Phase angle = _______°

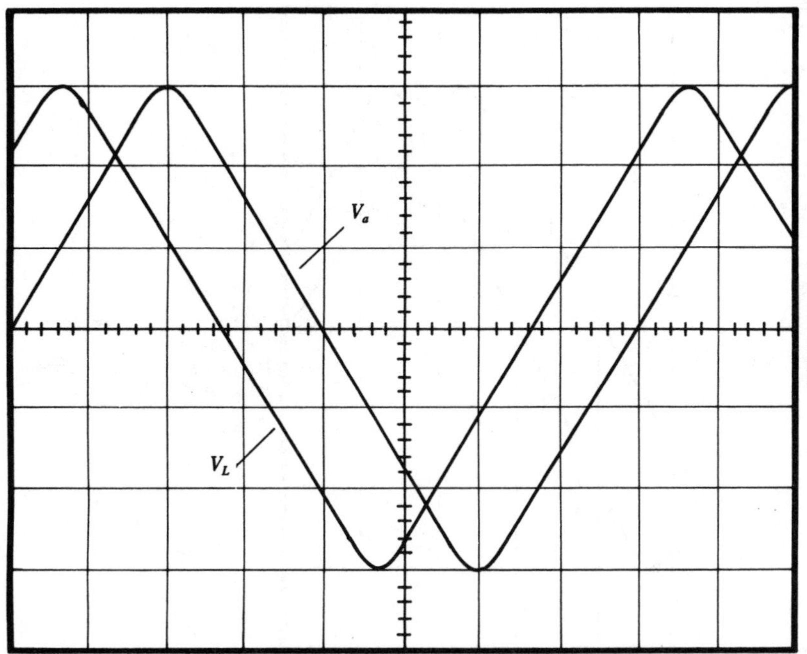

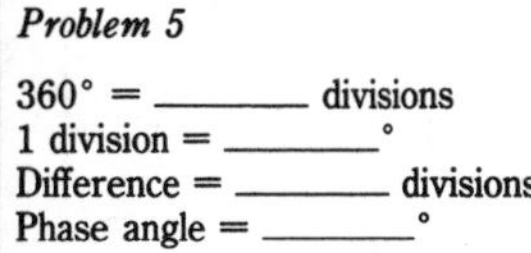

Problem 5

360° = ________ divisions
1 division = ________°
Difference = ________ divisions
Phase angle = ________°

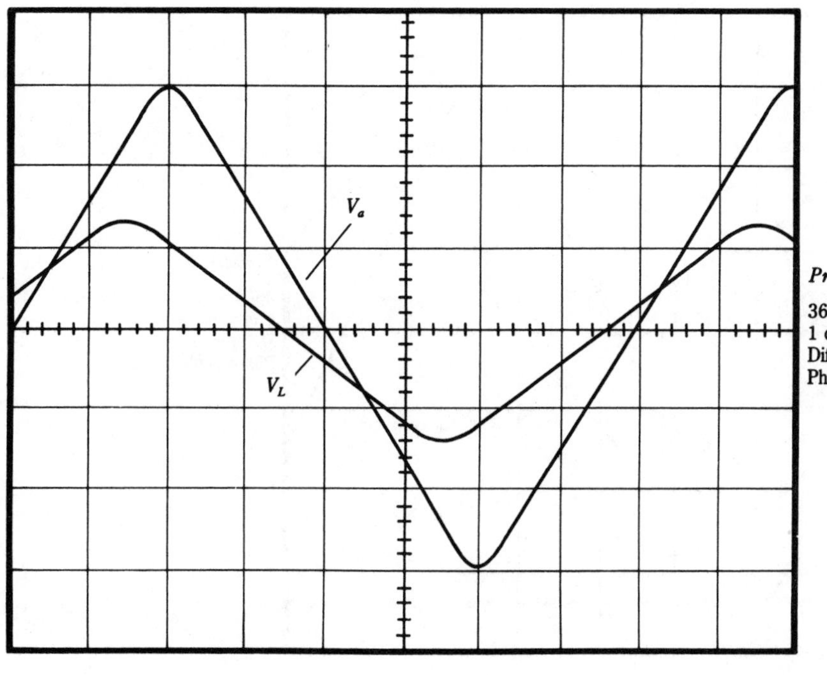

Problem 6

360° = ________ divisions
1 division = ________°
Difference = ________ divisions
Phase angle = ________°

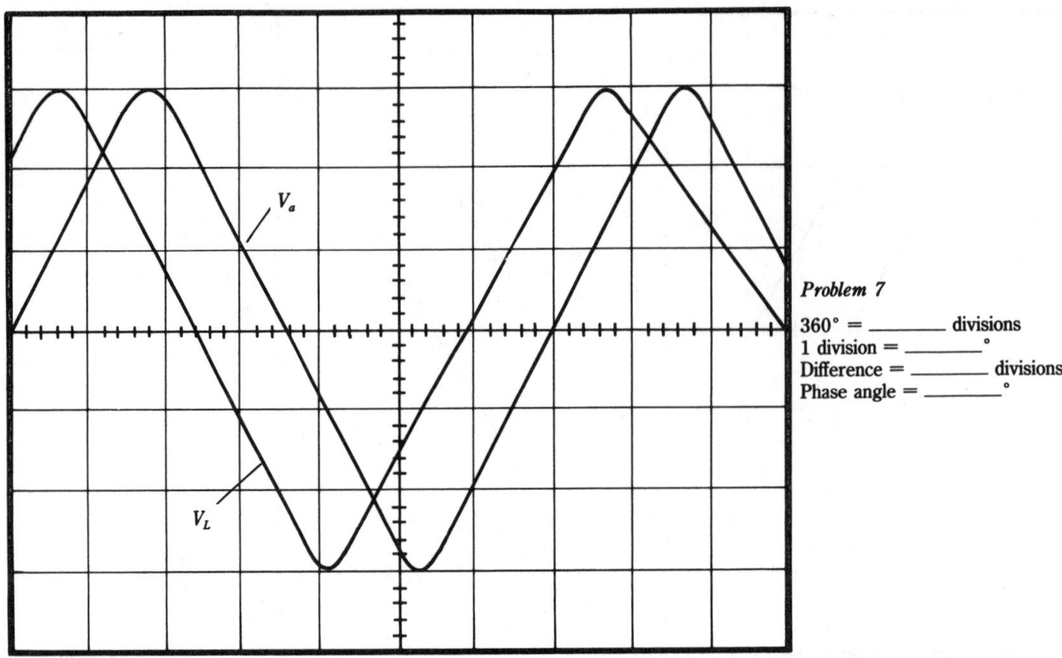

Problem 7

360° = ________ divisions
1 division = ________°
Difference = ________ divisions
Phase angle = ________°

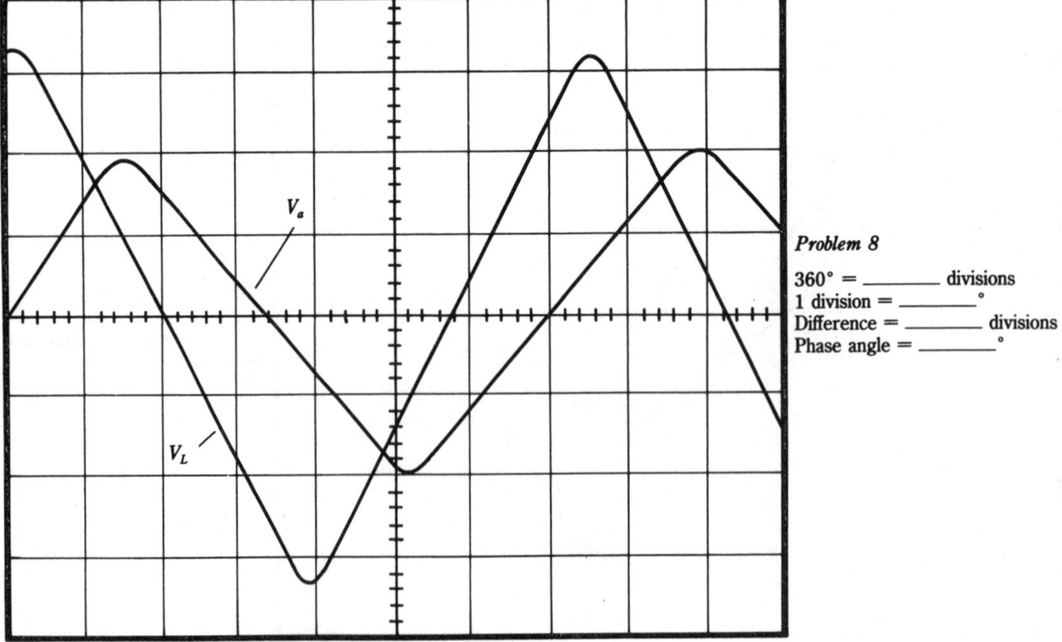

Problem 8

360° = ________ divisions
1 division = ________°
Difference = ________ divisions
Phase angle = ________°

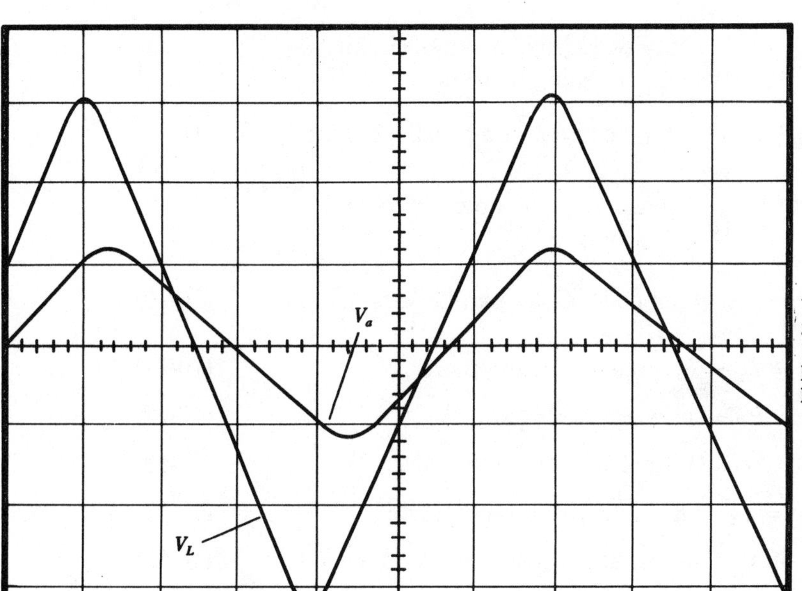

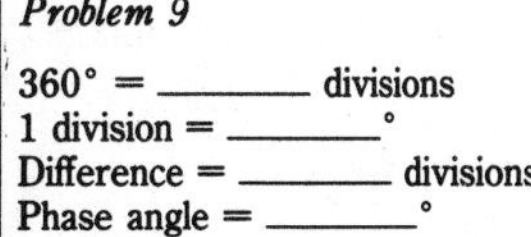

Problem 9

360° = ________ divisions
1 division = ________°
Difference = ________ divisions
Phase angle = ________°

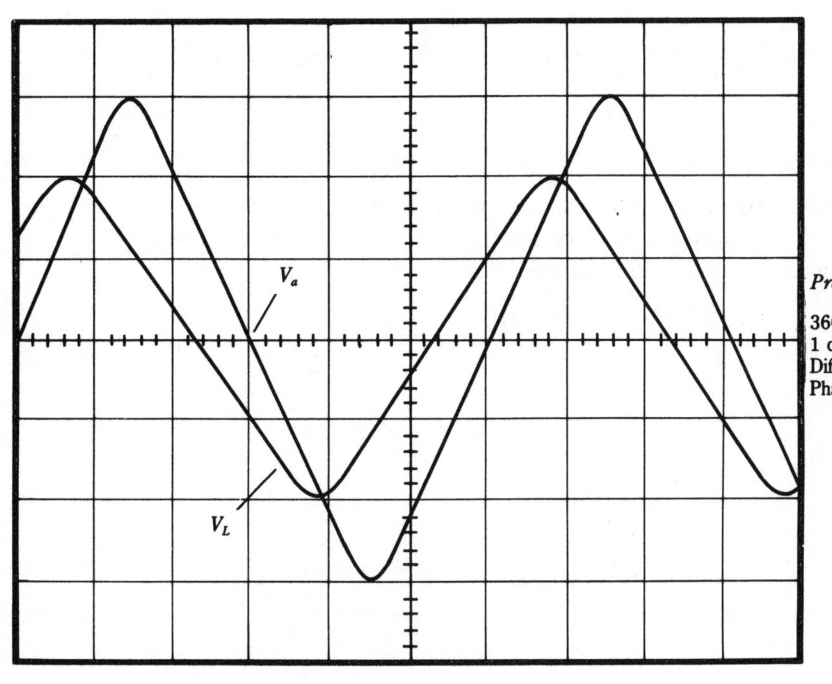

Problem 10

360° = ________ divisions
1 division = ________°
Difference = ________ divisions
Phase angle = ________°

$$\Theta = \tan^{-1}\frac{X_L}{R} \quad \text{phase angle of a series circuit, using resistance} \tag{10-7}$$

$$I_T = \sqrt{I_R^2 + I_L^2} \quad \text{total current of a parallel circuit,} \tag{10-8}$$

$$\Theta = \tan^{-1}\frac{I_L}{I_R} \quad \text{phase angle of a parallel circuit} \tag{10-9}$$

$$Z = \frac{V_a}{I_T} \quad \text{total impedance of a parallel circuit} \tag{10-10}$$

$$P_R = I_R \times E_R \quad \text{real—or true—power, W} \tag{10-11}$$

$$P_R = I^2 \times R \quad \text{real power, W} \tag{10-12}$$

$$P_R = VI\cos\Theta \text{ (total voltage)} \quad \text{real power, W} \tag{10-13}$$

$$P_X = I_X \times E_X \quad \text{reactive power, VARS} \tag{10-14}$$

$$P_X = VI\sin\Theta \text{ (total voltage)} \quad \text{reactive power, VARS} \tag{10-15}$$

$$P_A = VI \text{ (total voltage)} \quad \text{apparent power, VA} \tag{10-16}$$

$$PF = \cos\Theta \quad \text{power factor, no units} \tag{10-17}$$

$$PF = \frac{R}{Z} \text{ (no units)} \tag{10-18}$$

$$PF = \frac{I_R}{I_T} \quad \text{applies to parallel circuits, power factor (no units)} \tag{10-19}$$

Practice problems

Each of the problems are drawings representing an oscilloscope screen display with each drawing, find: a) number of divisions per cycle, b) degrees per division, c) differences between waveforms, and d) phase angle.

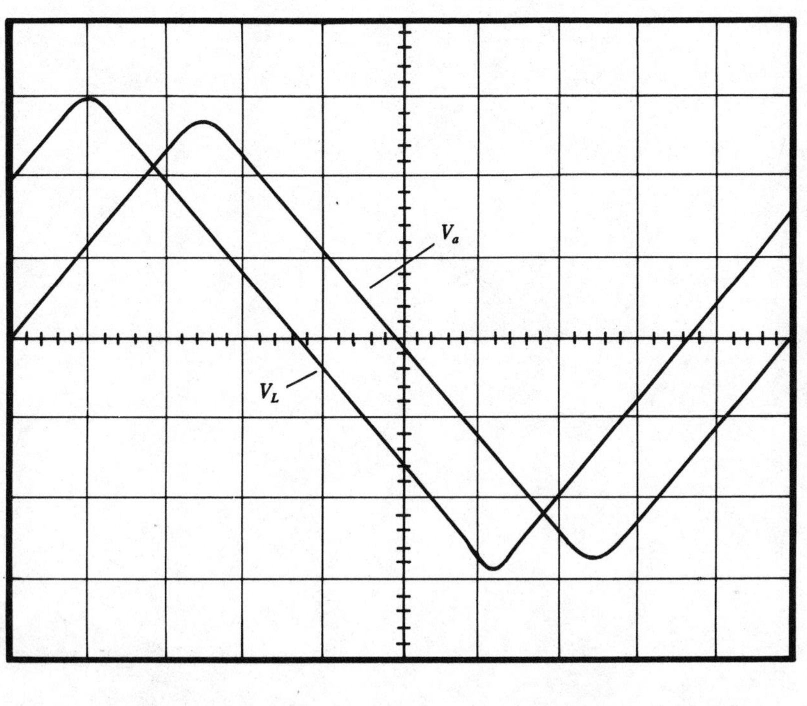

Problem 1

360° = ________ divisions
1 division = ________°
Difference = ________ divisions
Phase angle = ________°

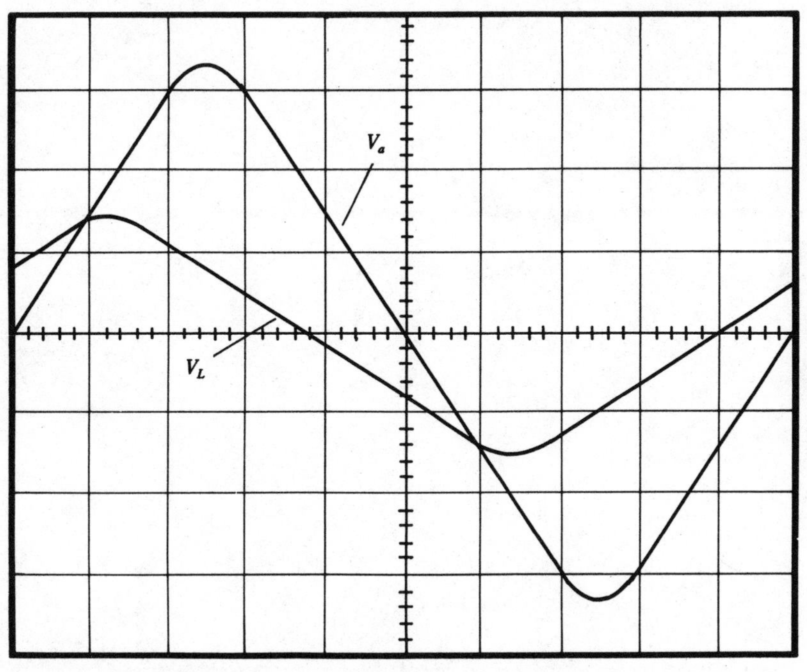

Problem 2

360° = ________ divisions
1 division = ________°
Difference = ________ divisions
Phase angle = ________°

11
Capacitive reactance

THE PREVIOUS CHAPTER DISCUSSES INDUCTIVE REACTANCE AS BEING AN ac resistance. This same general statement also applies to a circuit containing a capacitor with an ac power source.

It will be demonstrated that many of the statements made about inductive circuits will also apply to capacitors. The big difference will be that capacitive reactance has an inverse relationship to frequency and capacitance, also the phasor triangles will be drawn opposite those of the inductor circuits.

Capacitive reactance

$$X_C = \frac{1}{2\pi fC} \tag{11-1}$$

The factor 2π is a constant term in this formula.
f is the frequency measured in hertz.
C is capacitance measured in farads.
X_C is capacitance reactance measured in ohms.

Notice from the formula, the capacitive reactance varies inversely with the frequency and capacitance value. An inverse relationship means that the X_C is smaller with larger values of either f or C.

- Capacitive reactance, X_C, is indirectly related to frequency and capacitance.

Sample calculations of capacitive reactance

The following sample calculations show how to use the capacitive reactance formula. When using a reciprocal formula, such as this one, perform the arithmetic in the denominator first, then use $1/x$ on the calculator to take the reciprocal.

If $C = 0.22\ \mu F$ and $f = 2500$ Hz, find X_C.

$$X_C = \frac{1}{2\ \pi\ fC} \qquad (11\text{-}1)$$

$$= \frac{1}{2 \times \pi \times 2500\text{ Hz} \times 0.22\ \mu\text{F}}$$

$$= X_C = 289\ \Omega$$

If $X_C = 100\ \Omega$ and $f = 10$ kHz, find the capacitance.

$$X_C = \frac{1}{2\ \pi\ fC}$$

$$C = \frac{1}{2\ \pi\ f\ X_C}$$

$$= \frac{1}{2 \times \pi \times 10{,}000\text{ Hz} \times 100\ \Omega}$$

$$= 0.159\ \mu\text{F}$$

If $X_C = 100\ \Omega$ and $C = 15\ \mu F$, find the frequency.

$$X_C = \frac{1}{2\ \pi\ fC} \qquad (11\text{-}1)$$

$$f = \frac{1}{2\ \pi\ C\ X_C}$$

$$= \frac{1}{2 \times \pi \times 15\ \mu\text{F} \times 100\ \Omega}$$

$$= 106\text{ Hz}$$

Practice problems

Find the value of the unknown quantity using the information given.

1. $f = 1000$ Hz, $C = 0.01\ \mu F$; find X_C.
2. $X_C = 1.59$ kΩ, $f = 1000$ Hz; find C.
3. $X_C = 15.9\ \Omega$, $C = 1\ \mu F$; find f.
4. $X_C = 15.9$ kΩ, $f = 1000$ Hz; find C.
5. $X_C = 159\ \Omega$, $C = 1\ \mu F$; find f.
6. $f = 10$ kHz, $C = 150\ \mu F$; find X_C.
7. $f = 0$ Hz (dc), $C = 10\ \mu F$; find X_C.
8. $f = 3$ MHz, $C = 0.002\ \mu F$; find X_C.
9. $f = 1$ kHz, $C = 5\ \mu F$; find X_C.
10. $f = 1000$ MHz, $C = 1$ pF; find X_C.

Series or parallel capacitive reactance

Whenever two or more of the same type of reactance are connected together, the total value of the reactance is calculated the same as resistors in a dc circuit.

$$X_{Cs} = X_{C1} + X_{C2} + X_{C3} \ldots \text{series } X_C \qquad (11\text{-}2)$$

$$\frac{1}{X_{Cp}} = \frac{1}{X_{C1}} + \frac{1}{X_{C2}} + \frac{1}{X_{C3}} \ldots \text{parallel } X_C \qquad (11\text{-}3)$$

When it is necessary to calculate voltage drops, with only reactance in the circuit, follow Ohm's Law. This only applies when there is only reactance in the circuit, not resistance and reactance.

Figure 11-1 shows a series circuit with two capacitors connected to a 10 Vac power supply. The largest voltage drop is across the largest capacitive reactance. Keep in mind, however, the largest capacitive reactance results from the smallest value of capacitance.

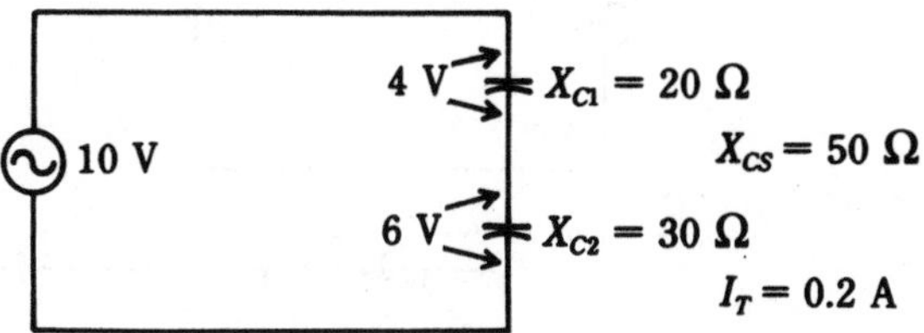

11-1 Series capacitive reactance.

Figure 11-2 shows a parallel circuit with two capacitors connected in parallel across and ac power supply. Note that the larger current flows through the smallest value of reactance.

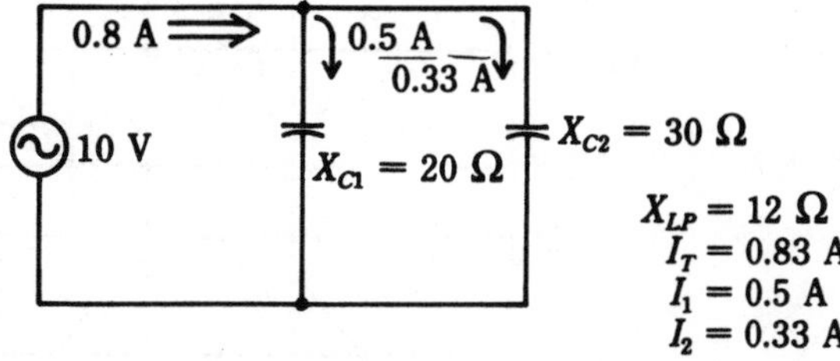

11-2 Parallel capacitive reactance.

Because the two sample circuits, Figs. 11-1 and 11-2, contain only capacitance and no resistance or inductance, the values of voltage drops and current can be calculated based on Ohm's Law. If the circuit contains any other component, it is necessary to use vector addition.

Series X_C and R

The current is the same throughout the series circuit, regardless of the types of components contained in the circuit.

In a circuit containing capacitance, the capacitor must charge and discharge. This charging, discharging process causes a delay or phase shift in the circuit. The current in the resistor is considered the reference point. The resistive voltage drop will be the same phase as the circuit current. The current in the capacitor will be the same also since it is a series circuit. The capacitive voltage drop, however, will be 90 degrees out of phase with the current. This will be a lagging voltage or a leading current.

- In a series circuit, I_C leads V_C, V_R leads V_C always by 90 degrees.

Refer to Fig. 11-3 for the sine wave analysis of a series circuit containing capacitance and resistance. Notice how the resistor voltage is plotted in phase with the current (Fig. 11-3) where the capacitive voltage is shown as lagging the resistor

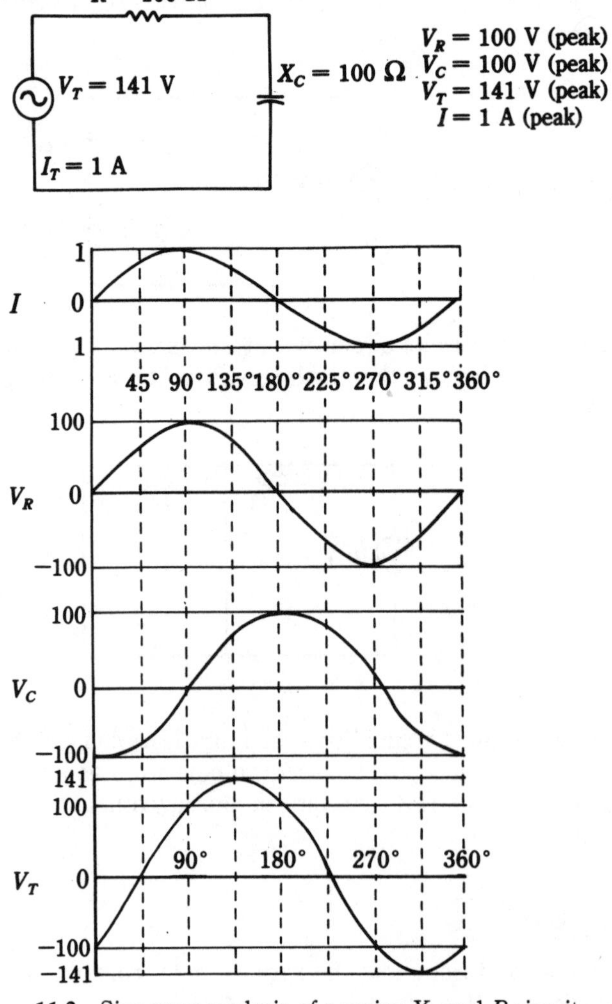

11-3 Sine wave analysis of a series X_C and R circuit.

voltage by 90 degrees. The total voltage is the vector sum of these two. Figure 11-4 shows the vectors with the capacitive voltage drop plotted straight down. The line drawn down represents a phase angle of −90 degrees. The circuit phase angle (also called phase shift or operating angle) is a negative angle. Compare this drawing to the same drawing for an inductive circuit (Fig. 10-5).

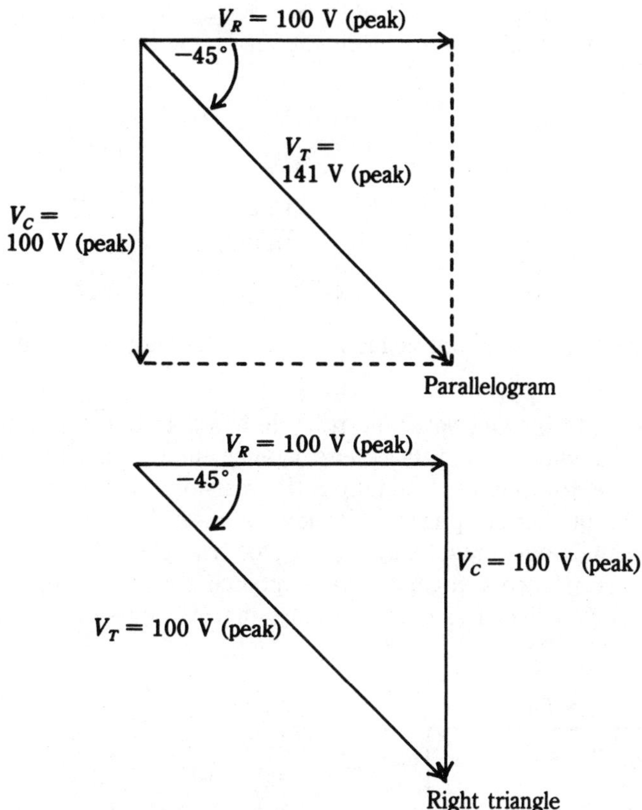

11-4 Vector analysis of a series X_C and R circuit.

- The voltage vectors for inductance and capacitance are opposite.
- The operating angle of a circuit with a capacitor in series with a resistor will be between 0 degrees and −90 degrees, depending on circuit values.

Calculating the triangle

Figure 11-4 is the vector analysis of Fig. 11-3.

V_R is plotted on the horizontal 0 degrees axis.
V_C is plotted on the vertical −90 degrees axis.
V_T is plotted to connect the two vectors.

Calculate the magnitude, length, of V_T.

$$= \sqrt{V_R^2 + V_C^2} \quad \text{total voltage in a series circuit} \qquad (11\text{-}4)$$

$$= \sqrt{100^2 + 100^2}$$

$$= 141 \text{ V}$$

Calculating the operating angle of the series circuit, in other words, the hypotenuse, V_T, of the voltage triangle forms the angle, Θ.

$$\Theta = \tan^{-1} \frac{V_C}{V_R} \qquad (11\text{-}5)$$

$$= \tan^{-1} \frac{100 \text{ V}}{100 \text{ V}}$$

$$= -45°$$

Note The angle will come out negative on the calculator if V_C is used as a negative voltage.

In the example shown above, the triangle has a 45 degree angle. This could have been predicted by the fact that the resistance and the capacitive reactance are equal. Therefore, the voltage drop across the resistor and capacitor would be the same value with 90 degrees phase difference between them. Having the two sides of the right triangle equal results in a 45 degree triangle.

When the resistance and capacitive reactance are not equal, it is possible to predict if the angle will be less than or greater than 45 degrees as demonstrated by Fig. 11-5.

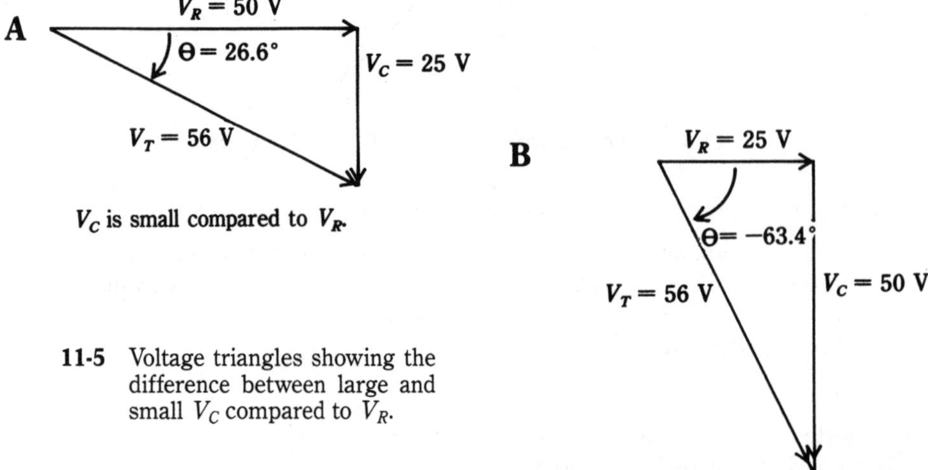

11-5 Voltage triangles showing the difference between large and small V_C compared to V_R.

The first triangle has the resistance larger than the reactance, therefore, the phase shift is a negative angle less than −45 degrees. The second triangle has the resistance smaller than the capacitive reactance, therefore, the operating angle will have a negative angle, with a magnitude greater than −45 degrees.

Calculating the impedance in a series circuit

The calculations for the impedance of a series circuit with a reactive component is the same as calculating the total resistance of a dc circuit. Impedance is the total ac resistance of a circuit. It will have an angle equal to the operating angle of the circuit. In other words, the operating angle can be found using either the voltage triangle or the impedance triangle.

Figure 11-6 shows a series circuit with its impedance and voltage triangles. Notice both triangles are plotted in the negative direction, which is opposite the triangles for an inductive circuit. See Fig. 10-7 to compare the inductive circuit.

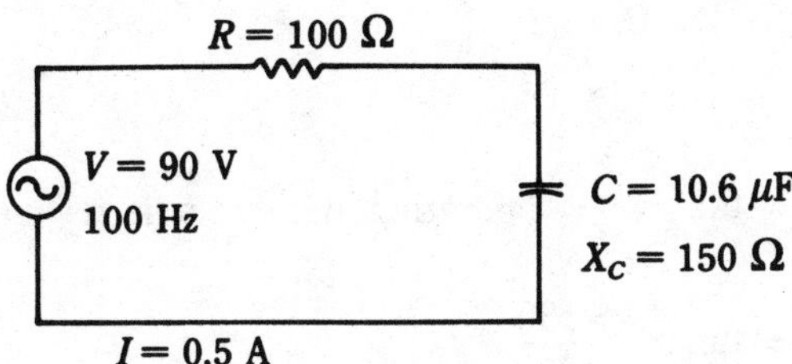

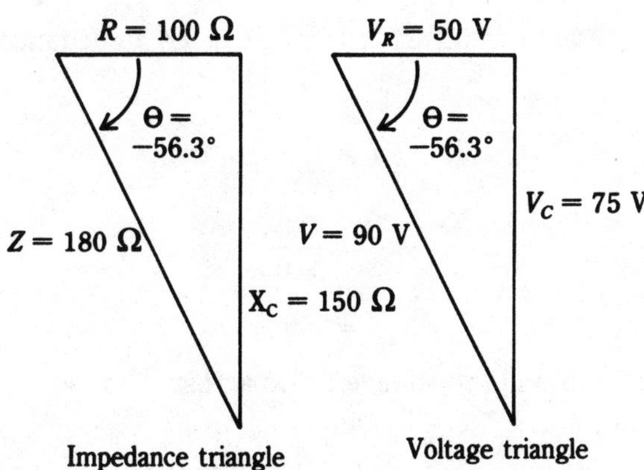

11-6 Solving a series *RC* circuit using phasor triangles.

Calculating a complete series circuit

Figure 11-6 given values: $C = 10.6\ \mu F$, $f = 100$ Hz, $V = 90$ V, $R = 100\ \Omega$. Find: X_C, Z, I, V_R, and V_C.

Calculate the X_C.

$$X_C = \frac{1}{2\,\pi\,fC} \tag{11-1}$$

$$= \frac{1}{2 \times \pi \times 100\text{ Hz} \times 10.6\,\mu\text{F}}$$

$$= 150\ \Omega$$

Calculate impedance (Z) using the impedance triangle. Plot R on the horizontal and X_C down.

$$Z = \sqrt{R^2 + X_C^2} \quad \text{impedance of a series circuit} \tag{11-16}$$

$$= \sqrt{100^2 + 150^2}$$

$$= 180\ \Omega$$

Calculate the phase angle, Θ.

$$\Theta = \tan^{-1}\frac{X_C}{R} \quad \text{operating angle of a series circuit} \tag{11-17}$$

$$= \tan^{-1}\frac{>150}{100}$$

$$= = \tan^{-1}\frac{>150}{100}$$

Calculate the total current using the calculated impedance and the applied voltage.

$$I = \frac{V}{Z} \tag{2-1A}$$

$$= \frac{90\text{ V}}{180\ \Omega}$$

$$= 0.5\text{ A}$$

Use the current to find the voltage drops across the resistor and capacitor.

$$\begin{aligned} V_R &= I \times R \\ &= 0.5\text{ A} \times 100\ \Omega \\ &= 50\text{ V} \end{aligned} \tag{2-1}$$

$$\begin{aligned} V_C &= 0.5\text{ A} \times 150\ \Omega \\ &= 75\text{ V} \end{aligned}$$

Refer to Fig. 11-6 to see the voltage drops plotted in the voltage triangle. The hypotenuse of the voltage triangle should calculate to be equal to the voltage given as the applied voltage.

$$V = \sqrt{V_R^2 + V_C^2} \tag{11-14}$$
$$= \sqrt{50^2 + 75^2}$$
$$= 90 \text{ V}$$

Summary for series circuits

- Calculate the capacitive reactance if it is not already given.
- Calculate the impedance, Z, using the impedance triangle.

$$Z = \sqrt{R^2 + X_C^2} \tag{11-14}$$

- Calculate the operating angle, Θ, based on the impedance triangle. The angle can also be found using the voltage triangle.

$$\Theta = \tan^{-1} \frac{X_C}{R} \tag{11-17}$$

- Calculate the total circuit current using the applied voltage and the impedance.
- Use the current to calculate the voltage drops across each component. Use Ohm's Law.
- Compare the voltage drops to the given applied voltage by using vector addition with the voltage triangle. Theta can also be found using the voltage triangle, rather than the impedance triangle.

Practice problems

Use the two schematics to calculate the answers to the problems.

Use schematic A for problems 1 and 2. Frequency is 60 Hz. Find: C_T, X_{C1}, X_{C2}, X_{C3}, and X_{CT}.

1. $C_1 = 1\ \mu F$, $C_2 = 0.01\ \mu F$, $C_3 = 0.1\ \mu F$
2. $C_1 = 47\ \mu F$, $C_2 = 150\ \mu F$, $C_3 = 100\ \mu F$

Use schematic B to answer problems 3 through 8. V_a is 100 V. Find: Z, I, V_R, V_C, Θ, and draw the impedance triangle.

3. $R = 100\ \Omega$, $X_C = 100\ \Omega$
4. $R = 25\ \Omega$, $X_C = 50\ \Omega$
5. $R = 10\ \Omega$, $X_C = 100\ \Omega$
6. $R = 75\ \Omega$, $X_C = 5\ \Omega$

7. $R = 75\ \Omega$, $X_C = 25\ \Omega$
8. $R = 0\ \Omega$, $X_C = 50\ \Omega$

Use schematic B to answer problems 9 and 10. Find: R, X_C, Z, V_a, Θ, and draw voltage triangle.

9. $V_R = 20$ V, $V_C = 40$ V, $I = 0.25$ A
10. $V_R = 25$ V, $V_C = 15$ V, $I = 0.333$ A

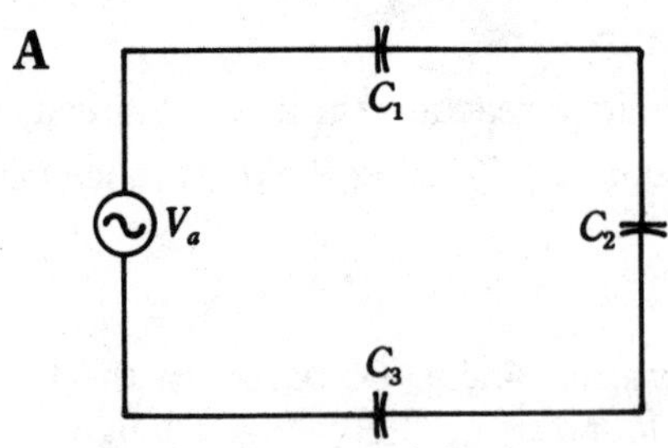

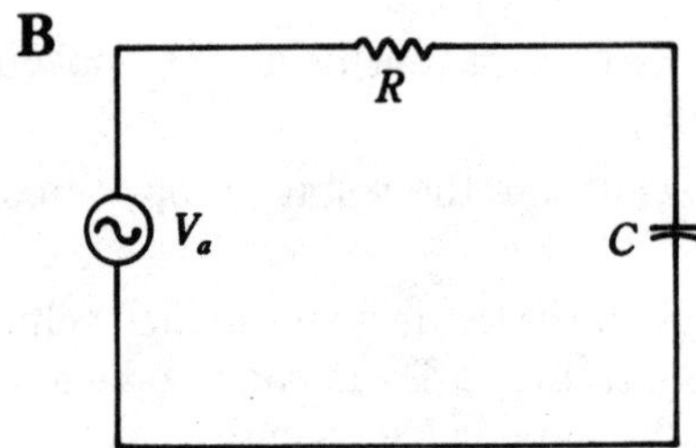

Parallel X_C and R

The rules for a parallel circuit state that the voltage throughout a parallel circuit is the same and that the current will divide to the individual branches. Because the voltage is the same throughout the parallel circuit, there cannot be any phase shift. There can, however, be a phase shift in the current to the individual components.

Because a capacitor causes the voltage to have a lagging relationship, it makes sense then that the current will be a leading current when compared to the current of a resistor in parallel.

- In a parallel circuit, I_C leads I_R by 90 degrees.

Figure 11-7 shows the sine wave analysis of a circuit with equal values of R and X_C. The equal values have been chosen to simplify the drawings, using a 45-degree phase angle. Notice the resistive current is in phase with the applied voltage; you would expect that because resistors do not cause any phase shift. The capacitive current is then shown leading the resistive current by 90 degrees.

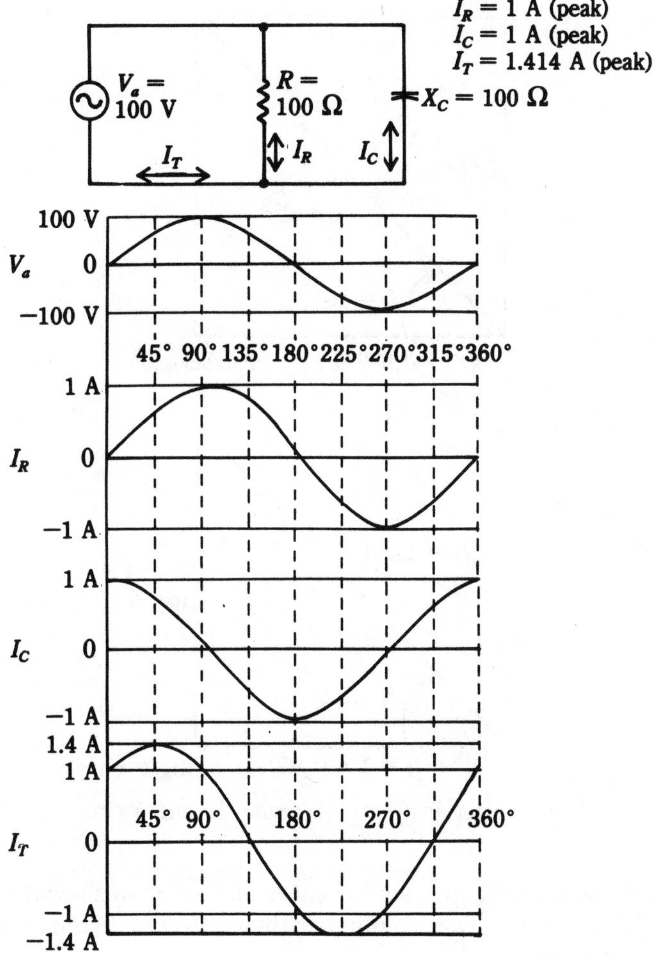

11-7 Sine wave analysis of a parallel X_C and R circuit.

Because the two currents are out of phase, the total current must be between the two currents, with its phase shift. Each point along the total current line is found by simply adding the individual points along the sine waves of the two other currents. The resultant total current will have a 45-degree phase shift, as shown, and this phase shift will be leading the applied voltage. The phase angle is therefore considered a positive angle.

Figure 11-8 is the vector analysis of Fig. 11-7. Figure 11-8A is the parallelogram drawing of the vectors and Fig. 11-8B is the right triangle method of drawing the vectors. The only difference between the two drawings is the point at which the capacitor current vector has its starting point. In the parallelogram, it starts at the same point as the resistor vector. In the triangle, it starts at the tail end of the resistor vector. The triangle method is the preferred method in this book, although it is a matter of personal preference.

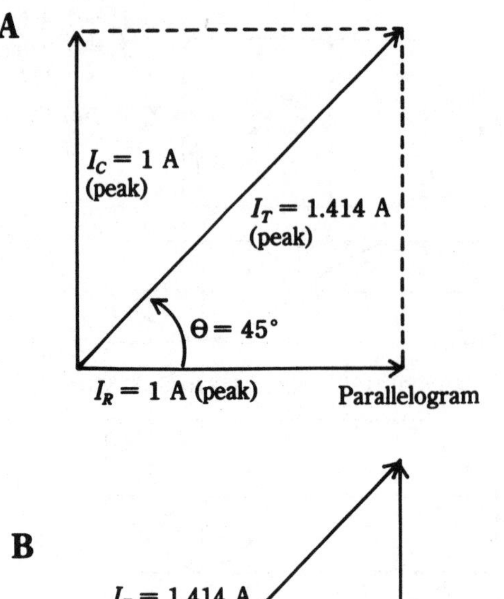

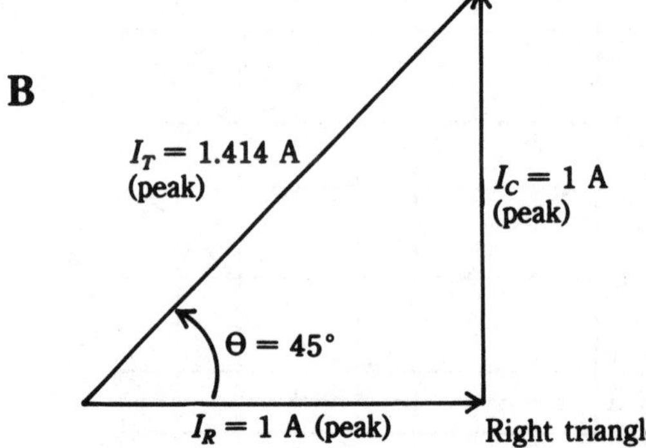

11-8 Vector analysis of a parallel X_C and R circuit.

Figure 11-8 shows the capacitor current is plotted up to show the leading condition. Resistive current is always plotted on the horizontal, to the right. The phase angle is a positive angle.

Calculating the hypotenuse of the right triangle is the square root of the sum of the squares formula and the phase angle is the inverse tangent function as follows:

$$I_T = \sqrt{I_R^2 + I_C^2} \quad \text{total current for a capacitor in parallel with a resistor} \tag{11-8}$$

$$\Theta = \tan^{-1} \frac{I_C}{I_R} \quad \text{phase angle of an RC parallel circuit} \tag{11-9}$$

The impedance is determined by using Ohm's Law with the total current and the applied voltage.

$$Z = \frac{V}{I_T} \quad \text{impedance of a parallel RC circuit} \tag{11-10}$$

Figure 11-9 shows two different triangles to demonstrate the difference between an I_C larger than I_R and an I_C smaller than I_R. The larger phase angle, Fig.

A

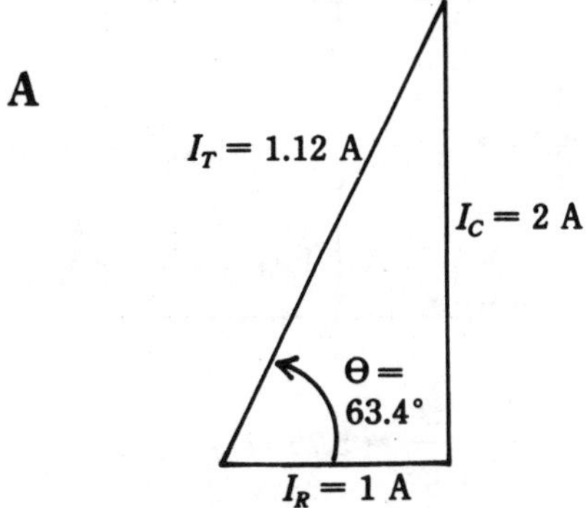

I_C is large compared to I_R.

B

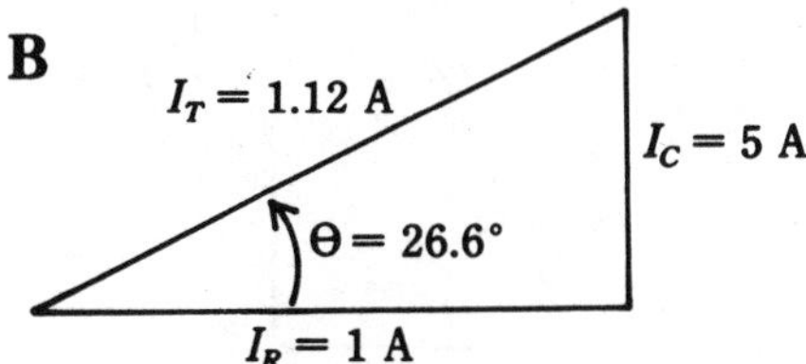

I_C is small compared to I_R.

11-9 Current triangles showing the difference between large and small I_C compared to I_R.

11-9A, is caused by a larger amount of I_C. The smaller phase angle results when the resistive current is larger.

To continue the analysis of the larger and smaller currents shown in Fig. 11-9, consider which one would have the larger or smaller value of capacitance. Take the larger capacitive current. A larger current means the value of the capacitive reactance must be smaller. A smaller value of reactance means the value of capacitance is larger, assuming there has been no change in frequency. That is, the capacitance value stays the same and the frequency is caused to vary. The larger the frequency, the smaller the capacitive reactance. Therefore, the two triangles shown in Fig. 11-9 could represent a circuit where the values of resistance and capacitance are not changed, only the frequency is changed. Figure 11-9B would be the current triangle for a parallel RC circuit with a low frequency and Fig. 11-9A would be the same circuit for a higher frequency.

Calculating a parallel circuit

Figure 11-10 shows a parallel RC circuit and the current triangle used in performing the calculations. It is very important to keep in mind that the total current is the vector sum of the branch current; it is not simply an arithmetic sum.

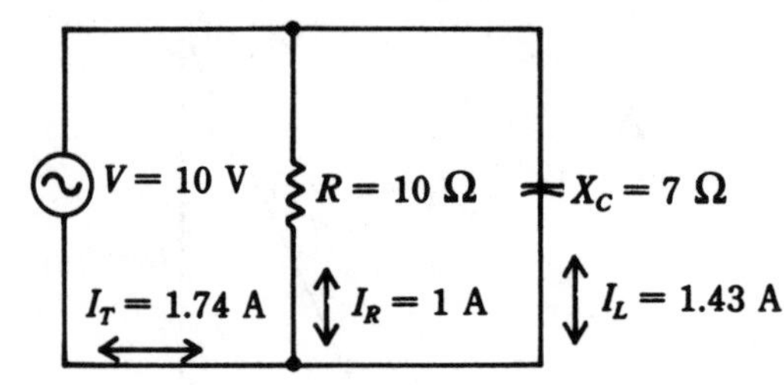

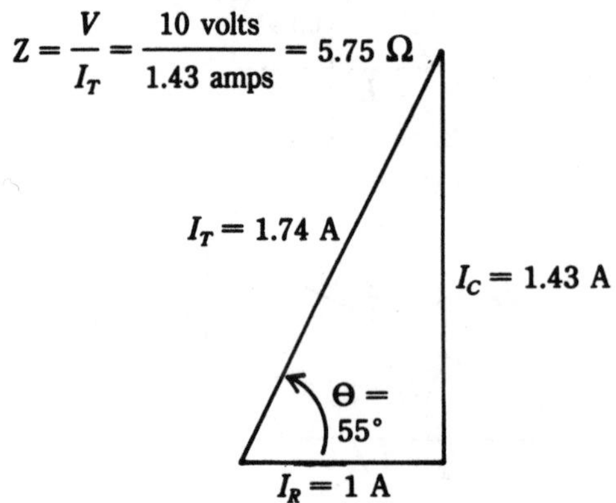

11-10 Using the current triangle to solve a parallel circuit with capacitance and resistance.

In a parallel circuit, the applied voltage is the same across all branches of the parallel circuit.

Figure 11-10, given: $V = 10$ V, $R = 10\ \Omega$, $X_C = 7\ \Omega$. Find: I_R, I_C, I_T, Θ, and Z. Calculate the resistive current, I_R.

$$I_R = \frac{V}{R} \tag{2-1A}$$

$$= \frac{10\text{ V}}{10\ \Omega}$$

$$= 1\text{ A} \quad I_R \text{ is plotted in the triangle at 0 degrees}$$

Calculate the capacitive current, I_C.

$$I_C = \frac{V}{X_C}$$

$$= \frac{10\text{ V}}{7\ \Omega}$$

$$= 1.43\text{ A} \quad \text{plot at } +90 \text{ degrees in the triangle}$$

Calculate the hypotenuse of the current triangle, the total current, using the Pythagorean theorem.

$$I_T = \sqrt{I_R^2 + I_C^2} \qquad (11\text{-}18)$$

$$= \sqrt{1^2 + 1.43^2}$$

$$= 1.74 \text{ A}$$

Calculate the operating angle, Θ.

$$\Theta = \tan^{-1} \frac{I_C}{I_R}$$

$$= \tan^{-1} \frac{1.43 \text{ A}}{1 \text{ A}}$$

$$= 55 \text{ degrees}$$

Calculate the total impedance, Z, using the calculated total current and the given applied voltage.

$$Z = \frac{V}{I_T} \qquad (2\text{-}1B)$$

$$= \frac{10 \text{ V}}{1.74 \text{ A}}$$

$$= 5.75 \ \Omega$$

Summary for parallel circuits

- Calculate the current for each of the parallel branches, using Ohm's Law.
- Calculate the total current using the branch currents to form the current triangle.
- Calculate the operating angle based on the current triangle. The angle will be positive for a capacitive circuit.
- Calculate the total impedance using the total current and the applied voltage.

Practice problems

Use the schematics to calculate the answers to the problems.

Use schematic A to answer problems 1 and 2. Frequency is 100 Hz. Find: C_T, X_{C1}, X_{C2}, X_{C3}, and X_{CT}.

1. $C_1 = 1\ \mu F$, $C_2 = 2\ \mu F$, $C_3 = 3\ \mu F$
2. $C_1 = 50\ \mu F$, $C_2 = 100\ \mu F$, $C_3 = 75\ \mu F$

A

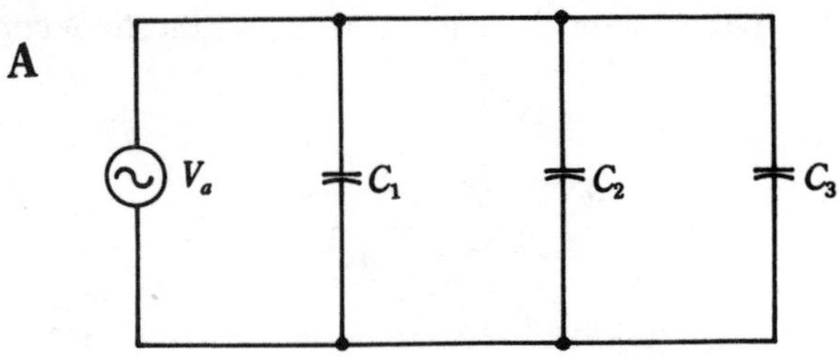

B

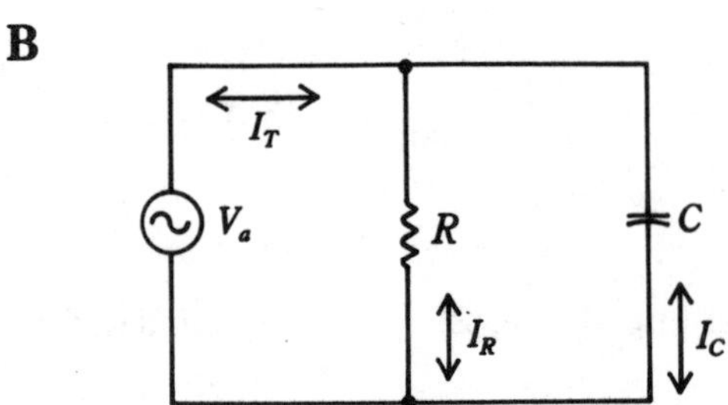

Use schematic B to answer questions 3 through 8. $V_a = 100$ V. Find: I_R, I_C, I_T, Z, and Θ.

3. $R = 100\ \Omega$, $X_C = 100\ \Omega$
4. $R = 25\ \Omega$, $X_C = 50\ \Omega$
5. $R = 10\ \Omega$, $X_C = 100\ \Omega$
6. $R = 75\ \Omega$, $X_C = 5\ \Omega$
7. $R = 75\ \Omega$, $X_C = 25\ \Omega$
8. $R = 5\ \Omega$, $X_C = 5\ \Omega$

Use schematic B to answer questions 9 and 10. $V_a = 100$ V. Find: R, X_C, I_T, Θ, Z, and draw current triangle.

9. $I_R = 1$ A, $I_C = 3$ A
10. $I_R = 25$ mA, $I_C = 10$ mA

Power in a reactive circuit

Power in a reactive circuit is the same whether it is an inductive circuit, as in chapter 8, or it is a capacitive circuit. In fact, all the calculations are the same. Because it is a fairly important concept, the key points are shown here as they were in chapter 8.

- Real power (also called true power) is the power dissipated in pure resistance, unit is watts.

- Reactive power is the power dissipated in pure reactance, unit is *VARS* (volt-ampere-reactive).
- Apparent power is the power of a circuit containing both resistance and reactance. It is calculated by: $I \times E$ (multiplying total current by the applied voltage), unit is *VA* (volt-ampere).
- Power factor is a ratio of the real power to the total power. It is a pure number, with no units. It will always be between 0 and 1. It is calculated by taking the cosine of the operating angle, cos Θ.

Note Sample calculations and a summary of formulas can be found in chapter 8.

Practice problems

Use the two schematic diagrams shown to calculate the answers to the problems.

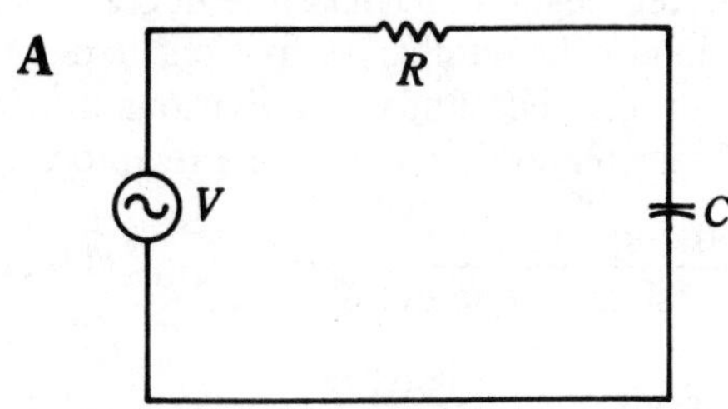

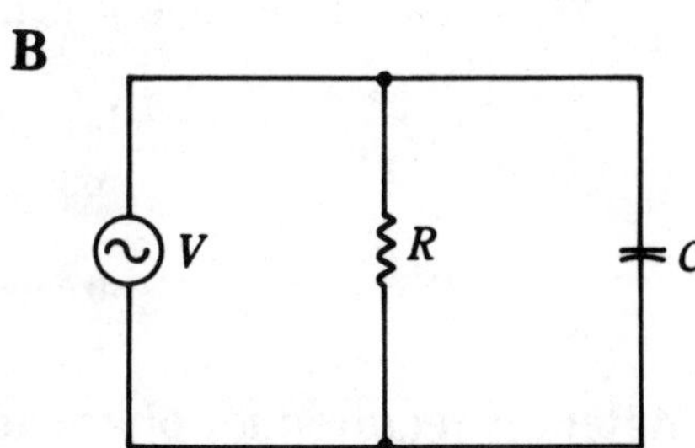

Use schematic A for problems 1 through 5. V = 100 V. Find: Z, I, V_R, V_C, Θ, P_R, P_X, P_A, and PF.

1. R = 500 Ω, X_C = 500 Ω
2. R = 100 Ω, X_C = 250 Ω
3. R = 1000 Ω, X_C = 750 Ω
4. R = 1000 Ω, X_C = 500 Ω
5. R = 10 Ω, X_C = 20 Ω

Use schematic B for problems 6 through 10. V = 100 V. Find: I_R, I_C, I_T, Z, Θ, P_R, P_X, P_A, and PF

6. R = 500 Ω, X_C = 500 Ω
7. R = 100 Ω, X_C = 250 Ω

8. $R = 1000\ \Omega$, $X_C = 750\ \Omega$
9. $R = 1000\ \Omega$, $X_C = 500\ \Omega$
10. $R = 10\ \Omega$, $X_C = 20\ \Omega$

Measuring phase angle with an oscilloscope

The end of chapter 8 discusses how to make phase-angle measurements using an oscilloscope. To perform phase-angle measurements with inductors or capacitors is exactly the same with the exception being that the capacitor has a lagging phase angle (for voltage), and the inductor has a leading phase angle.

In order for this section not to be a repeat of the section in chapter 10, this section explains the calculations necessary for you to set up experiments in measuring the phase angle using an oscilloscope.

Keep in mind that the phase angle is found by first counting the number of divisions in one cycle and dividing that into 360 degrees to determine the degrees per division. Then multiply the degrees per division by the number of divisions to determine the phase angle. Find frequency by counting the number of divisions in one cycle and multiplying that by the time per division and then taking the reciprocal.

$$\frac{\text{degrees}}{\text{division}} = \frac{360\text{ degrees}}{\text{number of divisions in one cycle}} \tag{11-11}$$

$$\text{phase angle} = \text{number of divisions} \times \frac{\text{degrees}}{\text{division}} \tag{11-12}$$

$$\text{period} = \frac{\text{number divisions in}}{\text{one cycle}} \times \frac{\text{time per division}}{\text{(scope setting)}} \tag{11-13}$$

$$\text{frequency} = \frac{1}{\text{period}} \tag{11-14}$$

Materials required for phase-angle measurements

Sine wave signal generator with variable frequency output:

100 Ω resistor, 1/4 W
1 μF capacitor, disc type
Dual-trace oscilloscope

Figures 11-11 through 11-16 show the series circuit connected to the oscilloscope and the resultant oscilloscope displays. The same circuit will be used with only the frequency of the variable sine wave generator being varied. The following calculations are to predict the required frequency needed to arrive at the approximate phase angle.

Each time the frequency is adjusted, the voltage across the capacitor should change its relative amplitude in comparison to the applied voltage. The applied voltage amplitude should change only slightly. As the capacitor voltage changes, adjust the oscilloscope to try and maintain the two waveforms at approximately the same amplitude.

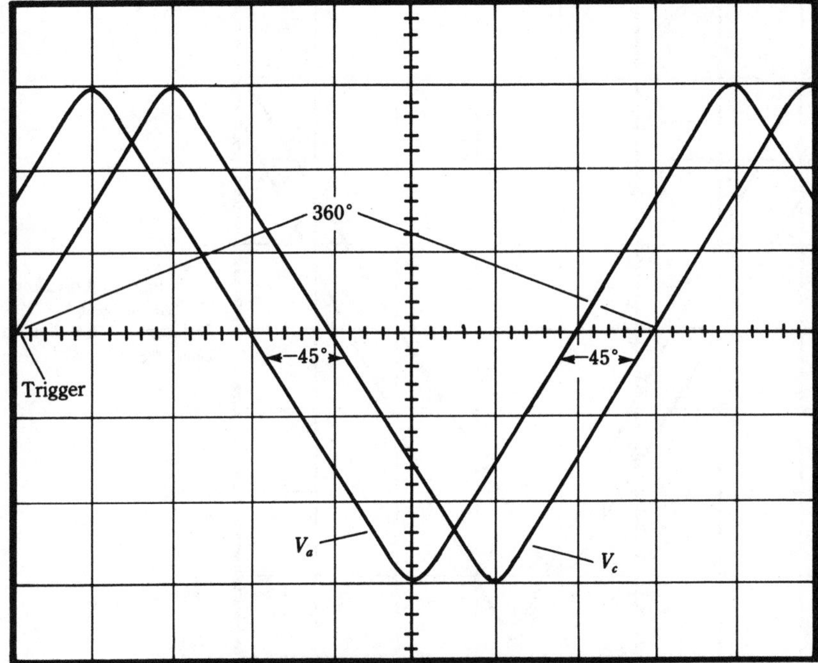

$R = X_C$
360° = 8 divisions
1 division = 45°
Phase angle = −45°

11-11 Scope triggered on the capacitor voltage.

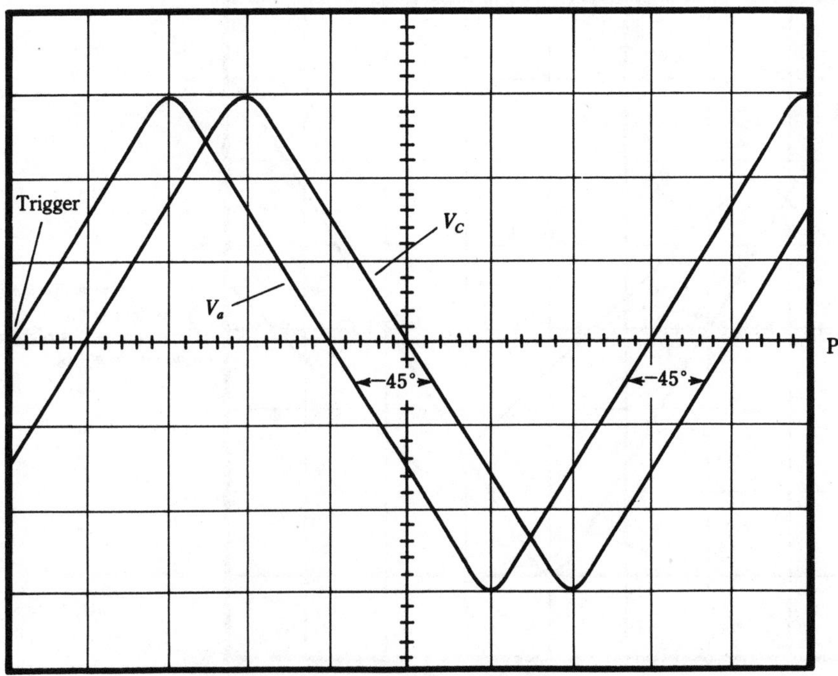

$R = X_C$
Phase angle = −45°

11-12 Scope triggered on the input voltage.

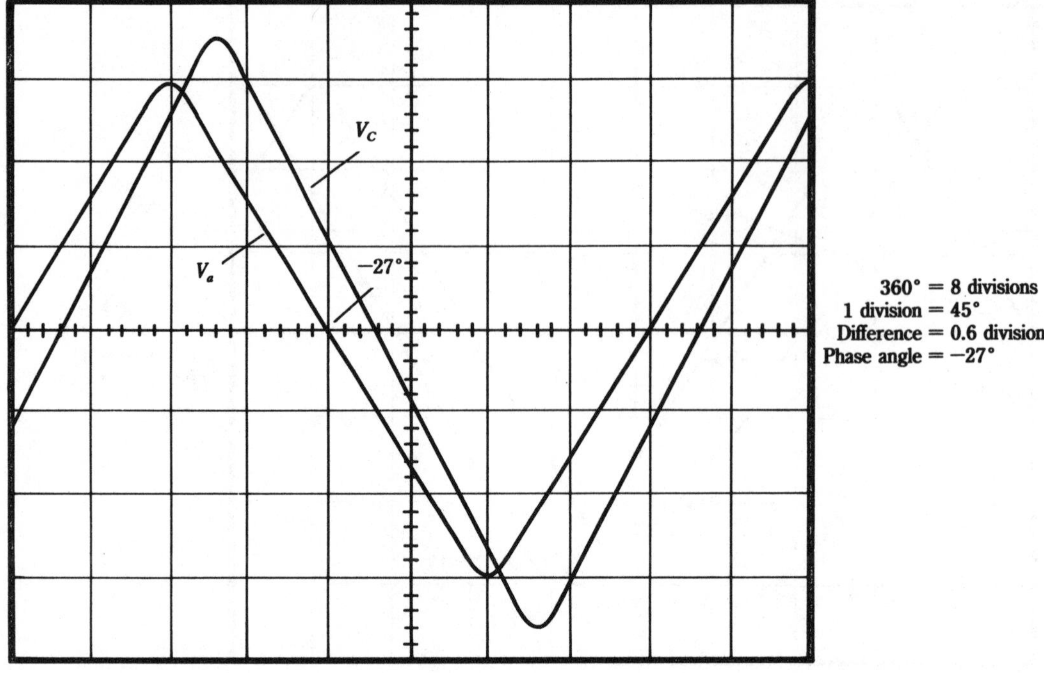

11-13 *R* larger than X_C.

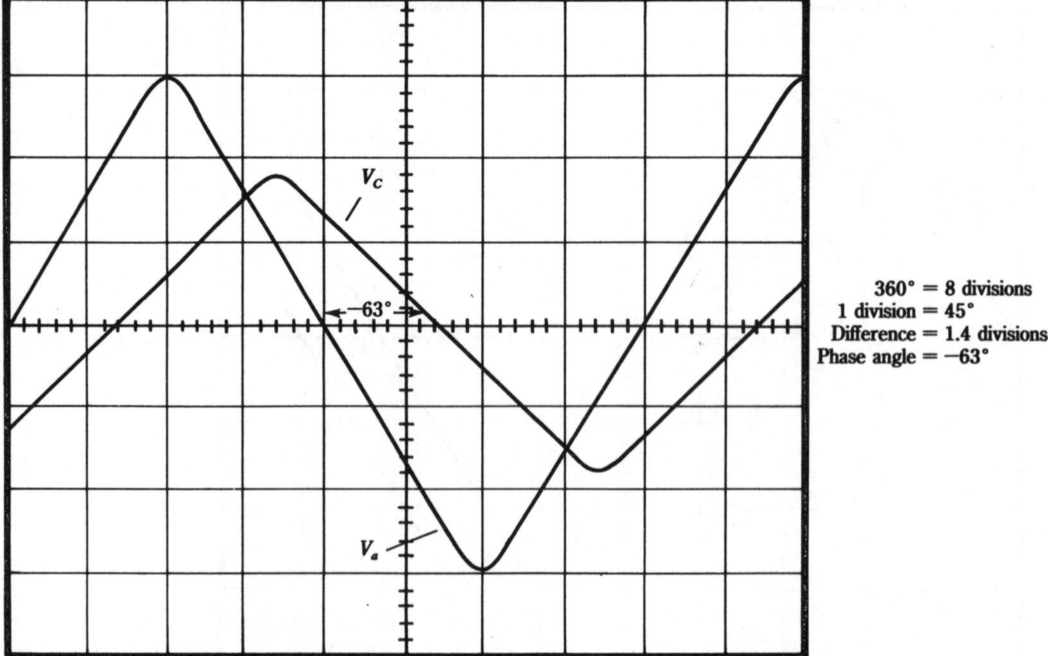

11-14 *R* smaller than X_C.

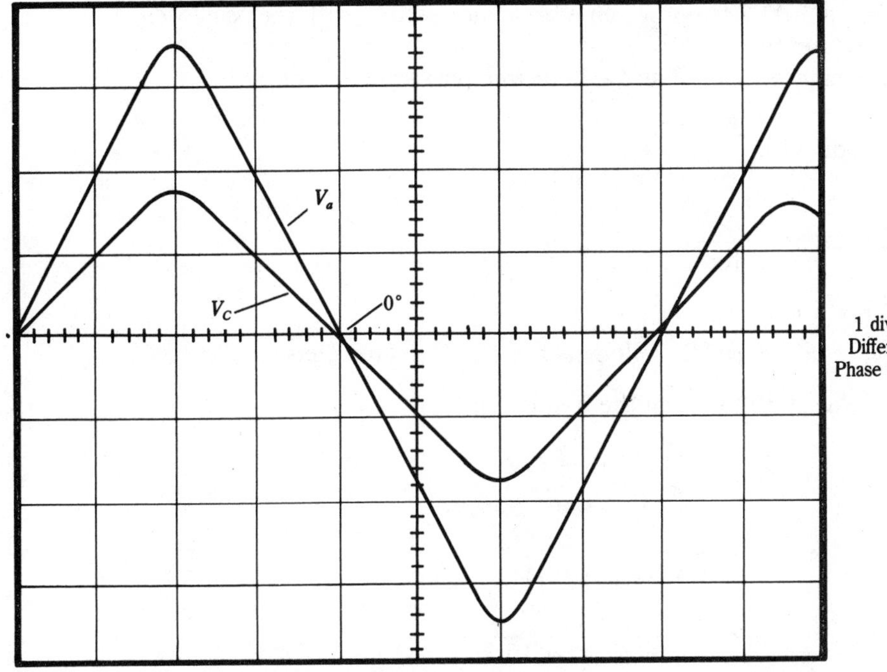

11-15 *R* much larger than X_C.

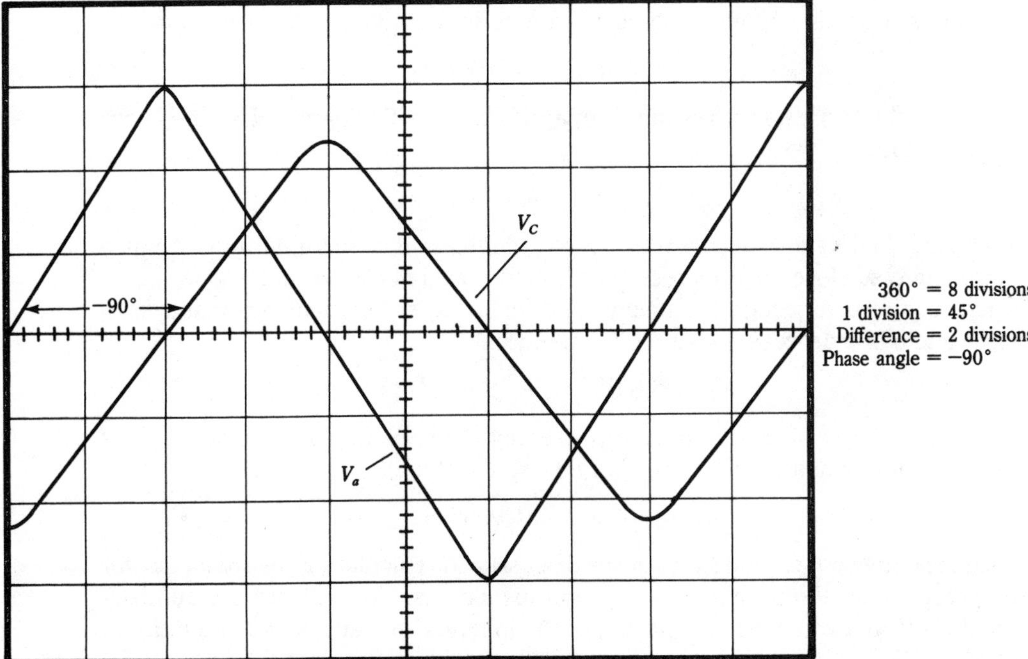

11-16 *R* much smaller than X_C.

Figures 11-11 and 11-12 are the same frequency setting with the oscilloscope being adjusted differently.

Find X_C. Resistance is 100 Ω and a 45-degree phase angle is desired.

$$\tan \Theta = \frac{X_C}{R}$$

$$\tan 45 \text{ degrees} = \frac{X_C}{100\ \Omega}$$

$$1 \times 100 = X_C$$

$$X_C = 100\ \Omega \quad \text{value needed for 45-degree phase angle}$$

If $C = 1\ \mu F$, X_C is 100 Ω. Find the frequency.

$$X_C = \frac{1}{2\ \pi\ fC} \quad \text{therefore;}$$

$$f = \frac{1}{2\ \pi\ C\, X_C}$$

$$= \frac{1}{2 \times \pi \times (1 \times 10^{-6}) \times 100}$$

= 1600 Hz (approximate frequency needed for 45-degree phase shift)

Figure 11-13 shows approximately a 27-degree phase shift between the two sine waves. Use the outline for the calculations shown above to determine:

$$X_C = 51\ \Omega \qquad f = 3120 \text{ Hz}$$

Figure 11-14 shows a phase angle of approximately 63 degrees. Use the above outline to find:

$$X_C = 193\ \Omega \qquad f = 825 \text{ Hz}$$

Figure 11-15 shows a phase angle of 0 degrees. It is possible to arrive at a phase angle so close to 0 degrees that it could not be seen on the scope, but it would require a very high frequency. The following calculations are made for a phase angle of 5 degrees, using the above outline.

$$X_C = 8.8\ \Omega \qquad f = 18{,}000 \text{ Hz}$$

Figure 11-16 shows a 90-degree phase angle, the results for the calculations are shown for an actual phase angle of 88 degrees.

$$X_C = 2860\ \Omega \qquad f = 56 \text{ Hz}$$

It is pointed out several times in the discussion on calculating the frequency for the phase angle, above, that these are approximations. Not only are the numbers rounded off, but you need to be aware of the inherent tolerances and other circuit inaccuracies.

Sometimes, when working with a calculator and numbers, it is easy to forget that the components that build the actual circuit are far from accurate by the standards of the calculations made beforehand. However, calculations are very important for the technician because they are used to predict circuit outcomes and therefore determine if a circuit is performing the way it was intended.

Chapter summary

Capacitive reactance, along with inductive reactance are the two forms of ac resistance. When capacitors are connected in series or parallel, with no resistance in the circuit, the individual reactances can be mathematically combined using the same formulas as are used for pure resistance.

When a capacitor is connected in series with a resistor, there will be a resulting phase shift between the voltage drops across each component. The phase shift will show the capacitive voltage to lag the resistive voltage.

When a capacitor and resistor are connected in parallel, the currents to the individual branches will have a resultant phase angle. The current through the capacitor will lead the current through the resistor.

Power in a reactive circuit must reflect the phase angle, or operating angle, of the circuit. There are three types of power: true power, reactive power, apparent power. Apparent will be the combination of both the true and apparent powers, and therefore the apparent power will always be the largest value. Power factor is the ratio of the resistive power to the total power of the circuit.

- Capacitive reactance, X_C, is indirectly related to frequency and capacitance.
- In a series circuit, I_C leads V_C, and V_R leads V_C always by 90 degrees.
- The voltage vectors for inductance and capacitance are opposite.
- The operating angle of a circuit with a capacitor in series with a resistor will be between 0 degrees and -90 degrees, depending on circuit values.
- In a parallel circuit, I_C leads I_R by 90 degrees.
- Real power is the power dissipated in pure resistance, unit is *W* (watts).
- Reactive power is the power dissipated in pure reactance, units is *VARS* (volt-ampere-reactive).
- Apparent power is the power of a circuit containing both resistance and reactance, unit is *VA* (volt-ampere).
- Power factor is the ratio of the real power to the apparent power. It is a pure number, with no units, between 0 and 1.

Summary of formulas

$$X_C = \frac{1}{2\pi fC} \quad \text{capacitive reactance} \tag{11-1}$$

$$X_{CS} = X_{C1} + X_{C2} + X_{C3} + \ldots \quad \text{series reactances} \tag{11-2}$$

$$\frac{1}{X_{CP}} = \frac{1}{X_{C1}} + \frac{1}{X_{C2}} + \frac{1}{X_{C3}} + \ldots \quad \text{parallel reactances} \qquad (11\text{-}3)$$

$$V_T = \sqrt{V_R^2 + V_C^2} \quad \text{total voltage in a series circuit} \qquad (11\text{-}4)$$

$$\Theta = \tan^{-1} \frac{-V_C}{V_R} \quad \text{operating angle using voltages} \qquad (11\text{-}5)$$

$$Z = \sqrt{R^2 + X_C^2} \quad \text{impedance in a series circuit} \qquad (11\text{-}6)$$

$$\Theta = \tan^{-1} \frac{-X_C}{R} \quad \text{operating angle using resistance} \qquad (11\text{-}7)$$

$$I_T = \sqrt{I_R^2 + I_C^2} \quad \text{total current in a parallel circuit} \qquad (11\text{-}8)$$

$$\Theta = \tan^{-1} \frac{I_C}{I_R} \quad \text{phase angle of a parallel circuit} \qquad (11\text{-}9)$$

$$Z = \frac{V}{I_T} \quad \text{impedance of a parallel circuit} \qquad (11\text{-}10)$$

Formulas 11-11 through 11-14 are used when making measurements on the oscilloscope:

$$\frac{\text{degrees}}{\text{division}} = \frac{\text{360 degrees}}{\text{number of divisions in one cycle}} \qquad (11\text{-}11)$$

$$\text{phase angle} = \frac{\text{number of divisions}}{\text{difference}} \times \frac{\text{degrees}}{\text{division}} \qquad (11\text{-}12)$$

$$\text{period} = \text{number of divisions} \times \text{time per division} \qquad (11\text{-}13)$$

$$\text{frequency} = \frac{1}{\text{period}} \qquad (11\text{-}14)$$

Practice problems

Find the requested information in each drawing.

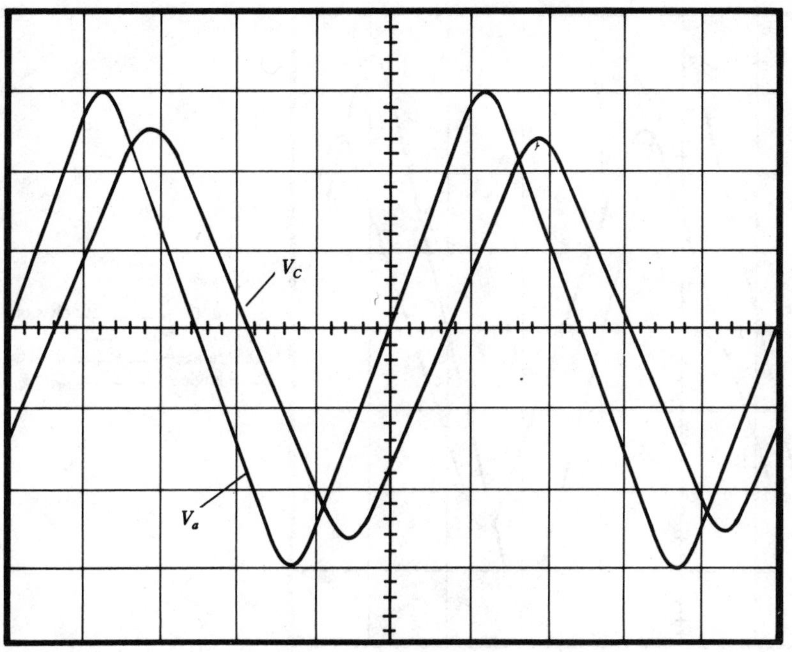

Problem 1

360° = _______ divisions
1 division = _______°
Difference = _______ divisions
Phase angle = _______°

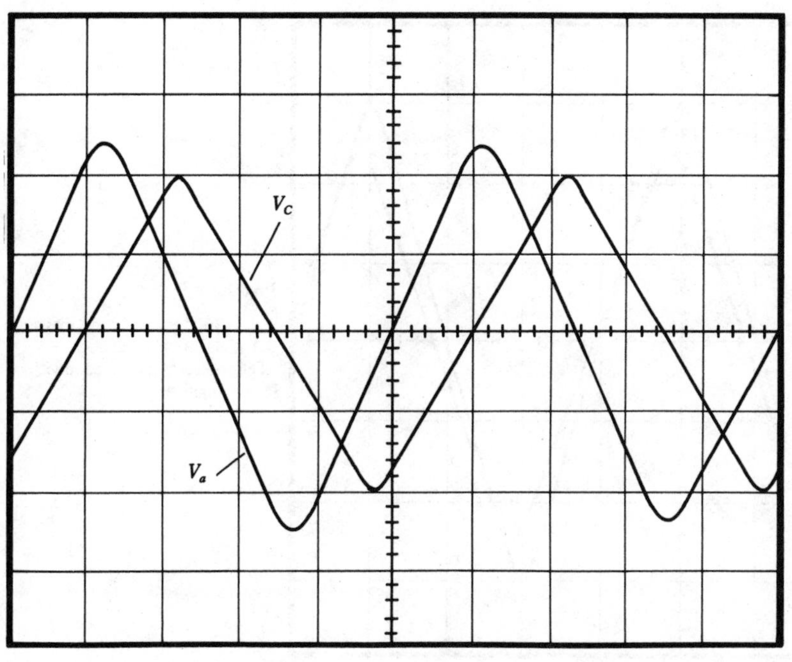

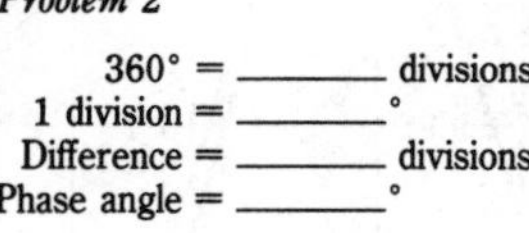

Problem 2

360° = _______ divisions
1 division = _______°
Difference = _______ divisions
Phase angle = _______°

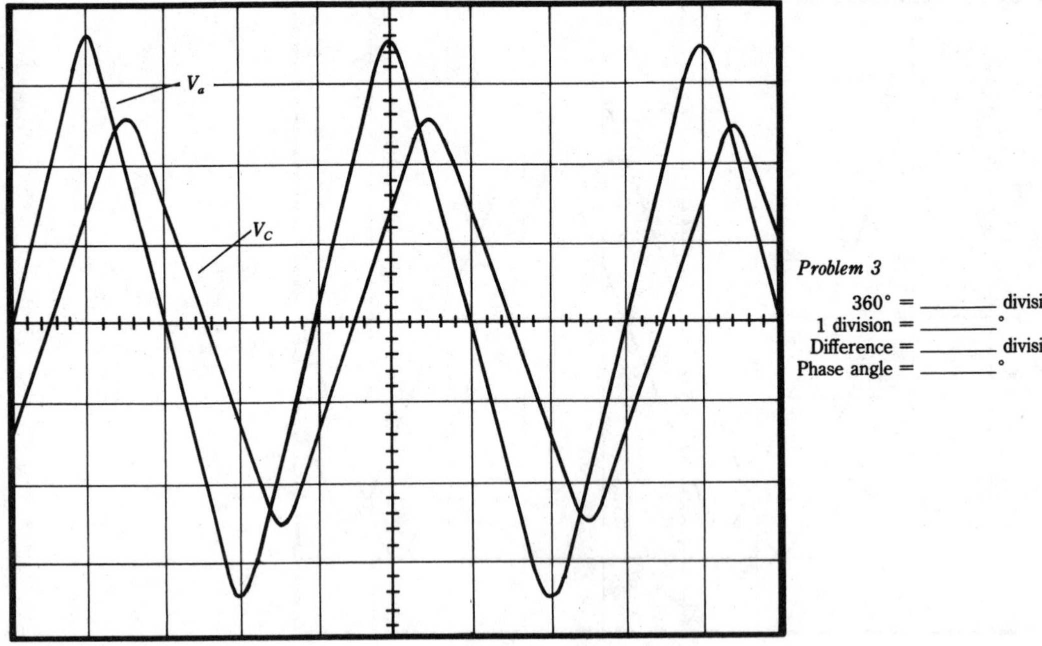

Problem 3

360° = ______ divisions
1 division = ______°
Difference = ______ divisions
Phase angle = ______°

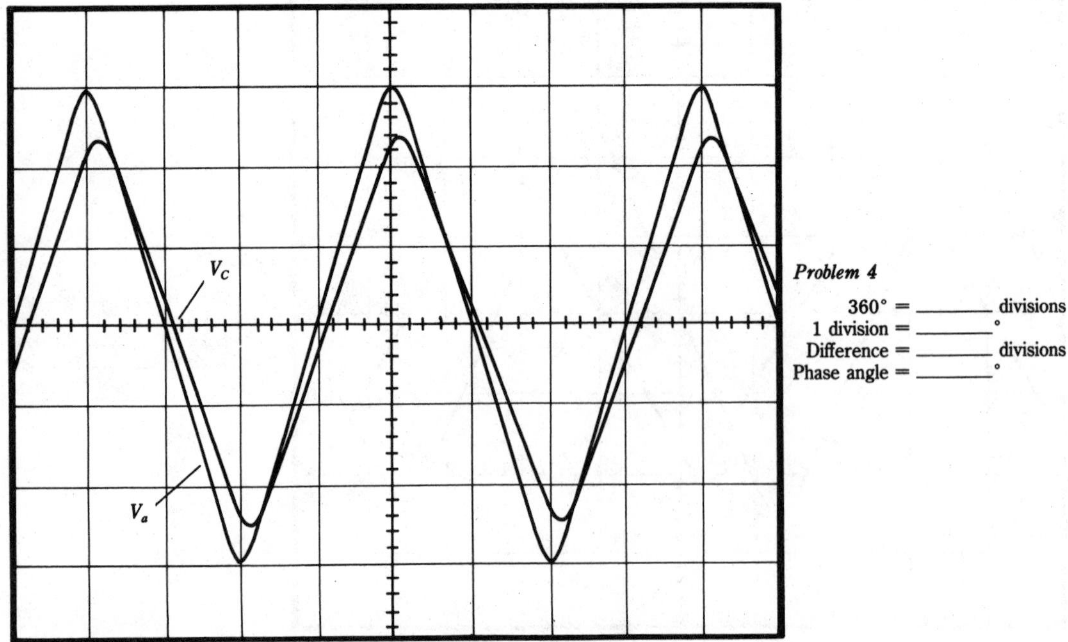

Problem 4

360° = ______ divisions
1 division = ______°
Difference = ______ divisions
Phase angle = ______°

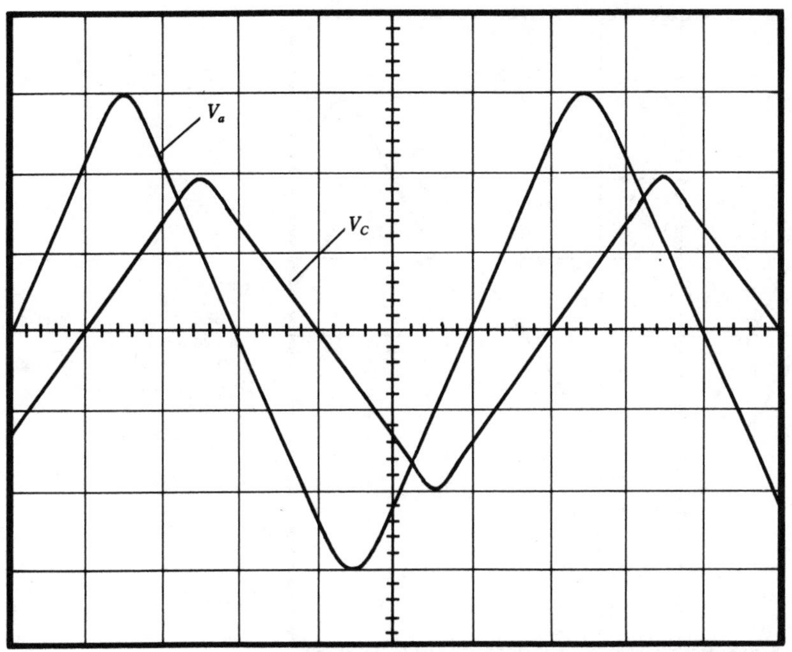

Problem 5

360° = ________ divisions
1 division = ________°
Difference = ________ divisions
Phase angle = ________°

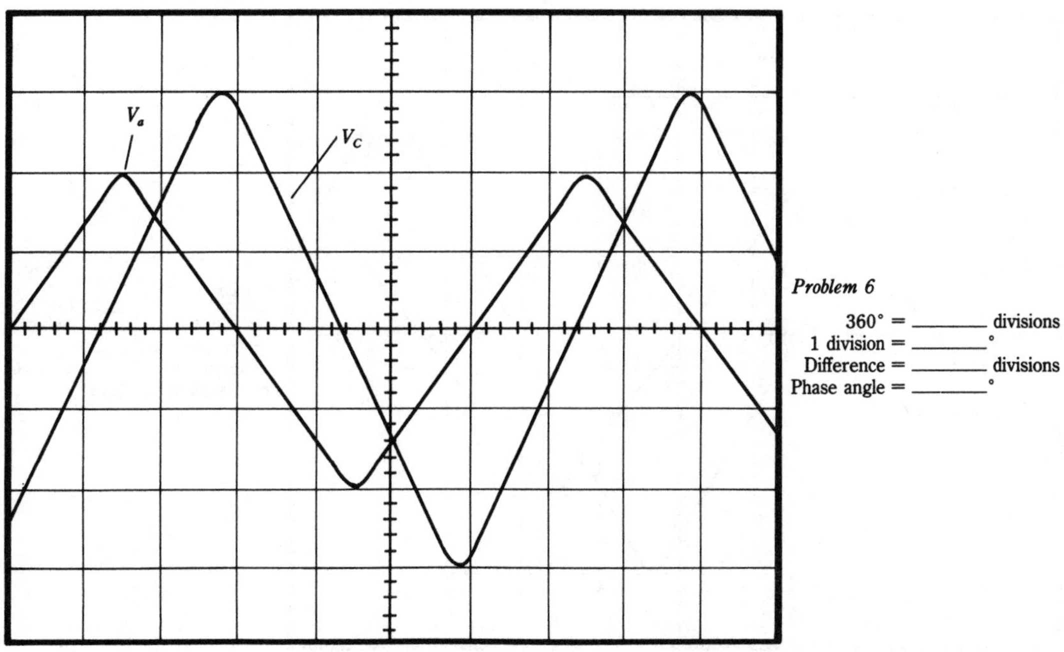

Problem 6

360° = ________ divisions
1 division = ________°
Difference = ________ divisions
Phase angle = ________°

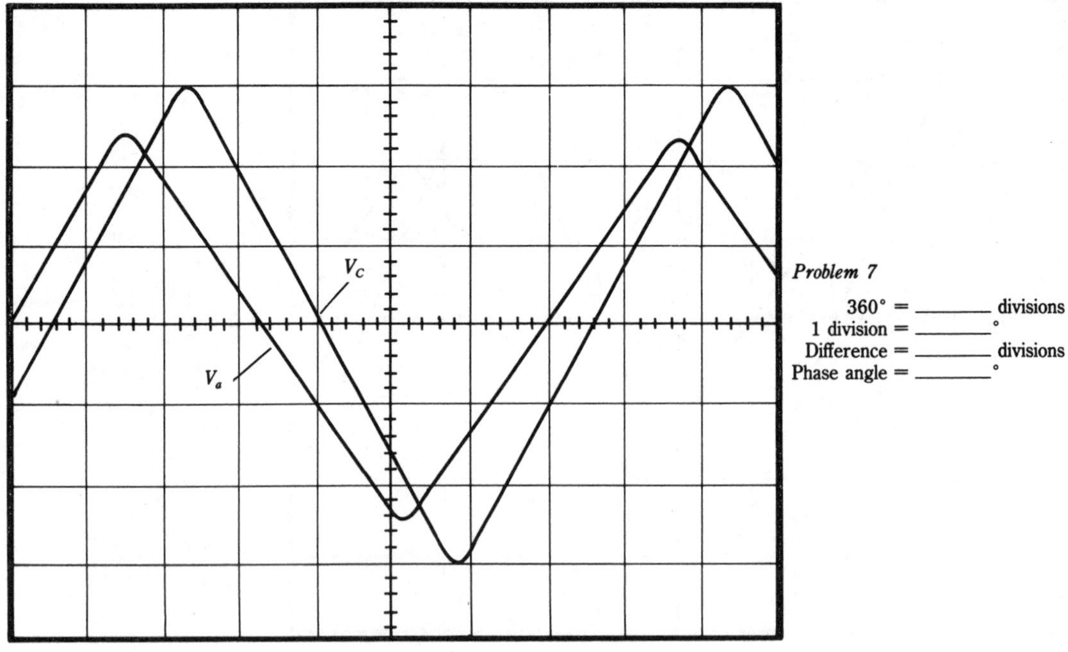

Problem 7

360° = ______ divisions
1 division = ______°
Difference = ______ divisions
Phase angle = ______°

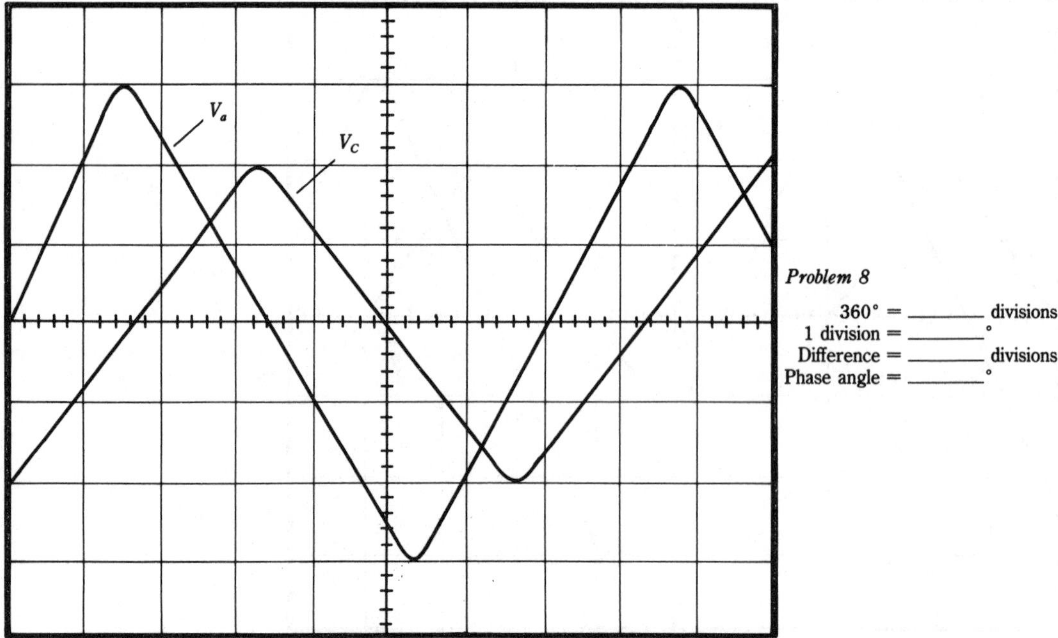

Problem 8

360° = ______ divisions
1 division = ______°
Difference = ______ divisions
Phase angle = ______°

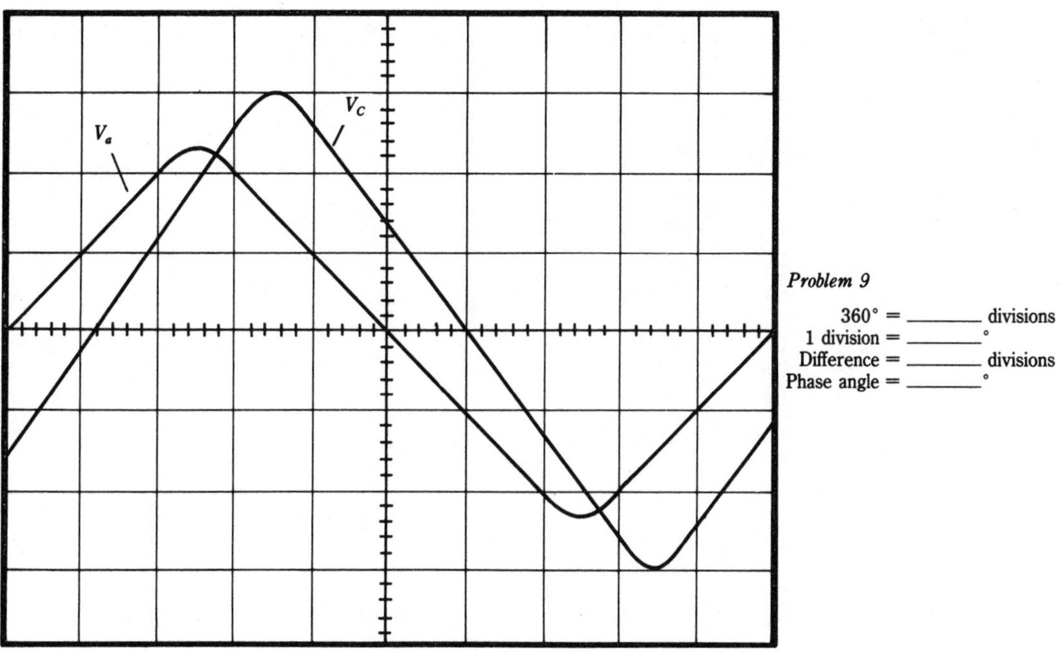

Problem 9

360° = ________ divisions
1 division = ________°
Difference = ________ divisions
Phase angle = ________°

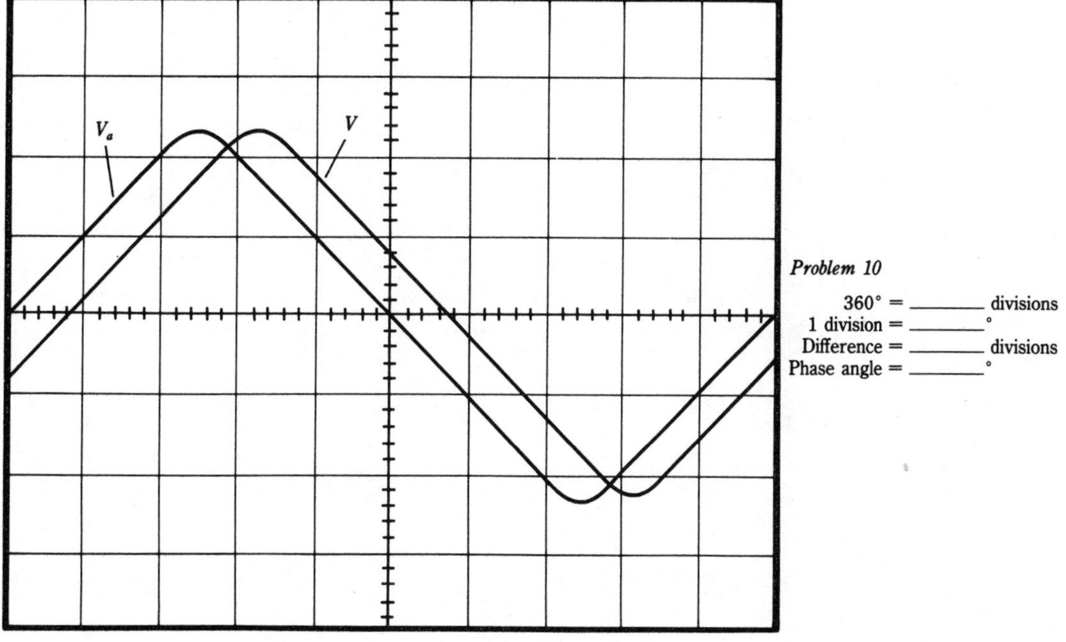

Problem 10

360° = ________ divisions
1 division = ________°
Difference = ________ divisions
Phase angle = ________°

12 ac circuits and the *j*-operator

THIS CHAPTER CONTAINS A DISCUSSION OF CIRCUITS WITH DIFFERENT combinations of resistance, capacitance, and inductance. Using vector addition to combine reactances to determine a net (or total) reactance for the combination is the primary concern of the chapter. The *j-operator* is a very useful mathematical tool that can be used to allow the calculator to do most of the work in performing calculations on complex circuits.

Net reactance

The *net reactance* is the reactance of a circuit that has more than one type of reactance; that is, both inductance and capacitance. In order to best determine the meaning of net reactance, draw the vectors of inductance and capacitance on the same graph. Notice in Fig. 12-1A that in a series circuit the vector for inductive reactance is plotted vertically up and the vector for capacitive reactance is plotted vertically down. That is, the vectors for inductive reactance and capacitive reactance are opposite. It can also be demonstrated that all of the vectors dealing with inductance and capacitance are opposite. A parallel circuit uses the current triangle (shown in Fig. 12-1B). In all cases, notice the relationship of the inductance vector and the capacitance vector.

- The vectors for inductance and capacitance are opposite.

A very interesting observation can be made from Fig. 12-1. In the series circuit, the opposite reactances partially cancel each other and result in a greatly reduced net reactance. The circuit current is determined by the net reactance, which results in an unusually high current. The individual voltage drops across each component are determined by using Ohm's Law with the circuit current and

A

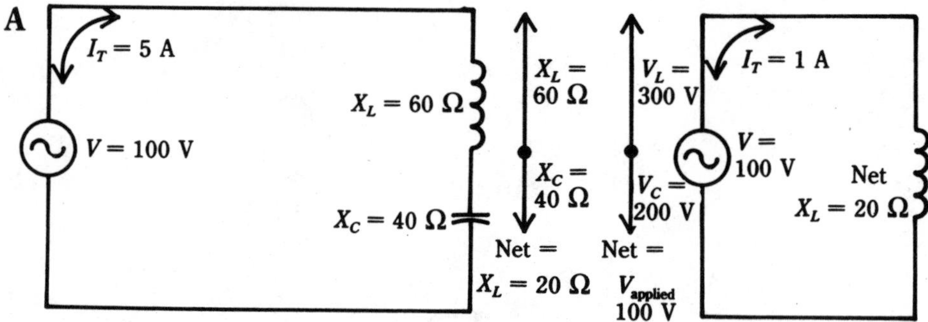

Series circuit to demonstrate net reactance.

B

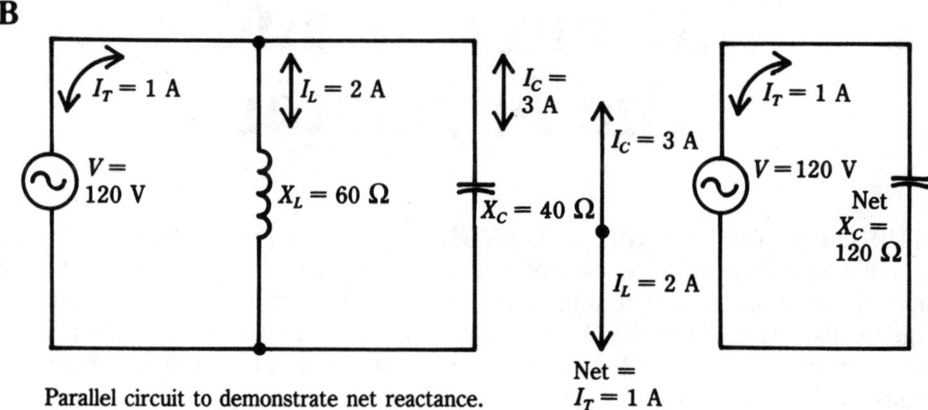

Parallel circuit to demonstrate net reactance.

12-1 Net reactance is demonstrated with a series circuit and a parallel circuit.

the individual reactances. The result is the voltage drop across the inductor and capacitor, each results in a voltage higher than the applied voltage.

The parallel circuit uses a current vector relationship. The total current is less than the individual branch currents. This is due to the fact that the individual branch currents are calculated using the applied voltage and the branch reactance. The branch currents are 180 degrees out of phase and will result in some cancellation in the total current.

In Fig. 12-1, on the right-hand side, an equivalent circuit showing the net reactance of the circuits is shown. This net reactance is based on the reactance vectors for the series circuit, and based on the total current in the parallel circuit.

The main purpose for using an equivalent circuit when dealing with ac circuits is that the net reactance can be shown as being either an inductor or a capacitor. The fact that the circuit has a net capacitance or inductance becomes very important when determining the characteristics of the circuit.

The above discussion, concerning Fig. 12-1, was all based on circuits that are only inductance and capacitance. In an actual circuit, it is also necessary to consider the effects of resistance. Chapters 10 and 11 show how to deal with the triangles for each of the individual components. In this chapter, the three components will be combined into one circuit. The rules will still be the same except that inductive reactance and capacitive reactance are opposite and must result in a net reactance to be used in the vector triangles.

It is possible for the inductive and capacitive reactances to completely cancel and leave the circuit with only a net resistance. This principle is used in resonant circuits.

Table 12-1. Vector relationships in a series and parallel *RLC* circuit.

Vector	Series RLC circuit	Parallel RLC circuit
X_L	+90° (up) ↑	
X_C	−90° (down) ↓	
R	0° (right) →	
Z	Hypotenuse of impedance triangle ◺ or ◸	$Z = \frac{V_a}{I_T}$
V_L	+90° (up) ↑	Same as applied voltage
V_C	−90° (down) ↓	Same as applied voltage
V_R	0° (right) →	Same as applied voltage
V_T	Hypotenuse of voltage triangle ◺ or ◸	Voltage is the same across all branches.
I_L	Same as total current	−90° (down) ↑
I_C	Same as total current	+90° (up) ↓
I_R	Same as total current	0° (right) →
I_T	Current is the same throughout a series circuit.	Hypotenuse of current triangle ◹ or ◿

Table 12-1 shows the relationship of the vectors used in ac circuits. To summarize:

- In a series circuit, use an impedance triangle and a voltage triangle.
- In a series circuit, X_L and V_L are +90 degrees, and X_C and V_C are −90 degrees.
- In a parallel circuit, use a current triangle.
- In a parallel circuit, I_C is +90 degrees and I_L is −90 degrees.
- Phase angle will be positive or negative depending on the net reactance in the circuit.

Analysis of a series RLC circuit

Refer to Fig. 12-2. There are six parts to this figure. Part A is the original circuit, containing resistance, capacitance and inductance. Part B shows the vectors of the

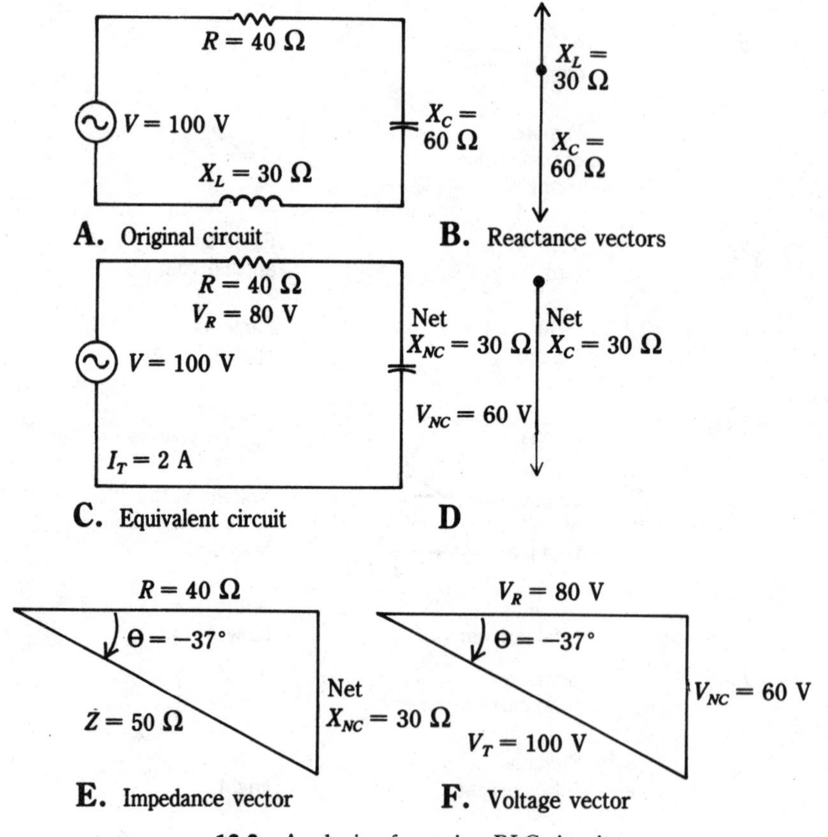

12-2 Analysis of a series *RLC* circuit.

reactance, not including the resistance vector. This is drawn this way to express the idea that X_L and X_C will partially cancel each other. Part C shows an equivalent circuit, resulting from the net reactance. Notice it is drawn as a capacitor, to show a net X_C. Part D is the vector of the net reactance. Part E is the impedance vector, using the resistance and the net reactance. Part F is the voltage vector showing the voltage drop across the resistor and the net reactance. The labeling on the net reactance is shown as X_{NC} and the voltage as V_{NC}. This labeling is used to ensure that it is clear that these values are for the net and not for the original value of capacitance.

Calculate the net reactance vector.

Step 1 Draw the vectors of X_L and X_C together with a connecting point to show opposite directions, as in Fig. 12-2B. Because the vectors go in opposite directions, use simple arithmetic to subtract the magnitudes (sizes) of the two vectors. $X_L = 30\ \Omega$ and $X_C = 60\ \Omega$. Using simple arithmetic, subtract the smaller from the larger. $60 - 30 = 30\ \Omega$ net. Figure 12-2A shows the resultant net X_C. Notice that the direction of the resultant vector is from the larger of the two original vectors.

Step 2 Draw the equivalent circuit showing the net reactance. The equivalent circuit shown in Fig. 12-2 is classified as a net capacitive circuit because the net reactance is capacitive.

Step 3 Draw the impedance and voltage triangles. Current for a series circuit is calculated using Ohm's Law.

- Series circuits are classified as inductive or capacitive by whichever is the larger reactance or voltage drop.

Analysis of a parallel RLC circuit

Refer to Fig. 12-3 for the drawings used in the analysis of a parallel circuit. Figure 12-3A is the original parallel circuit. In order to keep everything simple here, the circuit used has only one component in each branch. Keep in mind, it is possible for the branches to have more than one component forming series circuits, additional parallel circuits, or any combination.

Figure 12-3B shows the current vectors for the capacitive and the inductive branches. The resistive current is now shown here in order to show the relationship of these two vectors. Figure 12-3C is the equivalent circuit, showing the resistive branch unchanged and the second branch having a net capacitive reactance.

Figure 12-3D shows the vector of the net reactance. The magnitude, or length, of this vector is found by using simple subtraction of the inductive and capacitive currents. Direction is determined by the larger of the two vectors. Notice, the larger vector is the larger current, which is the smaller of the two reactances. Figure 12-3E is the current triangle.

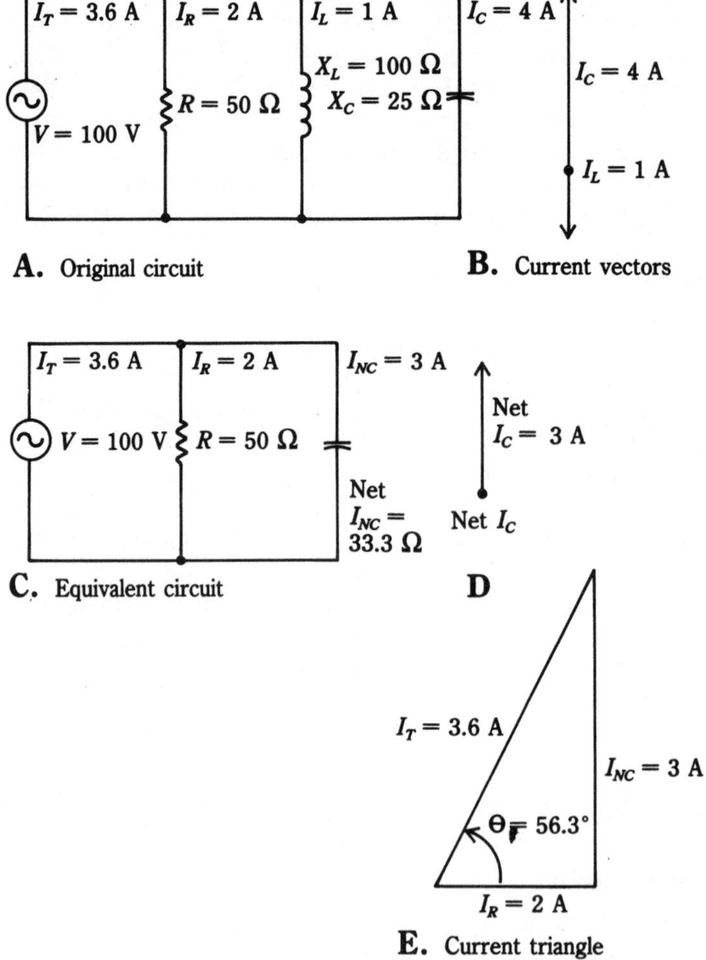

12-3 Analysis of a parallel RLC circuit.

Summary for analysis of a parallel RLC circuit

- Plot the vectors for the reactive currents.
- Subtract the reactive currents to determine the size and direction (inductive or capacitive) of the net current and therefore the net reactance.
- Draw the equivalent circuit.
- Draw the current triangle.

Note the following key points.

- Parallel circuits are classified as inductive or capacitive by whichever has the largest current. The largest current results from the smallest reactance.
- A circuit is classified as resistive only if the net reactance is zero.

Power in an RLC circuit

In chapter 10 the subject of power in inductive circuits is discussed. In chapter 11 the subject of power in capacitive circuits is discussed. This chapter deals with circuits that have both capacitance and inductance. Power is considered separately for each of the separate components. The points to remember are the definitions of the three types of power in an ac circuit.

True power is the power dissipated by the resistance of the circuit; the unit of power is the watt. Reactive power is the power dissipated by a reactive component, unit is *VARS* (volt-ampere-reactive). Apparent power is the power dissipated by a complete ac circuit, unit is *VA* (volt-ampere). Power factor is the ratio of true power to the total power of the circuit, no unit, it is a pure number.

Because the voltage of a series circuit or the current of a parallel circuit will have a phase angle, if a circuit contains both inductance and capacitance, the powers for these two components, because they are both in *VARS*, will be opposite and will have a cancelling effect, the same as would be seen with either the impedance or the current.

Practice problems

Use the schematics shown to calculate the answers to the problems. For each problem, find: net X, I_T, and Z; state if the circuit is a net C or L. Find angle Θ.

1. $X_{C1} = 25\ \Omega$, $R_1 = 25\ \Omega$, $X_{L1} = 100\ \Omega$, $R_2 = 25\ \Omega$
2. $X_{C1} = 200\ \Omega$, $R_1 = 50\ \Omega$, $X_{L1} = 50\ \Omega$, $R_2 = 50\ \Omega$
3. $X_{C1} = 10\ \text{k}\Omega$, $R_1 = 1\ \text{k}\Omega$, $X_{L1} = 5\ \text{k}\Omega$, $R_2 = 5\ \text{k}\Omega$
4. $X_{L1} = 10\ \Omega$, $X_{C1} = 10\ \Omega$, $R_1 = 10\ \Omega$, $X_{L2} = 20\ \Omega$, $X_{C2} = 30\ \Omega$, $R_2 = 5\ \Omega$, $R_3 = 5\ \Omega$
5. $X_{L1} = 200\ \Omega$, $X_{C1} = 100\ \Omega$, $R_1 = 50\ \Omega$, $X_{L2} = 200\ \Omega$, $X_{C2} = 100\ \Omega$, $R_2 = 75\ \Omega$, $R_3 = 75\ \Omega$
6. $R_1 = 10\ \Omega$, $X_{C1} = 10\ \Omega$, $X_{L1} = 20\ \Omega$
7. $R_1 = 100\ \Omega$, $X_{C1} = 200\ \Omega$, $X_{L1} = 100\ \Omega$
8. $R_1 = 1\ \text{k}\Omega$, $X_{C1} = 500\ \Omega$, $X_{L1} = 2\ \text{k}\Omega$
9. $X_{C1} = 20\ \Omega$, $R_1 = 10\ \Omega$, $X_{C2} = 20\ \Omega$, $X_{L1} = 40\ \Omega$, $X_{L2} = 50\ \Omega$
10. $X_{C1} = 50\ \Omega$, $R_1 = 20\ \Omega$, $X_{C2} = 40\ \Omega$, $X_{L1} = 20\ \Omega$, $X_{L2} = 10\ \Omega$

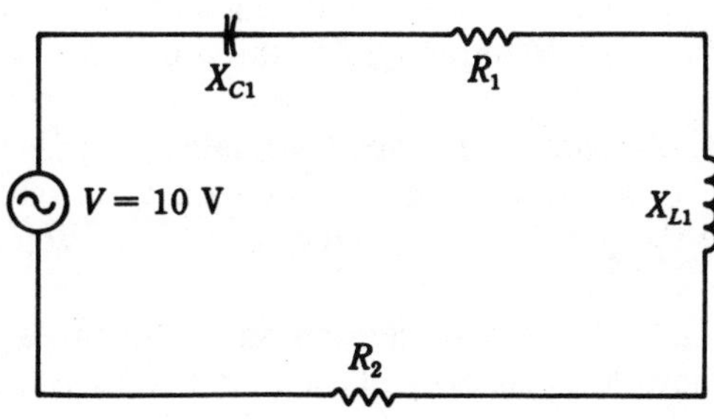

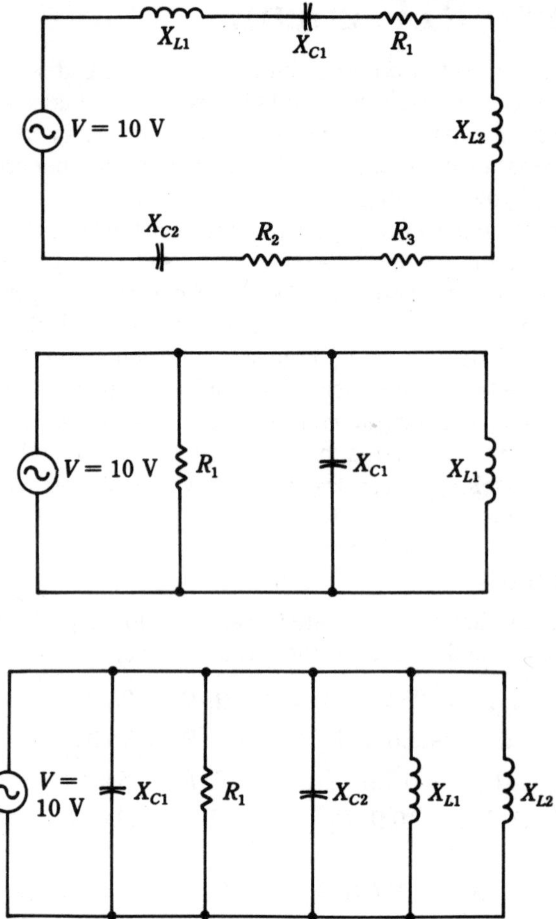

The *j*-operator

The *j*-operator is a mathematical tool used to simplify working with complex circuits. The *j*-operator concept can actually be used under any set of conditions dealing with vectors. Most scientific calculators are equipped with keys to allow the calculator to deal directly with numbers written in the *j*-operator format. When several vectors are to be combined, the use of trigonometric methods is somewhat time consuming, but the *j*-operator simplifies the process.

The standard, geometric x-y coordinate axis system is relabeled for use with the *j*-operator. Figure 12-4 shows how the coordinate system is relabeled. The horizontal axis, normally called the x axis, is now called the real axis. The vertical axis, normally called the y axis, is now labeled with $+j$ going up and $-j$ going down.

The real axis is labeled at both 0 degrees and 180 degrees. In electronics, the real axis is used to show the resistance of the circuit. Any point on the right of the

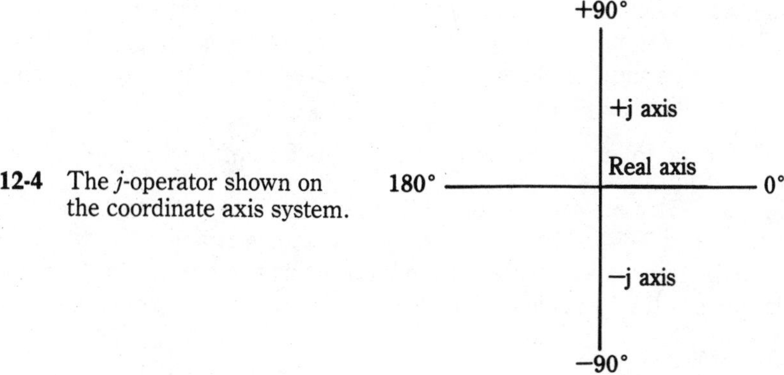

12-4 The *j*-operator shown on the coordinate axis system.

center point is positive, and any point on the left of the center point is negative. In electronics, points to the left cannot be considered for circuits with just resistance, capacitance, and inductance.

The j axis, both positive and negative, is of very important consideration in circuits. It is in these two directions, +90 degrees and −90 degrees, that you find the vectors for inductors and capacitors. It depends on whether the circuit is a series or a parallel circuit that determines the plotting of the vectors. For pure inductance or capacitance, it will always be 90 degrees in relation to the resistor vector. When dealing with *j*-operators and ac circuits, reactance is usually what is worked with. Even when the circuit is a parallel circuit, the reactance can be used with *j*-operators because it is possible to deal with it in terms of using the reciprocal formula necessary for parallel resistances.

Inductive reactance is called a $+j$ and capacitive reactance, is called $-j$. It is also possible to deal with currents in a parallel circuit, in which case, the inductance would be negative and the capacitance would be positive.

It is interesting to note that the *j* axis is often called the imaginary axis.

Complex numbers in rectangular form

A *complex number* is a number that contains both a real term and a *j* term. Complex numbers can be written in two forms; rectangular and polar.

The *rectangular form* of a complex number states the actual quantity of the real term and the actual quantity of the *j* term. In other words, the rectangular form tells the amount of resistance and the amount of net reactance. The vectors of these two quantities are plotted at right angles to each other, either up or down (+90 degrees or −90 degrees). A complex number, when dealing with resistance and reactance, states the amount of impedance in the circuit, in rectangular form. When working the mathematics of a circuit, sometimes it is necessary to use rectangular form, sometimes polar form.

- The rectangular form of a complex number states the actual quantities of the real and *j* terms.

When numbers are to be written in rectangular form, the real term is written first, followed by the *j* term. Because the *j* term can be either positive or negative, the plus or minus sign can be written as an addition or subtraction sign connecting the real term with the *j* term.

- The rectangular form is written with this format: real ± *j* term.

Figure 12-5 shows two simple series circuits, one with an inductor and one with a capacitor. With each circuit is the impedance triangle. Shown below each triangle is the impedance written out in both rectangular and polar forms. The rectangular form is the two sides of the triangle and the polar form is the hypotenuse.

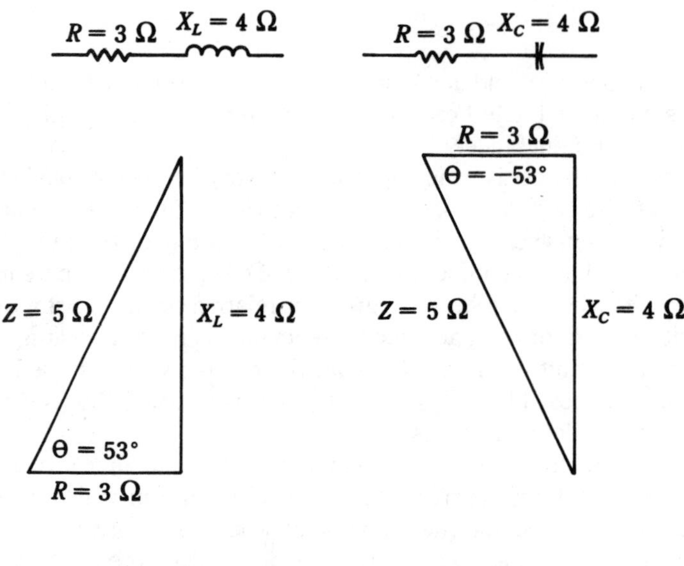

Rectangular form:	**(3 + j4)**	**(3 − j4)**
Polar form:	$5 \angle 53°$	$5 \angle -53°$

12-5 Examples of writing complex numbers from a circuit.

Complex numbers in polar form

The *polar form* of a complex number is a vector at some angle between 0 degrees and ± 90 degrees. In terms of an electrical circuit, that means the hypotenuse of the triangle is used to find either impedance, voltage, or current. The polar form also contains the operating angle, Θ.

- The polar form of a complex number is written in this format: $Z \angle \Theta$.
- When a number is written in polar form, the symbol $\angle$ means at an angle of.

- Rectangular form shows the horizontal (real) and vertical (imaginary) components.
- Polar form shows the resultant vector and its angle.

Consult the owner's manual of the calculator in use to determine the steps involved in converting from rectangular to polar and vice versa. If these calculator functions are not available, use the methods shown in chapters 10 and 11.

Calculations with complex numbers

Calculations with complex numbers have a set of rules that are unique to this type of mathematics. When working with circuits it is sometimes necessary to add or subtract and other times it is necessary to multiply or divide.

To understand when it might be necessary to use more than one operation, consider an example of a series parallel circuit. When working the parallel portion, total resistance is found by using the reciprocal formula, which involves both addition and division. Then the equivalent resistance of the parallel branches is added to the series resistance. Then, to find the total current, or voltage drops in the circuit, Ohm's Law is necessary.

Rules for calculations with complex numbers

Rule 1 Add or subtract in rectangular form.

Rule 2 When adding or subtracting in rectangular form, add/subtract the real terms together and add/subtract the j terms together.

Rule 3 Multiply or divide in polar form.

Rule 4 When multiplying in polar form, multiply the magnitudes (the Z term) and add the angles.

Rule 5 When dividing in polar form, divide the magnitudes (the Z term) and subtract the angles.

Rule 6 When taking the reciprocal of a number, it is to be done in polar form, because it is division. Take the reciprocal of the magnitude and change the sign of the angle.

Series ac circuits

Whenever it is necessary to solve any series circuit, keep in mind the basic rules of series circuits. In particular, remember that current is the same throughout a series circuit and the voltage drops across each component will add to the applied voltage.

When dealing with ac circuits, the reactive components, inductance and capacitance, should always be changed to their reactances, if not already given in that form. Using the rules for j-operators, inductors have an X_L of $+j$ and capacitors have an X_C of $-j$.

In a series circuit, resistances and reactances, are added to find the total resistance of the series circuit. Notice that the $+j$ for X_L and the $-j$ for X_C will take care of the fact that the reactances are opposite and in one step produce the total resistance and the net reactance of the circuit. This addition is performed in rectangular form and the final result will be in the form of: $R \pm jX$.

The calculations shown below are for the series circuit shown in Fig. 12-6. To solve a series circuit, using complex numbers, it is usually enough to find the total impedance, total current and the voltage drops of each component.

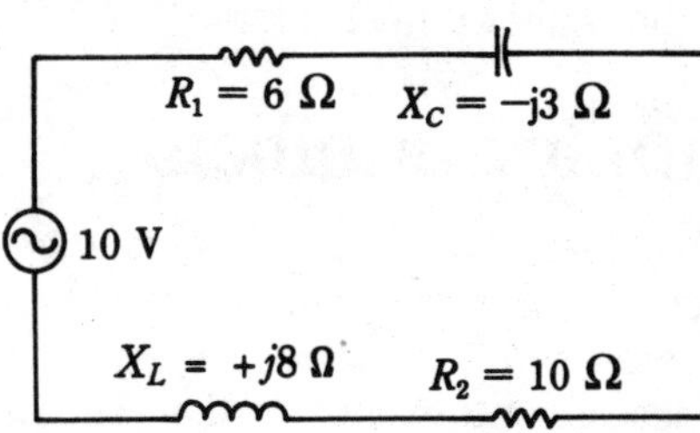

12-6 Sample series ac circuit.

Because this is a series circuit, the individual resistances will all add to form the total resistance. The reactive components will combine to produce a net reactance.

Calculations for Fig. 12-6

Step 1 Add the resistances for the real term of Z_T.

$$R_1 = 6\ \Omega$$
$$R_2 = 10\ \Omega$$
$$R_1 + R_2 = 6 + 10 = 16\ \Omega \text{ real term}$$

Step 2 Add the reactances, algebraically (subtract) to produce the net reactance or the j term.

$$X_C = -j3\ \Omega$$
$$X_L = +j8\ \Omega$$
$$X_C + X_L = -j3 + +j8 = +j5\ \Omega \text{ net reactance}$$

Step 3 Combine the real term and the j term to form the complete total impedance, written in rectangular form.

$$Z_T = R \pm jX$$
$$Z_T = (16 + j5)\ \Omega$$

Step 4 Change the Z_T in step 3 to polar form to make it ready to use to find I_T.

$$Z_T = (16 + j5) \text{ (rectangular form)}$$
$$Z_T = 16.76 \angle 17.35\ \Omega \text{ (polar form)}$$

Step 5 Use the applied voltage and total impedance with Ohm's Law to find the total current. Because this is a division problem, it must be worked using the numbers in polar form. The applied voltage is shown with no specific angle. Because it needs an angle for the calculations, an angle of zero degrees will be assigned.

$$I_T = \frac{V}{Z_T} \quad \text{Ohm's Law formula to be used} \tag{2-1A}$$

$$= \frac{10 \quad 0 \text{ V}}{16.76 \quad 17.35\ \Omega} \quad \text{substitute values}$$

$$= 0.596 \quad -17.35 \text{ A} \quad \text{solve by dividing the magnitudes and subtracting the angles, bottom from top}$$

Step 6 The voltage drops across each component can be found by using Ohm's Law and the total current. Because Ohm's Law is a multiplication problem, each component must be shown in polar form. This is best done by first writing in rectangular form and then changing to polar form.

$R_1 = (6 + j0)$ rectangular form (zero shows no reactance)
$= 6 \angle 0$ polar form

$X_C = (0 - j3)$ rectangular form (zero shows no resistance)
$= 3 \angle -90$ polar form

$R_2 = (10 + j0)$ rectangular form
$= 10 \angle 0$ polar form

$X_L = (0 + j8)$ rectangular form
$= 8 \angle 90$ polar form

Step 7 Find the voltage drops of each component using the polar form of each of the values found in Step 6, with the total current found in Step 5. $E = IR$

$E_{R1} = (0.596 \angle -17.35 \text{ amps})(6 \angle 0\ \Omega) = 3.576 \angle -17.35 \text{ V}$
$E_{XC} = (0.596 \angle -17.35 \text{ amps})(3 \angle -90\ \Omega) = 1.788 \angle -107.35 \text{ V}$
$E_{R2} = (0.596 \angle -17.35 \text{ amps})(10\ 0\ \Omega) = 5.96 \angle -17.35 \text{ V}$
$E_{XL} = (0.596 \angle -17.35 \text{ amps})(8 \angle 90\ \Omega) = 4.768 \angle 72.65 \text{ V}$

Step 8 Notice that simply trying to add the voltage drops to be equal to the applied voltage does not work. The reason is that the voltage drops must be added in rectangular form. Change each of the voltage drops to rectangular form.

$E_{R1} = 3.4 - j1.07$
$E_{XC} = -0.53 - j1.7$
$E_{R2} = 5.68 - j1.78$
$E_{XL} = 1.42 + j4.55$

Step 9 To add the rectangular numbers from Step 8, add the column of j terms separately, then the column of real term separately.

$V_T = 9.97 + j0$ rectangular form
$V_T = 9.97 \angle 0$ volts polar form for comparison with the applied voltage

Power can also be found for each component by using the power formula, $P = I \times E$. Power would be apparent power if it is used with the total voltage and total current. It would produce true power when total current is used with the voltage drop across a resistor and it would be reactive power if total current is used with the voltage drop of a reactive component.

Parallel ac circuits

When dealing with a parallel circuit, remember that voltage is the same across all parallel branches and the current divides to each branch.

Figure 12-7 shows a parallel circuit with two branches. Recall from working with dc circuits that there are several ways to work out a parallel circuit. Notice in this sample circuit, each branch is composed of a series circuit. It is probably best to work each branch individually as a series circuit, trying to find branch currents. The individual branch currents can then be added to find the total current and that will produce the total circuit impedance.

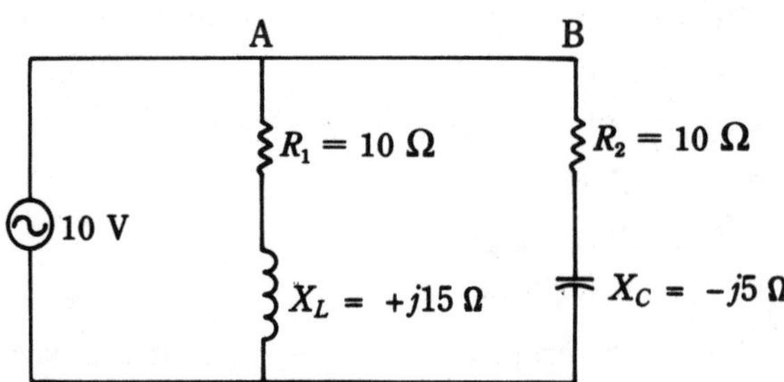

12-7 Sample parallel ac circuit.

Solving Fig. 12-7

Step 1 Find each branch current.

$$I = \frac{E}{R} \qquad \text{(2-1A)}$$

$$I_A = \frac{10 \angle 0 \text{ V}}{(10 + j15)\ \Omega} \quad \text{branch impedance needs to be changed to polar form}$$

$$= \frac{10 \angle 0 \text{ V}}{18\ 56.3\ \Omega} \quad \text{IA} = 0.556\ - \angle 56.3\text{A}$$

$$I_B = \frac{10 \angle 0 \text{ V}}{(10 - j5)\ \Omega} \quad \text{branch impedance needs to be changed to polar form}$$

$$= \frac{10 \angle 0 \text{ V}}{11.2 \angle\ 26.6\ \Omega} = 0.893 \angle\ 26.6 \text{ A}$$

Step 2 In order to add the individual branch currents to find the total current, the currents must be changed from polar form to rectangular form.

$$I_A = (0.308 - j0.46)\ \text{A (polar form)}$$
$$I_B = (0.798 + j0.399)\ \text{A (polar form)}$$

Step 3 Add the j terms separately, then add the real terms separately to find the total current.

$$I_T = (1.106 - j0.061)\ \text{A}$$

Step 4 Change to the polar form of the total current from the rectangular form for use in finding the total impedance.

$$I_T = 1.11 \angle -3.15\ \text{A}$$

Step 5 Find total circuit impedance using the applied voltage and the total current.

$$R = \frac{E}{I}$$

$$Z_T = \frac{10 \quad 0\ \text{V}}{1.11 \quad 3.15\ \text{A}} = 9\ 3.15\ \ \Omega$$

Complex ac circuits

The process of solving complex ac circuits using the j-operator, uses the same basic thought process involved in solving complex dc circuits. The parallel branches are solved first for the total impedance, and then the series resistance is added. This combination of parallel, then series, results in the total circuit resistance of the circuit. That is the same basic procedure that will be used with the following sample problem.

Refer to Fig. 12-8. There are three parallel branches and a complex series impedance. The method of finding the equivalent parallel impedance will be the reciprocal method.

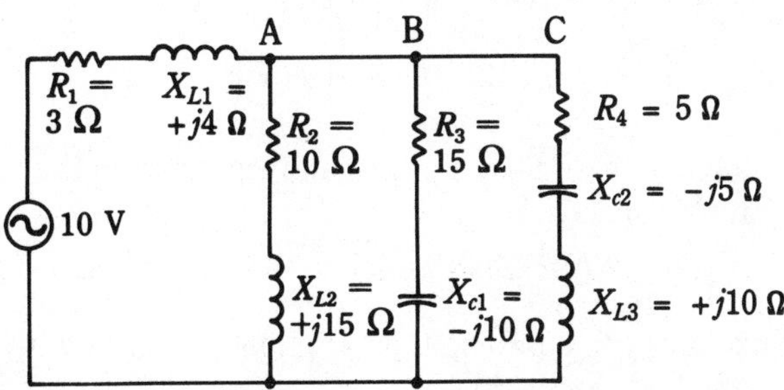

12-8 Sample complex ac circuit.

Step 1 Write the impedances for each individual branch in rectangular form, then change to polar form.

Branch A. $(10 + j15)\ \Omega = 18 \angle 56.3\ \Omega$
Branch B. $(15 - j10)\ \Omega = 18 \angle -33.6\ \Omega$
Branch C. $(5 + j5)\ \Omega = 7 \angle 45\ \Omega$

Step 2 Use the polar form of each branch resistance in the reciprocal formula to solve for total parallel impedance.

$$\frac{1}{R_T} = \frac{1}{R_1} + \frac{1}{R_2} + \frac{1}{R_3} \text{ formula}$$

$$\frac{1}{Z_{\text{Parallel}}} = \frac{1}{18\angle 56.3} + \frac{1}{18\angle 33.6} + \frac{1}{7\angle 45} \text{ substitute values}$$

To solve this type of equation, first take the reciprocal of the magnitudes (the numbers) and change the sign of the angles. For this step, it is necessary to change each reciprocal to a decimal.

$$\frac{1}{Z_P} = 0.05556\angle 56.3 + 0.05556\ 33.6 + 0.1428\angle 45$$

The formula now says to add these numbers. Remember, it is necessary to add in rectangular form. Therefore, change each of the polar forms to rectangular form.

$$\frac{1}{Z_P} = (0.0308 + j0.0462) + (0.0462 + j0.0307) + (0.1009 + j0.1009)$$

Now it is possible to perform the indicated addition. Add the real terms together, then add the j terms together.

$$\frac{1}{Z_P} = (0.1779 + j0.1164) \text{ (rectangular form)}$$

Now the equation indicates taking the reciprocal of both sides in order to change to Z_P rather than $1/Z_P$. In order to take the reciprocal, it is necessary to perform this in polar form. Therefore, change to polar form.

$$\frac{1}{Z_P} = 0.2125 \angle 33.19 \text{ (polar form)}$$

Now it is possible to take the reciprocal and end up with the value of the impedance of the parallel branches.

$$Z_P = 4.7 \angle 33.19 \text{ (parallel equivalent impedance)}$$

Step 3 It is necessary to combine the parallel impedance with the series impedance. Before doing this, it is necessary to change the parallel impedance to rectangular form.

$$Z_P = (3.9 + j2.57) \text{ rectangular form}$$

Write the series impedance in rectangular form so it can be added to the parallel impedance.

$$Z_S = (3 + j4)\ \Omega$$

Now, add the parallel impedance with the series impedance. This will be the total circuit impedance in rectangular form.

$$\begin{aligned} Z_T = Z_P + Z_S &= (3.9 + j2.57) + (3 + j4) \\ &= (6.9 + j6.57)\ \Omega \text{ (rectangular form)} \\ &= 9.52 \angle 43.59\ \Omega \text{ (polar form)} \end{aligned}$$

Practice problems

Calculate the answers to the problems using the *j*-operator and the circuit diagrams shown.

1. In the circuit shown, find the total impedance and the total current.

2. In the circuit shown, find the voltage drops across the resistor and capacitor, total current, and total impedance.

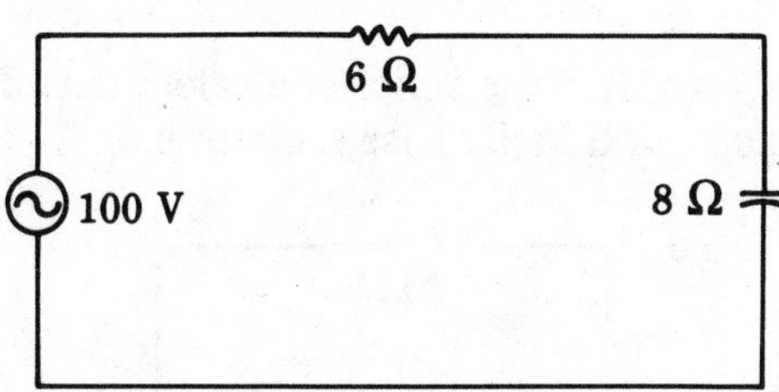

3. A circuit consisting of 175 Ω resistance in series with a capacitor of 5.0 μF is connected across a source of 150 V, 120 Hz. Determine the impedance of the circuit and the current through the circuit.
4. A circuit consisting of 100 Ω resistance, 0.35 H inductance, 13 μF capacitance is connected across a 220 V, 60 Hz voltage source. Find the voltage drops across each component, total current and total impedance.
5. In the circuit shown, find the total impedance and the total current.

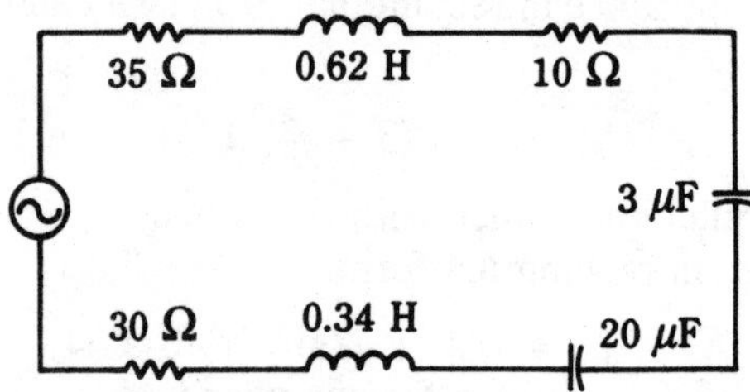

6. In the circuit shown, find the current through the resistor and inductor, total current and total impedance.

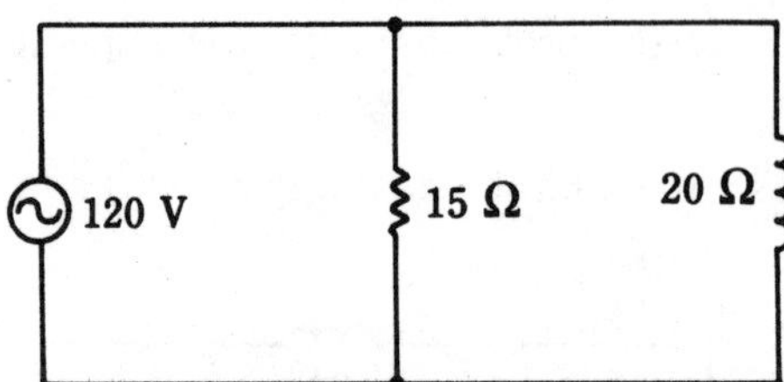

7. In the circuit shown, find the current through the resistor and capacitor, total current and total impedance.

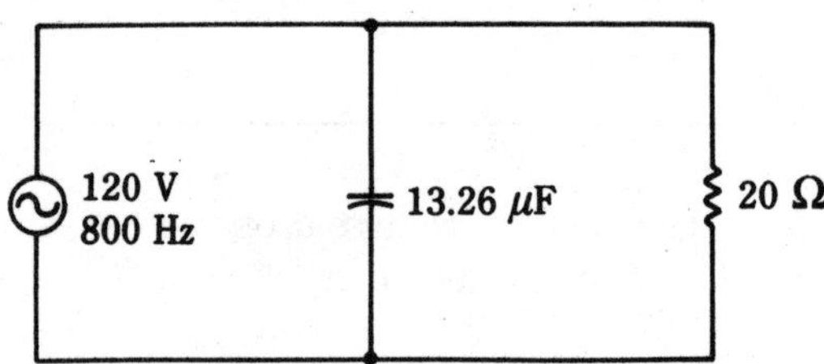

8. In the circuit shown, find the series equivalent impedance from point A to B with a frequency of 5 MHz. (Use reciprocals.)

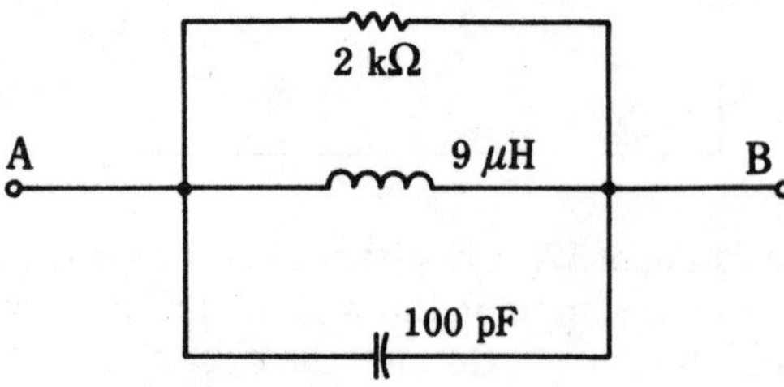

9. A parallel circuit contains two branches. The first branch has a 75-Ω resistor in series with a capacitive reactance of 30 Ω. The second branch has a 35 Ω resistor in series with an inductive reactance of 50 Ω. Find the total impedance.

10. A parallel circuit contains two branches. The first branch has an 80 Ω resistor in series with an inductive reactance of 26 Ω. The second branch has only a capacitive reactance of 100 Ω. Find the total impedance.
11. In the circuit shown, find the total impedance.

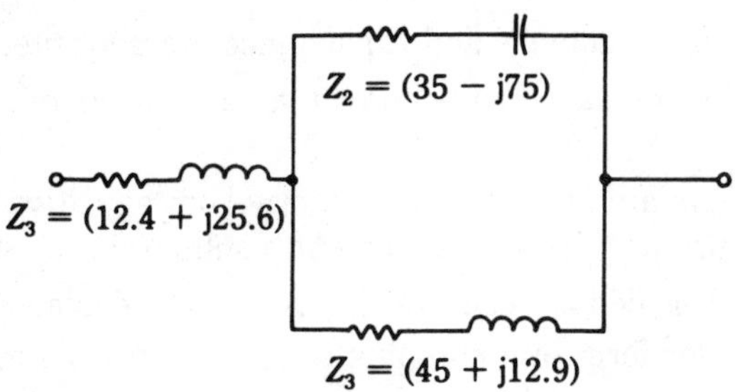

12. In the circuit shown, $Z_1 = (27.7 - j50)$, $Z_2 = (150 + j76.2)$, $Z_3 = 111.5 \angle 21$. Find the total impedance.

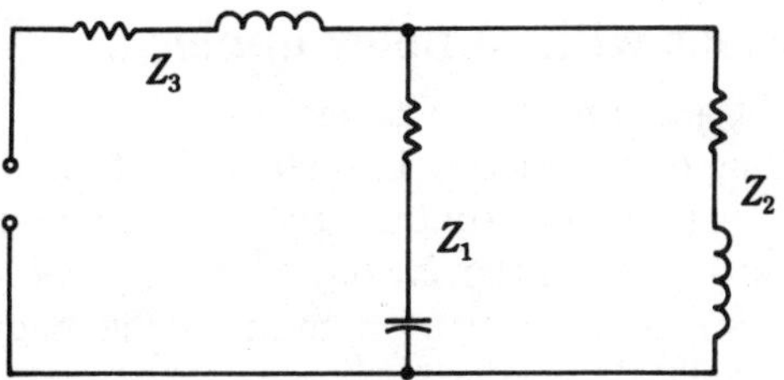

13. What is the equivalent impedance of two impedances $Z_1 = 151 \angle 4.07$ and $Z_2 = 50 \angle 53.1$ connected in parallel?
14. What is the equivalent impedance of two impedances $Z_1 = (73.8 - j34.4)$ and $Z_2 = (30 + j40)$ connected in parallel?
15. What is the equivalent impedance of two impedances, $Z_1 = 60.5 \angle 20$ and $Z_2 = (100 + j0)$ connected in parallel?

Chapter summary

Alternate current circuits that contain both inductance and capacitance, along with resistance are dealt with in a manner very similar to circuits with only one reactive component. The key to remember is the fact that the vectors representing capacitive reactance and inductive reactance are opposite in their direction. This results in at least a partial cancellation of the vectors. The result is called the net reactance.

Sometimes it is necessary to classify a circuit as being either inductive or capacitive. The reason for the classification is to be able to predict how the circuit will react in terms of a leading or lagging voltage or current.

The *j*-operator is a mathematical tool used to simplify the mathematics involved in solving complex ac circuits. When the *j*-operator is written in a number, the number then is called a complex number. Complex numbers can be written in either rectangular or polar form.

The following are considered the key points of the chapter.

- The vectors for inductance and capacitance are opposite.
- Series circuits are classified as inductive or capacitive by whichever is the larger reactance or voltage drop.
- Parallel circuits are classified as inductive or capacitive by whichever has the largest current. The largest current results from the smallest reactance.
- A circuit is classified as resistive only if the net reactance is zero.
- The rectangular form of a complex number states the actual quantities of the real and *j* terms.
- The rectangular form is written with this format; (real ± *j*-term).
- The polar form of a complex number is written in this format; $Z \angle \Theta$.

Rules for calculations with complex numbers

Rule 1 Add or subtract in rectangular form.

Rule 2 When adding or subtracting in rectangular form, add/subtract the real terms together and add/subtract the *j* terms together.

Rule 3 Multiply or divide in polar form.

Rule 4 When multiplying in polar form, multiply the magnitudes together (the *Z* term) and add the angles.

Rule 5 When dividing in polar form, divide the magnitudes (the *Z* term) and subtract the angles.

Rule 6 When taking the reciprocal of a number, it is to be done in polar form, since it is division. Take the reciprocal of the magnitude and change the sign of the angle.

13 Resonance

CREATING RESONANCE IS A VERY COMMON USE FOR AN INDUCTOR AND capacitor in the same circuit. A resonant circuit can be either series or parallel, and the characteristics of each are completely different. Some of the uses of a resonant circuit include the tuning of an antenna, tuning of a radio receiver, and filtering circuits. Resonant circuits can accept only certain frequencies, a group of frequencies, or it can reject certain frequencies—all depending on the application of the resonant circuit in the total circuit.

The resonance effect

In chapters 10 and 11, inductive reactance and capacitive reactance are discussed. Remember, inductive reactance will increase with an increase in frequency, and capacitive reactance will decrease with an increase frequency. Not only is the effect of frequency opposite, but all characteristics of the two reactances are opposite.

Once a set of values for inductance and capacitance have been selected, it is possible to calculate the frequency where both inductive reactance and capacitive reactance are equal. It is the point where the reactances are equal that is called the resonant frequency.

- At the resonant frequency, $X_L = X_C$.

Chapter 12 mentions that a circuit could be described as being either a capacitive circuit or an inductive circuit, depending on the net reactance. In a resonant circuit, the net reactance is zero.

Calculating the resonant frequency

By the definition given, the resonant frequency can be calculated by determining the frequency where the inductive reactance equals the capacitive reactance. Therefore:

$$X_L = 2\pi f L \text{ and} \qquad (10\text{-}1)$$

$$X_C = \frac{1}{2\pi f C} \text{ reactance formulas}$$

$$X_L = X_C \text{ definition of resonance} \qquad (11\text{-}1)$$

$$2\pi f L = \frac{1}{2\pi f C} \text{ substitute reactance formulas}$$

$$f^2 = \frac{1}{2^2\pi^2 LC}$$

$$f_r = \frac{1}{2\pi\sqrt{LC}} \text{ frequency at resonance} \qquad (13\text{-}1)$$

When the resonant frequency is calculated, it does not depend on it being a series circuit or a parallel circuit, the formula is still the same. The frequency of resonance depends only on the combined values of L and C.

Table 13-1 shows that the values of inductance and capacitance can be selected over quite a wide range of values. Once one of the two is selected, the other value is calculated so the reactance of the two is equal, and therefore, resonant at the desired frequency. The table is calculated for a resonant frequency of 1000 kHz, of 1 MHz.

Table 13-1. *LC* combinations resonant at 1000 kHz.

***L* (μH)**	***C* (pF)**	**Actual Value $X_L = X_C$ (Ω)**
23.9	1060	150
119.5	212	750
239	106	1500
478	53	3000
2390	10.6	15,000

The five different combinations shown in the table all have $X_L = X_C$, but there is a wide range of values that the reactance can have. Because there is such a wide range, what is the best way to select the values of the components? As a general rule, the inductor will have the deciding factor. Usually, with small values of inductance, the resistance of the wire used to make the inductor becomes smaller. The ratio of X_L to its coil resistance, R_s (or r_s) is called Q, or the quality of the coil. You will investigate the effects of Q later in this chapter. To conclude this discussion in simplified terms, for normal circuits, X_L is usually selected to be approximately 1500 Ω. A lower value means the resonant circuit does not have a shape of a bell curve (the curve of the output) and a higher value of X_L results in a circuit with a very sharp bell curve.

Calculating the value of *C* when resonant frequency is known

The sample shown here is selected from on of the points in Table 13-1.

$$f_r = \frac{1}{2\pi\sqrt{LC}} \tag{13-1}$$

$$f_{r2} = \frac{1}{4\pi^2 LC} \quad \text{square both sides to remove the square root sign}$$

$$C = \frac{1}{4\pi^2 L f_r^2} \quad \text{equation to solve for } C$$

Find the value of the capacitor to be used if the value of inductor is selected to be 239 μH, with a resonant frequency of 1000 kHz.

$$C = \frac{1}{4\pi^2 L f_r^2}$$

$$= \frac{5}{4\pi^2 \times 239\ \mu\text{H} \times (1000\ \text{kHz})^2}$$

$$= 106\ \text{pF}$$

Calculating the value of *L* when resonant frequency is known

The sample shown here is selected from one of the points in Table 13-1.

$$f_r = \frac{1}{2\pi\ \sqrt{LC}} \tag{13-1}$$

$$f_r^2 = \frac{1}{4\pi^2 LC} \quad \text{square both sides to remove the square root symbol}$$

$$L = \frac{1}{4\pi^2 C f_r^2} \quad \text{equation to solve for } L$$

Find the value of inductance to be used in a resonant circuit where the value of capacitance is 212 pF and the frequency of resonance is 1000 kHz.

$$L = \frac{1}{4\pi^2 C f_r^2}$$

$$= \frac{1}{4\pi^2 \times 212\ \text{pF} \times (1000\ \text{kHz})^2}$$

$$= 119.5\ \mu\text{H}$$

Effect above and below resonant frequency

When a circuit is at the resonant frequency, the effects of inductive reactance and capacitive reactance cancel each other, leaving a net reactance of zero, because they are equal and opposite. However, when the frequency drifts from resonance, the circuit will start to display characteristics of either an inductive circuit or a capacitive circuit because not all of the reactance will be canceled. The characteristics of inductive or capacitive depend on the circuit being either a series circuit, with the effects being determined by the voltage drops, or a parallel circuit, with the effects being determined by the current through each branch.

Even though the circuit configuration (series or parallel) determines the characteristics, it is possible to predict the actual values of inductive reactance and capacitive reactance, above and below the resonance frequency.

Table 13-2 has a resonant frequency of 1000 kHz. Reactance is calculated below resonant frequency and above resonant frequency. The values of reactance calculated in this table will apply to both series and parallel circuits.

Notice, below resonant frequency, the net reactance is larger in the capacitive side. Above the resonant frequency, the net reactance is larger on the inductive side. However, keep in mind that the effects of net reactance on the circuit depend on the circuit being series or parallel.

Table 13-2. Reactance at frequencies near 1000 kHz.

Frequency (kHz)	X (Ω)	X_C (Ω)	Net reactance (Ω)
600	900	2500	1600
800	1200	1875	675
1000	1500	1500	0 (resonance)
1200	1800	1250	550
1400	2100	1070	1030

Series resonance

Figure 13-1 shows a series resonant circuit. Notice, there is a resistor in series with the inductor and capacitor. This resistor is labeled r_s and is intended to represent the resistance of the inductor, resistance of the wires and any other resistance that may be present in the circuit. This r_s should be less than 10 Ω or the effect of the resonant circuit may be minimized to a point of uselessness.

By the definition of resonance, if $X_L = X_C$ and they cancel each other, the only remaining component in the circuit is the series resistor. It should make sense, then that if the reactances cancel and the only thing left is the resistor, the impedance of the circuit is very small. In fact, the impedance of the circuit is the remaining resistance.

The impedance is considered to be a minimum. If the impedance is minimum,

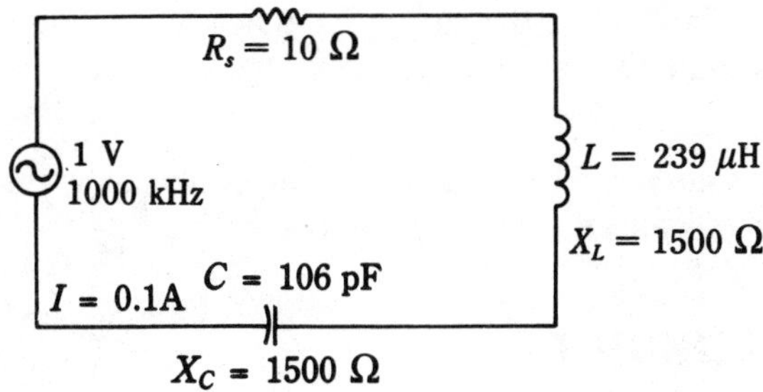

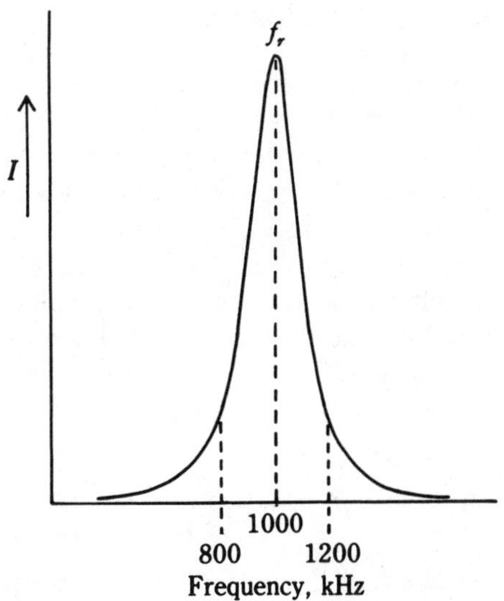

13-1 Series resonant circuit and the bell curve of current versus frequency.

the current flowing through a series resonant circuit would be large, restricted only by the series resistor.

- In a series resonant circuit, impedance is minimum and current is maximum.

The curve shown with the circuit in Fig. 13-1 is called the *bell curve*. The horizontal axis is the frequency, and the vertical axis is the current. Right at the resonant frequency is the peak of the curve. It is at this point that the current through the circuit is the maximum value. Notice, the curve drops off very steeply on either side of the resonant frequency. This drop off in current can also be seen in Table 13-2 by looking at the net reactance.

In a series circuit, the circuit is classified as either inductive or capacitive depending on the voltage drops of the individual components. The voltage drops depend on the relative sizes of the two components, that is to say the reactances.

Referring to Table 13-2, the capacitive reactance below resonant frequency is higher than the inductive reactance. Above resonant frequency, the inductive reactance is higher.

- In a series circuit, the circuit is net capacitive below resonant frequency and net inductive above resonant frequency.

Parallel resonance

Many of the items discussed earlier will still hold true in a parallel circuit because the circuit configuration does not effect such things as the resonant frequency, and the values of inductive or capacitive reactance.

The main change between the series circuit and the parallel circuit is the circuit conditions at resonance. In a series circuit, X_L equals X_C and cancel each other to leave the series resistance as the only component to limit current. In a parallel circuit, X_L equals X_C but it is the branch currents that cancel, leaving a path of no current flow. In a series circuit, the maximum current resulted in a minimum impedance. In a parallel circuit, the canceled branch currents make the circuit appear to be a very high impedance. In the parallel circuit, the higher reactance above or below frequency results in the lower value of branch current. The higher branch current determines the circuit characteristics. Therefore, below resonant frequency, the capacitive reactance is higher than the inductive reactance, this produces more current in the inductance branch. Therefore, the circuit is classified as an inductive circuit.

- In a parallel resonant circuit, impedance is maximum and current is minimum.
- In a parallel circuit, the circuit is net inductive below resonant frequency and net capacitive above resonant frequency.

Figure 13-2 shows a parallel resonant circuit with its respective bell curves. Notice in the circuit, the resistor r_s is placed in series with the inductor. This is to represent the resistance of the inductor windings. As long as the value is quite small, there is little or no effect on the resonant circuit.

The two bell curves show the comparison between the current versus frequency and the impedance versus frequency. The current curve is to show the sharp drop in current at the resonant frequency. The impedance curve shows a very sharp increase at the resonant frequency. If this parallel circuit is used in conjunction with another circuit, the impedance curve could also be used to represent the voltage dropped across the resonant circuit in comparison to another section of a circuit this might be connected to. For example, this could be connected as a filter in parallel with the applied voltage and the remainder of the circuit. The resonant frequency would see the filter as an open circuit. All frequencies that are not the resonant frequency would see the filter as a path to ground.

Using the same idea as an example for the series resonant circuit, if the series circuit were used as a filter, it would be connected in series with the load and volt-

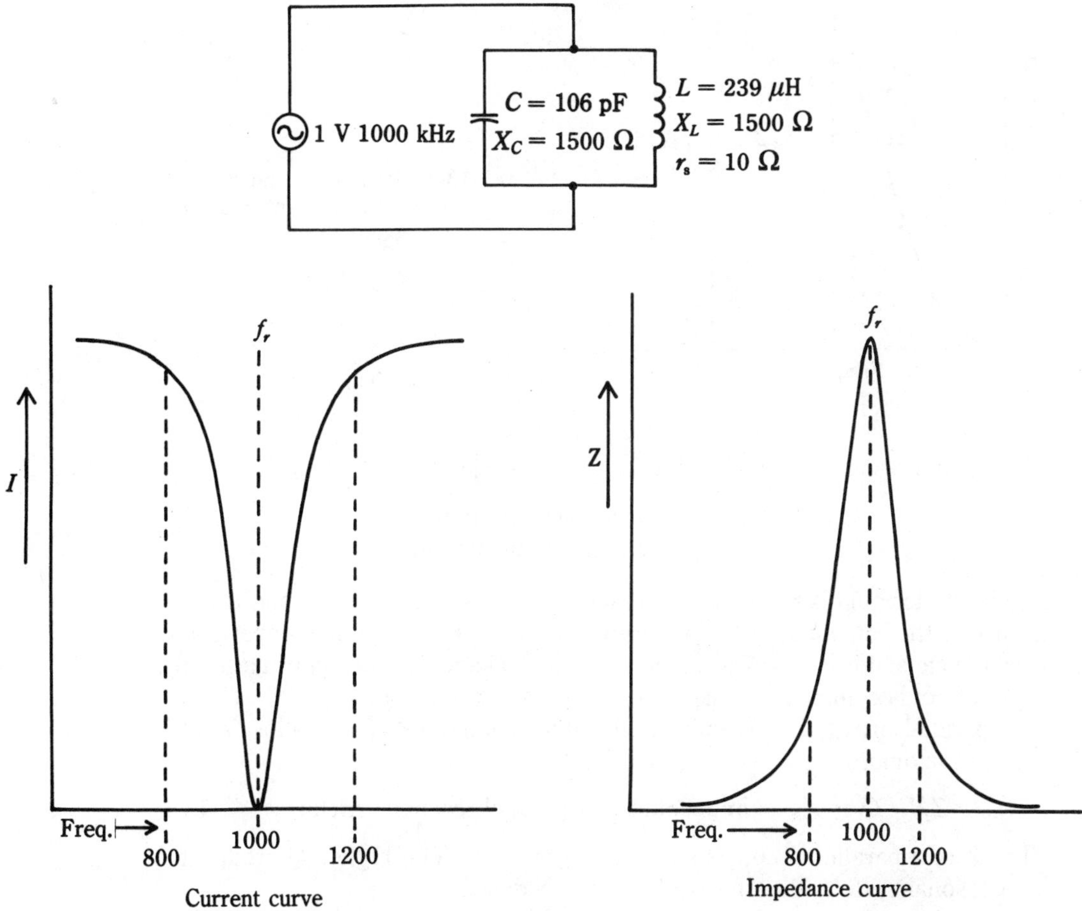

13-2 Parallel resonant circuit and the bell curves of current versus frequency and impedance versus frequency.

age source. At the resonant frequency, the resonant circuit would appear as only a piece of wire. At any other frequency, the circuit would have high impedance and would stop a signal from passing through by having too large of a voltage drop.

Series and parallel resonant circuit can be used in combination with each other to produce filters that fit specialized needs. Although filters are not the only application of resonant circuits, it is an easy one to describe.

Q of a resonant circuit

The *Q* of a resonant is a measure of the quality or figure of merit of the circuit. In general, the higher the circuit *Q*, the sharper the resonance effect, that is, a sharper bell curve.

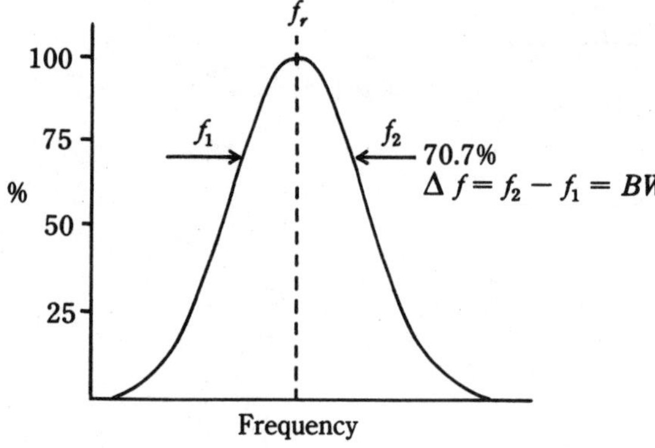

13-3 Bell curve showing the 70.7 percent response point.

The Q of a resonant circuit, either parallel or series can be calculated by:

$$Q = \frac{X_L}{r_s} \qquad Q \text{ of a resonant circuit. } Q \text{ is } r_s \text{ ratio; it has no units.} \qquad (13\text{-}2)$$

X_L in the formula is the inductive reactance, measured in ohms, and r_s is the resistance of the coil windings. This formula assumes r_s is small in comparison to X_L, and the capacitor is assumed not to have any leakage. If r_s becomes larger, the resonant effect becomes less sharp. The bell curve flattens out.

In a parallel circuit, the Q can be used to determine the Z of a parallel circuit is defined as maximum.

$$Z_P = Q \times X_L \qquad \text{impedance of a parallel resonant circuit} \qquad (13\text{-}3)$$

The Z of a parallel circuit is expected to be high. The higher Q produces a sharper resonant effect, therefore, a higher impedance.

Bandwidth of a resonant circuit

The bell curve discussed in this chapter is shown as having sloped sides to the curve. The resonant frequency curve allows calculations of the resonant frequency. When the resonant frequency is shown on the bell curve, however, you will see that there are frequencies that are close that have almost the same amplitude as the resonant frequency.

In order to make calculations concerning bandwidth, a standard has been set.

- Bandwidth is defined as the range of frequencies with a response of 70.7 percent (or more) of the maximum response. 70.7 percent of current (series) or 70.7 percent of voltage (parallel). Measured in frequency units.

The bandwidth is also known as the *half-power* points, or the −3dB points. Half-power and −3dB are equal to 70.7 percent of the maximum current or voltage.

Bandwidth can be represented by Δf or by BW. The first symbol best described exactly what the bandwidth is. The triangle in the first symbol is actually the Greek letter delta. It is used here to mean change in or range of and the small letter f stands for frequencies. When the two symbols are put together, Δf, it means change in frequencies or range of frequencies.

Refer to Fig. 13-3. This drawing is a simplified drawing of a bell curve. Its purpose is to demonstrate the bandwidth in terms of the 70.7 percent of the maximum amplitude and to show the bandwidth in terms of the Δf. Notice, there are points labeled f_1 and f_2. These two points are the frequencies at the 70.7 percent point. f_1 is the lower frequency and f_2 is the higher frequency. If f_1 is subtracted from f_2, the difference will be the range of frequencies between these two points, also called the bandwidth.

If the bell curve is perfectly symmetrical, the resonant frequency is found exactly in the middle, at the highest point on the curve. Making the assumption that the bell curve is perfectly symmetrical, the bandwidth can be divided in half. Half will be above resonant frequency and half will be below resonant frequency.

For example, in Fig. 13-3, if the resonant frequency, f_s is 100 kHz and the bandwidth is 20 kHz, then the points at the 70.7 percent level could be identified as 90 kHz and 110 kHz.

$$f_1 = f_r - \frac{BW}{2} \quad \text{substitute}$$

$$= 100\text{ kHz} - \frac{20\text{ kHz}}{2}$$

$$= 90\text{ kHz}$$

$$f_2 = f_r + \frac{BW}{2} \quad \text{substitute}$$

$$= 100\text{ kHz} + \frac{20\text{ kHz}}{2}$$

$$= 100\text{ kHz}$$

It is possible to calculate the bandwidth of a circuit in terms of the circuit values. The bandwidth is affected by the Q of the circuit. A circuit with a very high Q will have a very small bandwidth, steep bell curve. A circuit with a low Q will produce a very wide bandwidth or shallow bell curve.

$$BW = \frac{f_r}{Q} \quad \text{bandwidth in terms of circuit } Q \qquad (13\text{-}4)$$

$$BW = \Delta f = f_2 - f_1 \quad \text{bandwidth in terms of frequency} \qquad (13\text{-}5)$$

Refer to Fig. 13-4. There are three bell curves drawn. Each curve has the same resonant frequency. The three curves show the effect of circuit Q on the bandwidth. The curve with the Q of 80 has a steep bell curve, therefore, at the 70.7 percent points, the bandwidth is quite narrow. The middle curve, with a Q of 40 is

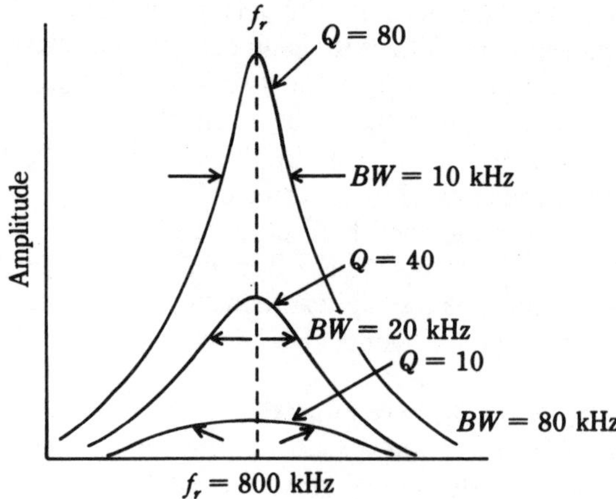

13-4 Three bell curves, each with the same resonant frequency and different Q.

not as steep of a bell curve because the resonant effect is not as great and the bandwidth is twice as wide as the top curve. The lowest curve on the graph shows a *Q* of 10. This curve has very little resonant effect and a very wide bandwidth. The curve for a pure resistance would be a flat line.

Chapter summary

The resonant circuit is useful for accepting, or rejecting certain frequencies. The key to any resonant circuit is the fact that the inductive reactance and capacitive reactances are equal, at resonant frequency.

The *Q* of a resonant circuit is an important factor, because it will determine the sharpness, or bandwidth of the resonant effect.

Bandwidth is defined as the 70.7 percent of maximum response. The points that have less than this response are considered too small to be of any practical use. If the circuit has a narrow bandwidth, the sides of the curve are so steep that the frequencies below the 70.7 percent points would be very hard to measure.

If the bell curve is perfectly symmetrical the bandwidth can be divided in half and the upper and lower frequencies (the 70.7 percent points) can be calculated.

- At the resonant frequency, $X_L = X_C$.
- In a series resonant circuit, impedance is minimum and current is maximum.
- In a series, the circuit is net capacitive below resonant frequency and net inductive above resonant frequency.
- In a parallel resonant circuit, impedance is maximum and current is minimum.

- In a parallel circuit, the circuit is net inductive below resonant frequency and net capacitive above resonant frequency.
- Bandwidth is defined as the range of frequencies with a response of 70.7 percent (or more) of the maximum response.

Summary of formulas

$$f_r = \frac{1}{2\pi\sqrt{LC}} \quad \text{resonant frequency} \tag{13-1}$$

$$Q = \frac{X_L}{r_s} \quad Q \text{ of a resonant circuit (no units)} \tag{13-2}$$

$$Z_P = Q \times X_L \quad \text{impedance of a parallel resonant circuit} \tag{13-3}$$

$$BM = \frac{f_r}{Q} \quad \text{bandwidth in terms of } \frac{f_r}{\text{circuit } Q} \tag{13-4}$$

$$BW = \Delta f = f_2 - f_1 \quad \text{bandwidth in terms of frequency} \tag{13-5}$$

Practice problems

1. Find the f_r for a series circuit with a 10 μF capacitor and an inductor of 16 H, with 5 Ω internal resistance.
2. Find the f_r for a parallel circuit with an L of 80 μH and C of 120 pF.
3. In a series resonant circuit, X_L is 1500 Ω and the internal coil resistance is 15 Ω. At the resonant frequency determine:
   a. How much is the Q of the circuit?
   b. How much is X_C?
   c. With a generator voltage of 15 mV, how much is I?
   d. How much is the voltage across X_C?
4. What value of L is necessary with a C of 100 pF for a resonant frequency of:
   a. 1 MHz?
   b. 4 MHz?
5. Calculate the lowest and highest values of a variable capacitor needed with a 0.1 μH inductor to tune through the commercial FM broadcast band of 88 MHz to 108 MHz.

14
Power supplies

POWER SUPPLIES ARE A VERY IMPORTANT PART OF ELECTRONIC SYSTEMS. After all, without a power supply of some sort, electrical circuits cannot operate. As an electronic technical, it is equally important to determine if a power supply is operating properly.

A power supply is simply a means of changing ac power to dc for use in the circuits. Very few circuits can operate without dc. The primary device used for converting ac to dc is the diode.

The PN junction diode

Diodes that are used for the purposes of rectifying ac to dc are of the PN junction type of diode. The PN junction is two pieces of semiconductor material joined together.

Semiconductor materials are doped to become either P-type or N-type materials. The *doping* is adding atoms of an impurity to the pure silicon or germanium crystals. It causes the semiconductor material to have either one extra electron one extra electron in orbit, or be missing one electron in its outer orbit. The extra electron, or missing electron is what makes the semiconductor material useful in the field of electronics.

- A semiconductor with an extra electron is called an N-type semiconductor.
- A semiconductor missing an electron is called a P-type semiconductor. The atom with a missing electron is referred to as a hole.

Figure 14-1A shows a PN junction diode. The point where the two pieces of semiconductor material are joined is called the junction. On the P-type side, there is a large number of holes. The majority are carriers of a P-type material. In the P-

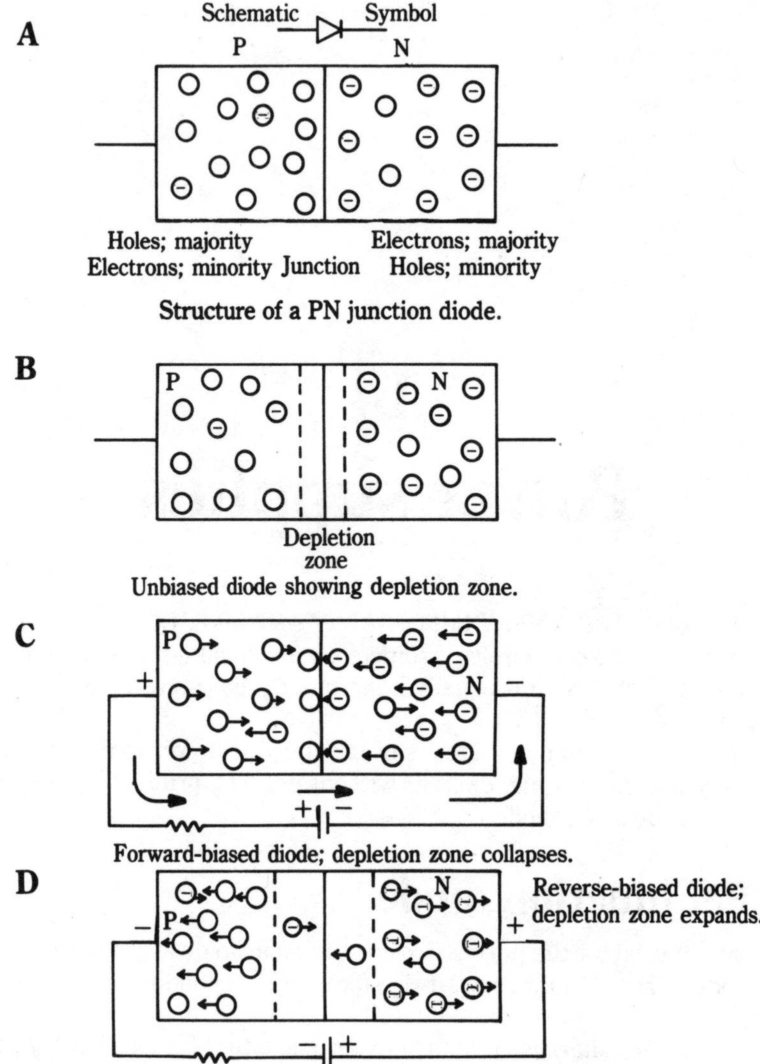

14-1 Characteristics of a PN junction diode.

material, there will be a very small number of minority carriers, electrons. Holes in the P-type material are represented as empty circles, to show they are an atom that is void one electron. They can be said to carry a positive charge.

- P-type semiconductors have a majority of holes, and a minority of electrons. Holes carry a positive charge.

Figure 14-1A also shows an N-type of material. The N-type has a majority of electrons and a minority of holes. The electrons of N-type material and holes of P-type material are considered to be current carriers of the semiconductor. The circles with a minus sign represent the electrons.

- N-type semiconductors have a majority of electrons, and a minority of holes. Electrons carry a negative charge.

Figure 14-1B shows the PN junction diode with a depletion zone. As soon as the P material and the N material are joined together, some of the electrons will join with some of the holes in the vicinity of the junction. The region is then called the depletion zone because it becomes void of any mobile carrier because when the holes and electrons join, the atom no longer is missing or have an extra electron.

- The depletion zone is depleted of any mobile carriers and acts like an insulation region.

In the area of the junction, the depletion zone will increase only to a certain point. It cannot expand any further because the attraction between the positive and negative charges is not strong enough without an outside voltage being applied to the semiconductor.

Figure 14-1C shows the PN junction diode with a forward bias applied to the semiconductor. A forward bias is a voltage that will cause the semiconductor to conduct current flow. Current flow is always assumed to be electron flow. However, keep in mind, inside a semiconductor material, the current flow is the majority carrier.

In order to have current flow, there must be a complete circuit, which means that for every electron that leaves the negative terminal of the battery, an equal number of electrons must return to the positive terminal of the battery.

In Fig. 14-1C, the electrons leave the negative terminal of the battery, travel to the N-type semiconductor material. These electrons enter the N-type material and become majority carriers. The negative force of the battery repels the electrons in the N material to travel towards the junction. The depletion zone will then start to collapse. While this is happening at the N material, the positive battery is attracting electrons from the P-type material. This attraction removes the electrons from the P-type material and causes them to travel to the battery. The positive force of the battery will also repel the positive holes, causing them to move toward the junction, collapsing the depletion zone.

With the depletion zone of the forward biased diode collapsed, the electron from the N-type material will enter the P-type material and recombine with the holes and be swept toward the positive terminal. The holes of the P-type material will enter the N-type and be recombined with the electrons and be swept toward the negative terminal. The result is, in the forward direction, the diode conducts current easily.

Even though the forward biased diode conducts easily, the junction still acts like a region of resistance. This is called the barrier potential which is 0.1 to 0.3 V for germanium and 0.5 to 0.7 V for silicon. This small voltage drop must be considered in all semiconductor circuits.

- A forward biased diode has negative connected to the N material and positive to the P material.

Figure 14-1D shows a reverse-biased diode. The positive terminal of the battery is connected to the N-type material and the negative terminal of the battery is connected to the P-type material. The positive on the N material attracts the electrons toward the battery, away from the junction. This causes the depletion zone to widen. The negative terminal of the battery attracts the holes toward the battery, away from the junction. This causes the depletion zone to widen. The depletion zone acts like a region of insulation and when it widens, it will not allow current to flow. Even though the depletion zone is quite wide, some electrons from the P material and some holes from the N material will travel across the junction. This is called minority current flow. Minority current flow is very small and is the only type of current in a reverse biased PN junction diode.

- A reverse-biased dioded has positive connected to the N material and negative to the P material.

Half-wave rectifier

The half-wave rectifier is an ideal example of an application for the PN junction diode. The half-wave rectifier is a power supply circuit made by using one diode.

Figure 14-2 shows a half-wave rectifier drawn to produce a positive output in Fig. 14-2A and a negative output in Fig. 14-2B. The two circuits are essentially the same with the exception of the diodes being in opposite directions. In order to simplify the discussion, Fig. 14-2A, the positive output is the one that is discussed.

In Fig. 14-2 the diode is connected to a transformer. With the half-wave rectifier circuit, this is really not necessary. However, most rectifier circuits will be con-

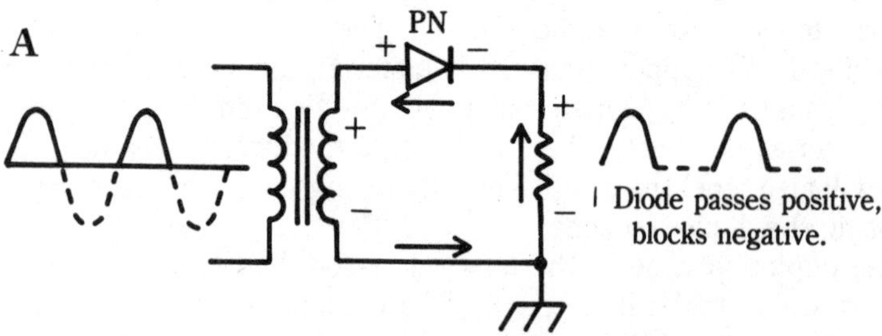

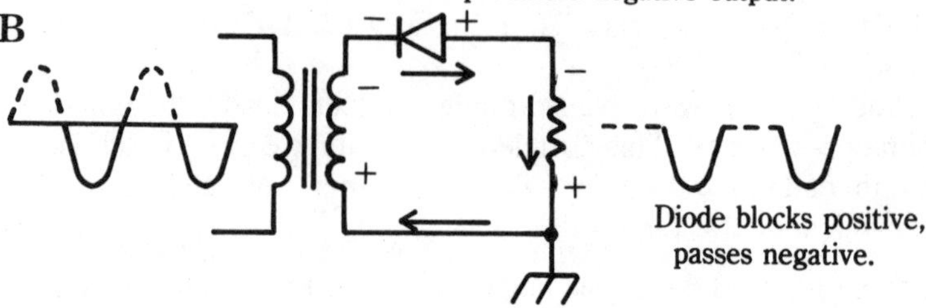

14-2 Half-wave rectifier.

nected to a transformer because that is an easy way to step down the ac input voltage.

The secondary of the transformer will have a sine wave exactly the same shape as the primary, with a different voltage. When discussing any rectifier circuit, the input sine wave is discussed in terms of positive half-cycle and negative half-cycle.

When the input sine wave is in the positive half-cycle, Fig. 14-2A, the diode will have a positive polarity on its P-type material. At the same time, the bottom of the transformer will be negative, which puts a negative polarity on the N-type material of the diode. The conditions for a forward biased junction are satisfied and the diode will conduct.

When the diode conducts the positive half-cycle of the input sine wave, there will be forward diode drop across the diode of either 0.3 V for Germanium (Ge) or 0.7 V for silicon (Si). The polarity of this voltage drop will be positive on the P side and negative on the N side. The resistor will drop the remainder of the voltage applied to the circuit. The voltage across the resistor will have a positive polarity on the top side and negative on the bottom side.

- When a rectifier diode is conducting, in the forward direction, the diode will have a voltage drop of 0.3 or 0.7 V, the remainder of the applied voltage is dropped across the load resistor.

When the input sine wave swings into the negative half-cycle, the P material will have a negative voltage applied to it and the N material will have a positive voltage applied to it. Note that the input sine wave polarity is always made in reference to the circuit ground, which is usually shown at the bottom of the transformer. Therefore, the negative half-cycle gives a transformer polarity of negative on the top and positive on the bottom of the transformer secondary. When the input sine wave is in its negative half-cycle, the diode is reverse biased and no current can flow. Minority current that can flow in the reverse direction is so small that it is considered zero.

- When the rectifier diode is reverse biased, there is no current flow. All of the applied voltage will be dropped across the diode and no voltage will be dropped across the load resistor.

When the two conditions of forward and reverse bias are both considered, note the arrows of both circuits (Figs. 12-2A and 14-2B) represent the current flow. Current flows only in one direction, which means the output voltage is considered dc. The name of this dc is *pulsating dc*. The dashed lines in the output waveform represent a period of time when there is no output. The input voltage is blocked during this time.

Full-wave rectifier

The full-wave rectifier circuit is shown in Fig. 14-3. The circuit uses two diodes instead of one, like the half-wave circuit. The primary advantage is the fact that the full-wave circuit will reproduce both halves of the input sine wave.

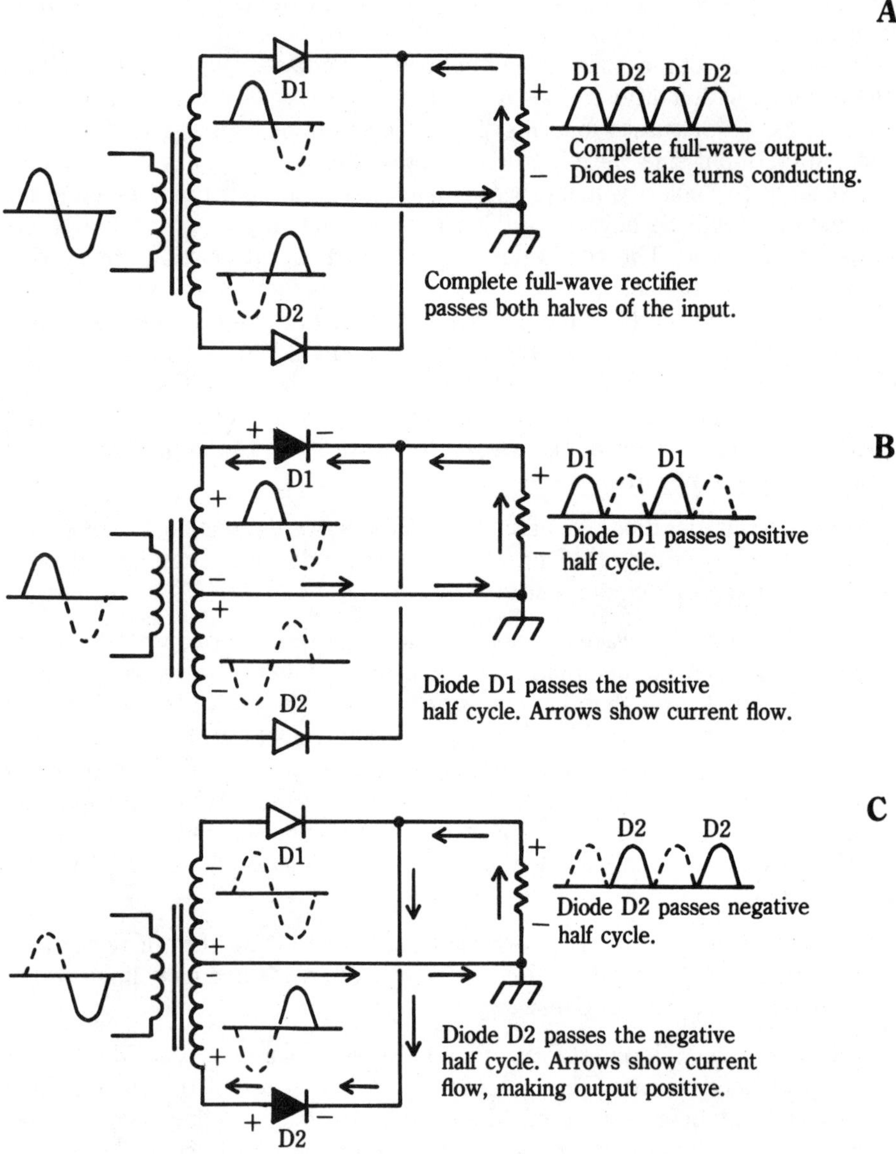

14-3 Full-wave rectifier.

The full-wave rectifier circuit requires the use of a center-tapped transformer. The function of the center-tapped transformer is to produce two sine waves, 180 degrees apart, from the one input sine wave. The center-tapped transformer can also be used as a step-up or as a step-down transformer. The center tap is connected to ground reference, which means the center tap is always zero volts. Figure 14-3B shows the secondary sine wave split into two sine waves 180 degrees apart.

When the input sine wave is in its positive half-cycle, as it is in Fig. 14-3B, the top part of the secondary becomes positive, in relation to the center tap. The bottom half of the secondary becomes negative, in relation to the center tap.

In Fig. 14-3C, the input sine wave is in its negative half-cycle and the secondary produces two sine waves 180 degrees apart, and the portion of the sine wave of interest is the second part, shown by the solid line. The polarity of the transformer reverses, putting positive at the bottom, with respect to the grounded center tap. The waveform looks like it is upside down. However, the wave is drawn correctly because it is in reference to the grounded center tap (imagine the bottom of the transformer to be drawn over the half of the circuit connected to the top of the transformer).

Figure 14-3A shows a complete full wave rectifier circuit. The solid lines drawn on the waveforms of the secondary show which part of the wave is in use with each diode. The dashed lines show the portion not in use at the time. Notice that each diode is responsible to produce only one half of the input sine wave. The center tap is a common return for current to the transformer. The arrows indicate only one direction of current flow. The output waveform drawn for Fig. 14-3A shows both halves of the sine wave, with the negative portion made positive. The labeling of D1 and D2 indicate which diode produced the output.

Figure 14-3B shows the operation of diode D1 in the circuit. When the input sine wave is in its positive half-cycle, the secondary has the polarity shown, with positive to D1 and negative to the center tap. The diode is forward biased and will allow current to flow to the load resistor. The diode will drop 0.3 V or 0.7 V, and the remainder of the voltage will be dropped across the load resistor. During the negative portion of the sine wave, diode D2 is reverse biased because the bottom of the transformer is negative. Current flow follows the arrows, from the center tap, through the load, through the diode D1 to the transformer.

Figure 14-3C shows the operation of diode D2. When the input sine wave is in its negative half cycle, it produces the polarity across the transformer as shown with a positive to diode D2 and a negative to the center tap. Diode D2 is forward biased and will allow current to flow to the load. The diode will have a voltage drop of 0.3 V or 0.7 V and the load will drop the remainder of the applied voltage. Notice the connection of diode D2 to the top side of the load resistor. It is this connection that causes the output to have both sides of the sine wave reproduced as positive. The polarity of the output is in relation to the ground reference point.

The full-wave rectifier circuit has the advantage over the half-wave circuit because it produces both halves of the input sine wave. This eliminates the period of time when there is an off condition produced with the half-wave rectifier circuit. The disadvantage is the fact that it does require a center-tapped transformer. Also, the center-tapped transformer produces one-half of its full secondary voltage between the tap and either side. This means the output voltage is only one-half of the voltage possible.

Bridge rectifier

The bridge rectifier is the most popular of modern power supplies. The reason is the fact that the circuit has some protection against failure. Also, bridge rectifiers are being manufactured as a single component for use in most low power circuits. The single unit makes the cost of building power supplies cheaper than other methods.

Figure 14-4 shows a bridge rectifier. There are four diodes used in this type of circuit. Two diodes take turns operating at a time.

Figure 14-4A shows the normal complete schematic. The output shown is the waveform of a full-wave rectifier, with both half-cycles being reproduced.

Figure 14-4B shows the bridge rectifier operating when the input sine wave is in its positive half-cycle. The positive will forward bias D2 and the negative of the transformer is connected to forward bias D3. To follow current flow, start from the bottom side of the transformer, which is now marked negative, to the junction labeled ac. Pass through diode D3, dropping 0.7 V to allow for diode drop. Continue to follow the current arrows to the junction marked negative, then proceed to the load. The bottom of the load will be negative, and the load will drop the majority of the voltage available to the circuit (all except 1.4 V diode drops). The large voltage drop across the load results in the load having the polarity as shown. Continue to trace current flow through the load, to the junction marked positive. Notice that diode D2 is forward biased, pass through D2 with a 0.7 V diode drop, to the junction marked ac and return to the transformer.

From the discussion of the previous paragraph, note that the output voltage for the positive half cycle of the input is from diodes D2 and D3.

Figure 14-4C shows the bridge rectifier as it operates on the negative half-cycle. Diode D1 and D4 are now forward biased. Follow the arrows shown in the drawing to trace the path of current to prove how the load once again will be positive.

The full-wave pattern is the result of the bridge rectifier. Because it uses two diodes at any one time, the reverse ratings of the diodes can be smaller than the conventional full-wave rectifier.

Calculating the rectifier output

The output of a rectifier circuit, without filtering, is called pulsating dc. It does qualify as a dc because it does not change direction. Therefore, if it is necessary to measure the output voltage, a dc voltmeter would be the proper measuring instrument to use.

A dc voltmeter will measure the average of the voltage applied to it. Even though a sine wave has its average located at 63.6 percent of peak, if a dc voltmeter were to be used with a sine wave, the result would be to measure zero.

Figure 14-5 compares the outputs of a half-wave, full-wave, and the input sine wave. In all three cases, the peak value is the same. The peak is measured from the zero reference line to the very peak of the wave. The values of rms and average apply only to the sine wave and not to the rectified outputs.

A. Complete bridge rectifier passes both halves of the input.

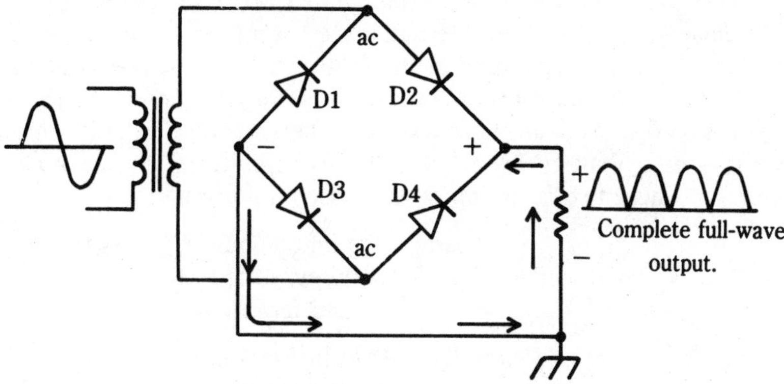

B. Diodes D2 and D3 pass the positive half cycle. Arrows show current flow.

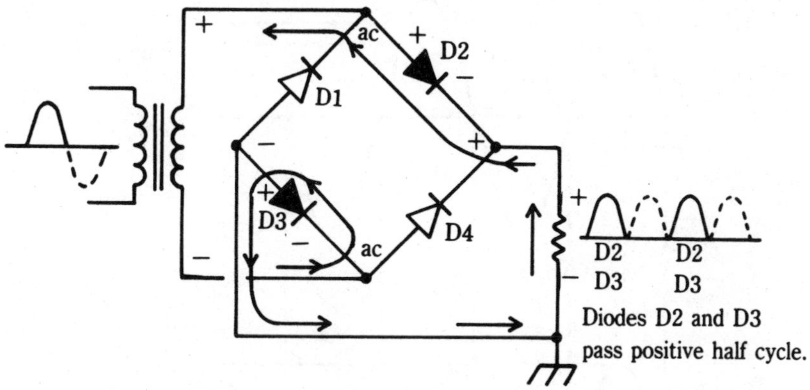

C. Diodes D1 and D4 pass the negative half cycle. Arrows show current flow, making output positive.

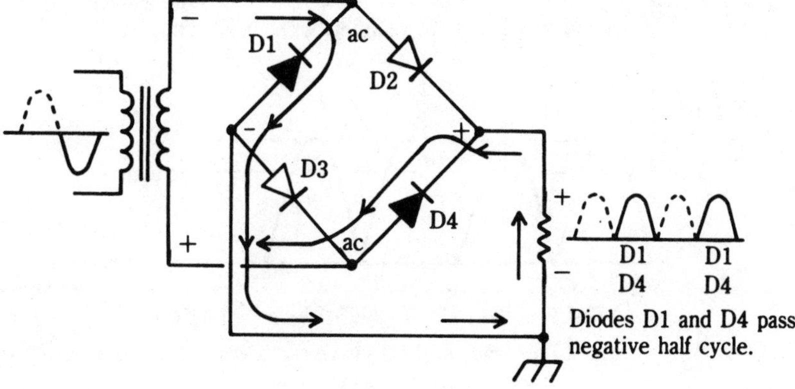

14-4 Bridge rectifier.

In Fig. 14-5, each of the rectified outputs shows a value called dc average. The average referred to there is the value of voltage that would be read on a dc voltmeter. The dc value is much lower on the half-wave output than it is on the full-wave output. In fact, the dc average value of the full-wave output is the same as the ac average of an equivalent sine wave. In other words, the dc average and the ac average both equal 63.6 percent of peak. Because the half-wave rectifier output has only one-half as many half-cycles as the full-wave circuit, the output must be one half the voltage. These two facts can be expressed in formulas.

$$Vdc = 0.626 \times \text{Peak} \qquad (14\text{-}1)$$

formula to find dc output voltage of a full-wave or bridge rectifier circuit

$$Vdc = \frac{0.636 \times \text{Peak}}{2} \qquad (14\text{-}2)$$

formula to find dc output voltage of a half-wave circuit

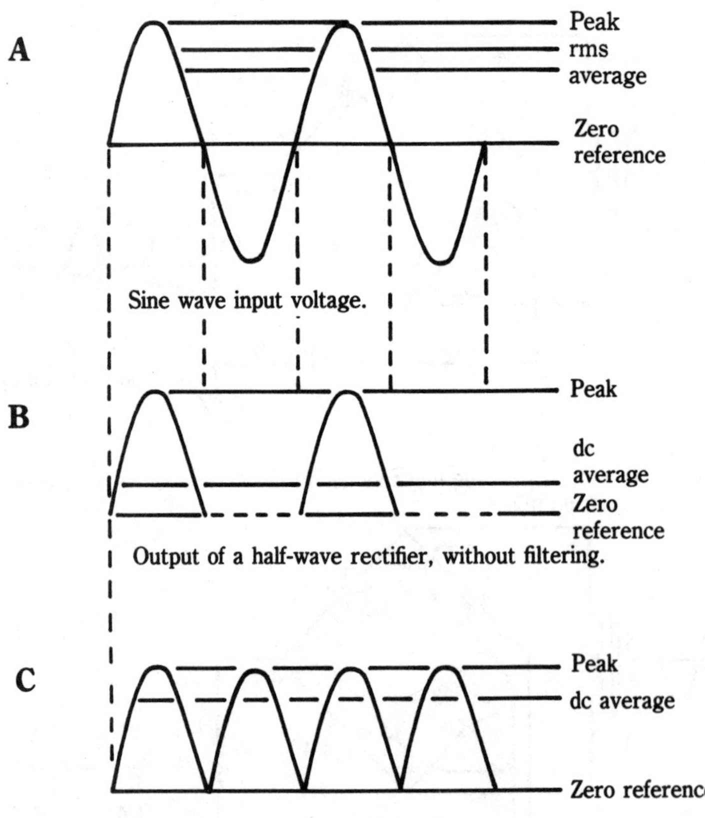

14-5 Waveforms of half-wave and full-wave rectifiers without filtering.

When dealing in terms of the peak, average, and rms, it is often more convenient to use rms because that is how most ac voltages are expressed. The following demonstrates a simple conversion from rms to average of a sine wave.

$$\text{Average} = 0.636 \times \text{Peak}$$
$$\text{rms} = 0.707 \times \text{Peak}$$

$$\text{Peak} = \frac{\text{rms}}{0.707}$$

$$\text{Average} = 0.636 \times \frac{\text{rms}}{0.707} \quad \text{substitute for peak}$$

$$\text{Average} = 0.9 \times \text{rms}$$

The discussion shown above leads to the following formulas:

$$\text{Vdc} = 0.9 \times \text{rms} \quad \text{formula to find dc output voltage of a full-wave or bridge rectifier circuit} \qquad (14\text{-}3)$$

$$\text{Vdc} = \frac{0.9 \times \text{rms}}{2} \quad \text{formula to find dc output voltage of a half-wave rectifier circuit} \qquad (14\text{-}4)$$

The next step in making calculations for the output of a rectifier circuit is to determine the frequency of the output waveform. Regardless of the frequency of the input sine wave, the output always follows the input exactly. The half-wave circuit will reproduce either the positive or negative half-cycles. The full-wave, or bridge, reproduces both half-cycles.

Frequency of a waveform is a measure of how often the waveform repeats itself. Notice from Fig. 14-5, the half-wave rectifier repeats itself after completing the positive voltage, then the off time, then it is back to the positive voltage. This is exactly the same period of time as the input sine wave. The full-wave waveform repeats itself each half-cycle of the input, therefore its frequency is twice as much as the input sine wave. The output frequency is called the ripple frequency and is important when considering the characteristics of a power supply.

$$\text{Ripple Freq. Half-Wave} = \text{Input Frequency} \qquad (14\text{-}5)$$

$$\text{Ripple Freq. Full-Wave} = 2 \text{ times Input Frequency} \qquad (14\text{-}6)$$

With the simple rectifier circuits that have been dealt with so far, there are only two calculations necessary; output voltage and output frequency. One consideration that is sometimes taken into account is the reverse voltage rating of the diodes, this is often called the *peak inverse voltage* (PIV). It is not unusual to find diodes with a PIV of 500 or even 1000 V. For this reason, the PIV is usually not an important consideration with modern circuits using relatively low voltages, usually under 100 V.

Filtering a rectifier circuit

The purpose of a filter on a rectifier circuit is to smooth the tremendous fluctuations in voltage. Many circuits cannot operate properly if there is any change in voltage. The filter is one step in making the dc voltage output a smooth, constant voltage to supply power to a circuit.

The filter discussed here is the simple capacitor filter. There are many kinds of filters available for power supplies, but the principle of operation is the same for all the filters.

With the simple capacitor filter, the principle characteristic of the capacitor is relied upon. That characteristic is the fact that a capacitor will charge and store a voltage. The stored voltage can then be discharged through a different path than it was charged from.

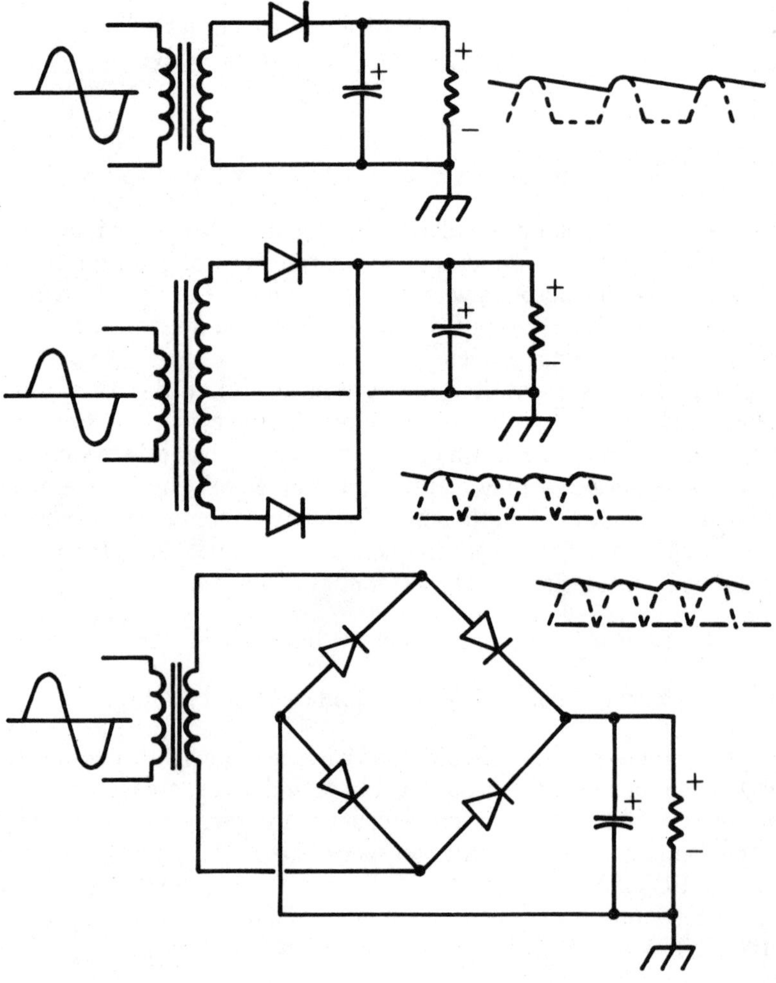

14-6 Rectifier circuits showing capacitor filter.

Figure 14-6 shows the three basic rectifier circuits, each with a simple capacitive filter connected. The capacitor is connected in parallel with the circuit, especially the load resistor. The rectifier diodes will charge the capacitor in the positive direction, then when the diode turns off, the capacitor will discharge through the load. Even though the load is represented by a resistor here, it could be any circuit that needs to be powered by a dc source.

Figure 14-7 shows what takes place with the filter in the circuit. Figure 14-7A is the input sine wave shown here as a figure of reference. Figure 14-7B is the waveform of a half-wave rectifier, and Fig. 14-7C is the waveform of a full-wave and bridge circuit.

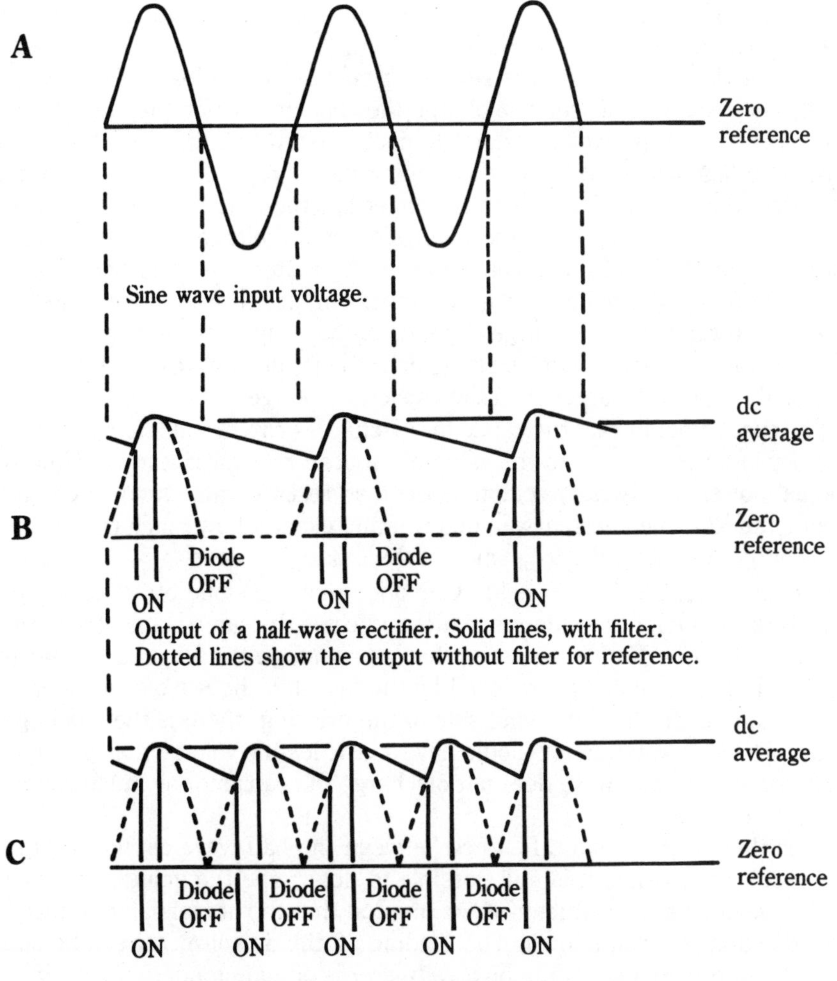

14-7 Waveforms of half-wave and full-wave rectifiers, with filtering.

Figure 14-7B, with the half-wave circuit waveform presents the best picture for an explanation of how the filter capacitor does its job. The dotted lines are for reference only, to show what the output would look like if there were no filter. The solid lines represent the actual voltage that is applied to the output of the rectifier circuit. The output voltage is a combination of voltage from the diodes and voltage from the capacitor. The figure assumes the circuit has been in operation for some period of time.

Looking at the left-hand side of Fig. 14-7B, notice that the on time is slightly delayed from the beginning of the waveform. This is also true for each of the other on times shown. It is during this on time that the diode is supplying current to the load. During the off times, the capacitor is supplying current to the load.

The only time the diode can turn on is when the diode has the proper forward bias applied. This will occur when the input sine wave is greater than the voltage across the capacitor. Once the diode turns on, it will follow the input voltage to peak. While the diode is on, it is supplying current to the load and, because the capacitor is in parallel, it will also charge the capacitor. Once the input sine wave reaches peak, the input voltage starts to decrease. At this moment, the capacitor has peak voltage across it. When the input voltage drops, the capacitor voltage is higher than the input and the diode no longer has the required 0.7 V forward bias to hold it on. Therefore, when the input sine wave voltage starts to decrease at peak, the diode turns off. The capacitor will then start to discharge through the load, maintaining the voltage at a constant, but decaying voltage. The rate at which the capacitor will discharge depends on how much current the load draws. The capacitor will continue to discharge until the input sine wave increases in voltage enough to become larger than the capacitor voltage.

The exact amount that the capacitor discharges can be calculated, if the value of the capacitor is known and the value of the load resistor is known. However, in an actual power supply, very seldom is the load resistor value known. Calculate it by using the RC time constant and determining exactly how much time is allowed for discharge, based on the frequency of the input sine wave.

Compare Fig. 14-7B to Fig. 14-7C. Notice that the slope of the capacitor discharge is the same for both figures, which indicates the same value of load resistor and filter capacitor. Notice, however, that even though the slope is the same, the length of discharge is greatly reduced by the fact that the waveform occurs twice as often. Notice, on the right-hand side of the drawing, there is the labeling of dc average. The half-wave circuit should have a little less dc voltage than the full-wave circuit because the full-wave does not discharge the capacitor as much as the half-wave circuit.

Even though there is a difference between the half-wave circuit and the full-wave circuit waveforms, it is still possible to determine the value of voltage that would be read on a dc voltmeter. Keep in mind that the actual value of load resistance is seldom, if ever, known. The reading on the dc voltmeter can be approximated. An approximate reading on a voltmeter is often enough to determine if the circuit is working correctly. The following formula is a figure between the dc average of a circuit without a filter and the peak value, because that is what the capacitor charges to. The formula will calculate the dc voltage of a circuit with a filter within 10 percent of the actual value that a voltmeter will read. This formula can be used for any circuit with simple capacitor filtering, full-wave or half-wave.

$$\text{Vdc} = 1.2 \times \text{rms} \quad \begin{array}{c}\text{approximation for dc voltage}\\ \text{for a rectifier with a filter}\end{array} \tag{14-7}$$

The rms value in the formula shown above is from the ac input voltage.

Sample half-wave rectifier circuits

With the following conditions, determine the dc output voltage and ripple frequency. Refer to Fig. 14-2 for the schematic diagram.

Alternating current input to transformer: 120 V/60 Hz (line voltage).
Transformer is a 5:1 step-down transformer.

Table 14-1. Summary of rectifier formulas.

	dc voltage without filter	dc voltage with filter	Ripple frequency
half wave	$V_{dc} = \dfrac{0.9 \times \text{rms}}{2}$	$V_{dc} = 1.2 \times \text{rms}$	same as ac input frequency
full wave	$V_{dc} = 0.9 \times \text{rms}$	$V_{dc} = 1.2 \times \text{rms}$	twice the ac input frequency
bridge	$V_{dc} = 0.9 \times \text{rms}$	$V_{dc} = 1.2 \times \text{rms}$	twice the ac input frequency

Calculate the ac applied to the rectifier. The rectifier is connected to the secondary of the transformer, not the primary.

$$\frac{\text{turns primary}}{\text{turns secondary}} = \frac{\text{voltage primary}}{\text{voltage secondary}}$$

$$\frac{5}{1} = \frac{120\ \text{V}}{V}$$

$$\text{voltage secondary} = \frac{120\ \text{V}}{5}$$

$$\text{ac voltage to rectifier} = 24\ \text{Vac rms}$$

Use the Vdc formula for no filter to determine the output voltage.

$$\text{Vdc} = \frac{0.9 \times \text{rms}}{2}$$

$$= \frac{0.9 \times 24\ \text{V}}{2}$$

$$= 10.8\ \text{Vdc dc voltage without a filter}$$

Calculate the dc voltage if the circuit had a simple capacitor filter.

$$\begin{aligned} \text{Vdc} &= 1.2 \times \text{rms} \quad \text{formula for circuits with a filter} \\ &= 1.2 \times 24 \text{ Vac} \\ &= 28.8 \text{ Vdc} \end{aligned}$$

Calculate the ripple frequency. Ripple frequency of a half-wave rectifier circuit is the same as the input frequency.

$$\text{Ripple Frequency} = 60 \text{ Hz}$$

Sample full-wave rectifier circuits

Refer to Fig. 14-3. With 120 V/60 Hz applied to the primary of a transformer with a turns ratio of 2:1, step down, secondary has a center tap; determine the dc output voltage with and without a filter. Also determine the ripple frequency.

When determining the ac voltage applied to the rectifier circuit, be cautious. The center-tapped transformer splits the voltage of the secondary in half. Therefore, only one-half of the secondary voltage is actually applied to the rectifier.

First using the turns ratio, determine the total secondary voltage, then divide that voltage in half.

$$\frac{\text{turns primary}}{\text{turns secondary}} = \frac{\text{voltage primary}}{\text{voltage secondary}}$$

$$\frac{2}{1} = \frac{120 \text{ V}}{x}$$

$$\text{Total secondary voltage} = 60 \text{ Vac}$$

With the total secondary voltage calculated, divide this figure in half to find the voltage to the rectifier.

$$\text{Vac to rectifier} = \frac{\text{total secondary voltage}}{2}$$

$$\text{Vac} = \frac{60}{2} = 30 \text{ Vac rms}$$

Use the Vdc formula to calculate the output voltage of this full-wave rectifier without a filter.

$$\begin{aligned} \text{Vdc} &= 0.9 \times \text{rms} \\ &= 0.9 \times 30 \text{ Vac} \\ &= 278 \text{ Vdc} \end{aligned}$$

Using the same value of rms applied to the rectifier, determine the value of the dc voltage with a simple capacitor filter.

$$\begin{aligned} \text{Vdc} &= 1.2 \times \text{rms} \\ &= 1.2 \times 30 \text{ Vac} \\ &= 36 \text{ Vdc} \end{aligned}$$

Determine the ripple frequency. Ripple of a full-wave is twice the input frequency.

$$\text{Ripple frequency} = 2 \times 60 \text{ Hz} = 120 \text{ Hz}$$

Sample bridge-rectifier circuits

Refer to Fig. 12-4. Determine the dc voltage with and without a filter. Also determine the ripple frequency. The ac input to the transformer is 120 V/60 Hz, and the transformer has a 8:1 turns ratio, step down.

Determine the ac voltage applied to the rectifier circuit. Because the entire secondary is used, the secondary voltage is simply calculated from the turns ratio.

$$\frac{\text{turns primary}}{\text{turns secondary}} = \frac{\text{voltage primary}}{\text{voltage secondary}}$$

$$\frac{8}{1} = \frac{120 \text{ Vac}}{x}$$

$$\text{Secondary volts} = 15 \text{ Vac rms}$$

Calculate the dc voltage without a filter.

$$\begin{aligned} \text{Vdc} &= 0.9 \times \text{rms} \\ &= 0.9 \times 15 \text{ Vac} \\ &= 13.5 \text{ Vdc} \end{aligned}$$

Calculate the dc voltage with a filter.

$$\begin{aligned} \text{Vdc} &= 1.2 \times \text{rms} \\ &= 1.2 \times 15 \text{ Vac} \\ &= 18 \text{ Vdc} \end{aligned}$$

Zener regulator

The voltage regulator is the section of a power supply circuit that regulates the voltage to compensate for any changes in load current or input voltage. Regulators range from the simplest being a zener diode to very complex used to regulate the voltage in a complex computer.

This section deals with only the zener diode in order to stay within the boundaries of the text.

Figure 14-8 is a schematic diagram of a simple zener regulator circuit. A power regulator circuit contains the zener diode, connected in reverse, and a series droppers resistor.

Figure 14-9 is the characteristic curve of a zener diode. In the forward direction the zener acts like any rectifier diode. Caution should be used in the forward direction, however, because the zener was not designed for forward operation. Its current capability in the forward direction is quite limited. Notice that the diode drop in forward direction is 0.7 V because most zeners are silicon.

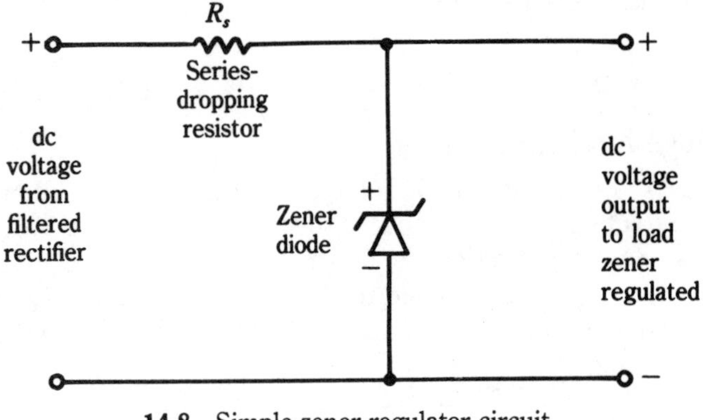

14-8 Simple zener regulator circuit.

Forward current
Zener diode forward characteristic curve is the same as a rectifier diode.
Zener voltage
Reverse voltage
Zener knee
Slight reverse current prior to zener knee.
0.7 V
Forward voltage
Reverse current
Zener region
Conducts current well while holding a constant voltage.
Very slight change in voltage in zener region.

14-9 Characteristic curve of a zener diode.

The zener diode in the reverse direction has a very unique characteristic curve. Starting from zero reverse volts, and increasing, there will be very small current flow, so small as to be considered zero. Each zener diode has a zener breakdown rating. This rating is the zener voltage in the reverse direction. When increasing the voltage, from zero, the reverse current will start to increase near the zener knee. Then, breakdown will be reached and there will be a large increase in current flow and the voltage across the zener will remain constant.

The reason there must be a series dropping resistor is that when the voltage is being increased in the reverse direction, during zener breakdown, the voltage not across the zener will be dropped across the resistor.

When the zener is operating in the zener region, there is a large amount of current through the zener, limited only by the series resistor and the applied voltage. Since the zener is conducting a significant amount of current and there is a voltage drop, there will be a considerable amount of power dissipated by the zener. For this reason, all zeners will have a zener voltage rating and a maximum wattage.

Zener diodes can have voltage ratings ranging from 2 V or so to quite high voltages. Wattage ratings can be from a few milliwatts to several watts. The higher the voltage rating, the more heat the zener must dissipate.

Block diagram of a power supply

Figure 14-10 shows a simple block diagram of a power supply. The block diagram is an easy tool to use because it shows the names of the circuits involved but is not limited to any one circuit inside the blocks. Almost every power supply uses a transformer. However, it really is not necessary with the half-wave and the bridge circuits. The transformer is shown because it is so common.

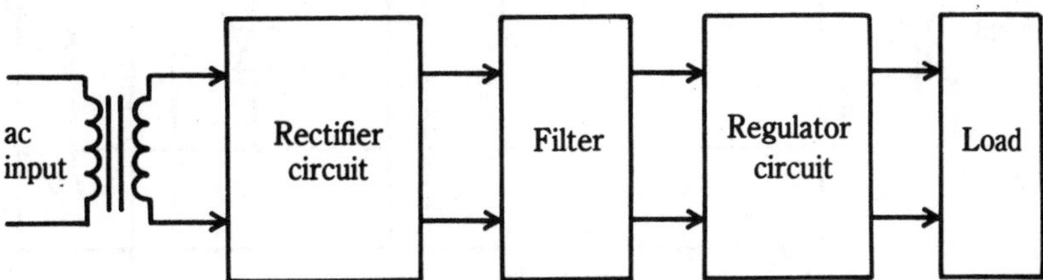

14-10 Block diagram of a power supply.

The first block of the diagram is the rectifier circuit. This could be any type of rectifier. Its purpose is to change the ac input to pulsating dc.

The second block of the diagram is the filter. Even though only the simple capacitor filter was discussed, there are many types of filters used with power supplies. All filters have the same function. Their function is to smooth out the fluctuating dc and to make it into as constant a voltage as possible.

The third block of the diagram is the regulator circuit. There are many types of regulators available, even though only the zener regulator will be discussed here. The function of the regulator circuit is to hold the output voltage constant under any changing load conditions or input conditions. Some regulators contain circuitry to turn off the power if the output voltage changes due to a failed regulator.

The last block of the diagram is the load. The load is the device that requires the dc power supply. The load could be a resistor, such as a heating element; it could be a radio, a tape player, record player, amplifier, a computer, or anything else that requires dc power to operate. It is quite normal for the load to vary its current requirements while maintaining the same voltage requirements. An example of this might be a tape player that is switched from the play mode to fast forward. The changing speeds of the motor will require different amounts of current.

Calculating a zener regulator

Usually when calculating the requirements of the regulator circuit, the load is the first thing that is known. The reason for knowing the load first is that the power supply would not even be needed if there wasn't a load to connect to it. Once the load is known, it is usually a simple matter of working the calculations backward in order.

Figure 14-11 shows the inside of the block diagram and is very useful when making the calculations for the zener regulator. The zener is in parallel with the load. That means that whatever voltage is required by the load is also the zener voltage. Select the zener to match the load.

- Whatever voltage that is required by the load, is also the zener voltage. Select the zener to match the load.

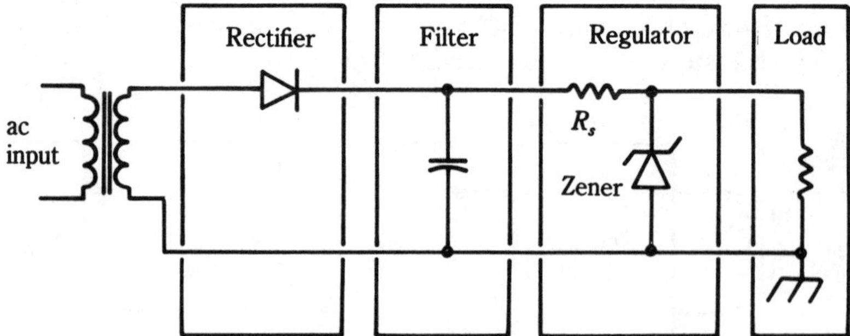

14-11 Inside the block diagram of a power supply.

R_S will be calculated, so you will come back to this. Keep in mind, R_S must drop whatever voltage the zener does not drop, coming from the filtered rectifier. A good rule to follow when determining the output voltage of the filter is to make it approximately double the zener voltage. That will give plenty of room for any fluctuations. Once the filter voltage has been approximated, then the output of the transformer can be approximated.

Sample zener regulator calculation

The load requires 12 V; maximum current is 1.5 A. With the information given, the zener is rated for 12 V. The zener is in parallel with the load. It must have the same voltage as the load. The next step is to determine the total current for the circuit. Notice how the total current is made up of both zener current and load current. Because the zener must always be operated beyond the zener knee, a good rule to follow is to allow a zener current of 10 percent of the maximum load current.

$$I_T = I_Z + I_L \quad \text{total current in a zener regulator circuit} \qquad (14\text{-}8)$$

Using the formula for total current, Formula 12-8, calculate the total current, allowing full load current and 10 percent of the maximum load current for the zener.

$$
\begin{aligned}
I_T &= I_Z + I_L \\
&= 1.5\text{ A} \\
&= 10\%\text{ of }1.5\text{ A}\quad\text{minimum zener current} \\
&= 0.15\text{ A}\quad\text{zener current when load is drawing full amount} \\
&= 0.15\text{ A} + 1.5\text{ A} \\
&= 1.65\text{ A}\quad\text{total circuit current}
\end{aligned} \tag{14-8}
$$

The load will seldom, if ever draw the full rated value of current. Whatever the load does not use, the zener will take the remainder. This way, the total current remains constant as long as the input voltage is a constant.

Next determine the voltage from the filtered rectifier. It is a good rule to use approximately double the zener voltage, to allow the fluctuations in the input voltage.

Zener voltage = 12 V, rectifier output = 24 V

With the value of voltage known across the zener, and the value of voltage across the filter known, the difference of these two values must be dropped across the series resistor, R_S.

$$
\begin{aligned}
V_{RS} &= \text{Vdc} - V_Z \\
&= \text{Vdc} - V_Z \\
&= 24 - 12 \\
&= 12\text{ V}
\end{aligned} \tag{14-9}
$$

Calculate the ohmic value of R_S. Total current will flow through R_S; therefore, the value of the resistor can be calculated using total current and the voltage dropped across the resistor.

$$R = \frac{E}{I} \tag{2-1A}$$

$$R_S = \frac{12\text{ V}}{1.65\text{ A}}$$

$$= 7.27\ \Omega\quad\text{calculated value of } R_S$$

Note The calculated value of R_S will have to be adjusted to a value that is possible to purchase at a parts store.

Using the calculated value of R_S and calculated total current; calculate the wattage rating of the resistor.

$$
\begin{aligned}
P &= I \times E \\
P &= 1.65\text{ A} \times 12\text{ V} \\
P_{RS} &= 19.8\text{ W}
\end{aligned} \tag{2-2}
$$

Using the calculated values, calculate the value of the zener wattage. Use total current because the worst condition for the zener is when the load is not using any current.

$$P = I \times E$$
$$P_Z = 1.65\ \text{A} \times 12\ \text{V} \qquad (2\text{-}2)$$
$$= 19.8\ \text{W}$$

Next in designing the zener regulator circuit is to determine the value of secondary voltage that will supply the calculated dc voltage across the filter. Remember, the filter voltage was calculated using the Vdc formula:

$$\text{Vdc} = 1.2 \times \text{rms}$$

$$\text{rms} = \frac{\text{Vdc}}{1.2}$$

$$= \frac{24\ \text{V}}{1.2}$$

$$= 20\ \text{Vac}$$

Note Primary of the transformer will be as needed, usually 120 V/60 Hz.

Selecting the actual zener circuit values

Even though the zener regulator circuit is such a simple circuit, it demonstrates the problems that are encountered when designing a working circuit.

When the calculations for a circuit are made, it is most common to start off at the load. The calculated values give a starting point, but that is not the final value. Once the calculations have been made, the easiest place to start with in adjusting the values is the transformer and work towards the load.

Transformers come in almost any value of secondary voltage, if round numbers are selected. The hard part in selecting a transformer is that larger values of secondary current will result in physically larger transformers. Most transformers will come with a center tap, so any type of rectifier circuit can be used. The actual output voltage of the transformer will not be critical in the final circuit, provided there is enough voltage.

The next step in the design is the rectifier circuit. The bridge rectifier has many advantages over the other types, and is usually selected for low power applications.

The filter capacitor can be selected over a wide range of values. It should be an electrolytic capacitor. It will have a voltage rating and a capacitor rating. The voltage rating must be a minimum of the peak value of the output of the rectifier. A good rule to follow is to double the necessary voltage rating. As far as the capacitor rating is concerned, 100 μF is a satisfactory value, but the higher, the better.

The next step in adjusting the values of the zener circuit is to select the zener. The load is usually selected first and will therefore determine the actual value of the zener. However, if that value cannot be found, it is possible to place zeners in

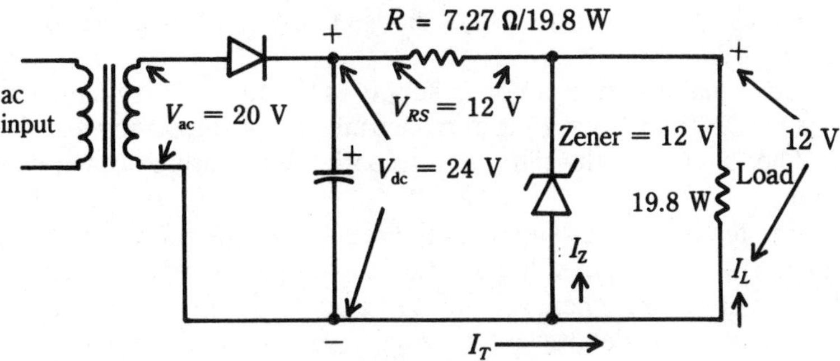

14-12 Complete zener circuit, showing calculated values.

series to arrive at a new value. For example, two 6 V in series equals 12 V. Do not try to connect them in parallel; it does not work that way.

The only component left in the circuit is the hardest one to choose. R_S determines the total current of the circuit. It must be selected in such a way as to ensure enough current will flow to the load and still have a small amount of current to keep the zener in the zener region. If the value of R_S is increased beyond the calculated value, there may not be enough current for all conditions. If the value of R_S is made smaller than the calculated value, it will increase the total current of the circuit and will therefore demand the zener handle more current. The zener must be able to handle all of the total current when the load is disconnected. Usually, the zener current rating will be selected high enough to allow for extra current. So, R_S should be selected slightly lower in value than the figure actually calculated. As far as the wattage rating is concerned, make sure the wattage rating is higher than needed. Resistors can be placed in series or parallel to make up any combination of resistance or power.

Chapter summary

The power supply is the part of an electronic system that is often forgotten, but it is a very important part when it is considered to be the thing that makes the electronic device operate.

Changing ac power to dc is what the power supply is used for. Its primary component for this purpose is the rectifier diode. The rectifier diode is made up of two pieces of semiconductor material, one N-type and the other P-type. The two types of semiconductor materials joined together form the PN junction. It is this PN junction that allows the diode to conduct only in one direction.

The three types of rectifier circuits discussed in this chapter are the half-wave, full-wave, and bridge circuits. These three circuits are the circuits used in most modern power supplies, with the bridge being the most favorite. The rectifier circuit is the circuit responsible for the actual changing of ac to dc. The different configurations of circuitry allow the use of diodes in different manners to accomplish the task.

The second major part of a power supply is the filter. This chapter discussed only the capacitor filter because it demonstrates all of the principles of any filter available. The function of the filter is to smooth the pulsating dc that comes from the rectifier.

The next step in a power supply is the regulator. The regulator is responsible for holding a constant output voltage to the load. A power supply without a regulator would vary in voltage whenever the load conditions of current were to change. Some devices would be destroyed by a supply voltage that is too low or too high. This chapter discusses the zener regulator and demonstrates the calculations that are involved in the zener circuit. Even though it is such a simple regulator circuit, the calculations can become quite complex. Zener circuits are only suitable for low power applications. Other circuits can be chosen for loads using over 5 amps.

The following are the key points of this chapter.

- A semiconductor with an extra electron is called an N-type semiconductor.
- A semiconductor missing an electron is called a P-type semiconductor. The atom missing an electron is called a hole.
- P-type semiconductors have a majority of holes, and a minority of electrons. Holes carry a positive charge.
- N-type semiconductors have a majority of electrons, and a minority of holes. Electrons carry a negative charge.
- The depletion zone is depleted of any mobile carriers and acts like an insulation region.
- A forward-biased diode has negative connected to the N material and positive connected to the P material.
- A reverse-biased diode has positive connected to the N material and negative to the P material.
- When a rectifier diode is conducting, in the forward direction, the diode will have a voltage drop of 0.3 V or 0.7 V; the remainder of the applied voltage will be dropped across the diode load resistor.
- When the rectifier diode is reverse biased, there is no current flow. All of the applied voltage will be dropped across the diode and no voltage will be dropped across the load resistor.
- Whatever voltage is required by the load in a simple zener regulator, is also the zener voltage. Select the zener to match the load.

Summary of formulas

$$\text{Vdc} = 0.636 \times \text{peak} \quad \text{output voltage of a full-wave or bridge rectifier circuit} \tag{14-1}$$

$$\text{Vdc} = \frac{0.636 \times \text{peak}}{2} \quad \text{output voltage of a half-wave rectifier circuit} \tag{14-2}$$

$$\text{Vdc} = 0.9 \times \text{rms} \quad \text{output voltage of a full-wave or bridge rectifier circuit} \qquad (14\text{-}3)$$

$$\text{Vdc} = \frac{0.9 \times \text{rms}}{2} \quad \text{output voltage of a half-wave rectifier circuit} \qquad (14\text{-}4)$$

$$\text{ripple frequency half-wave} = \text{input frequency} \qquad (14\text{-}5)$$

$$\text{ripple frequency full-wave} = 2 \times \text{input frequency} \qquad (14\text{-}6)$$

$$\text{Vdc} = 1.2 \times \text{rms} \quad \text{approximation for dc output voltage for a rectifier with a filter, all configurations} \qquad (14\text{-}7)$$

$$I_T = I_Z + I_L \quad \text{total current in a zener regulator circuit} \qquad (14\text{-}8)$$

$$V_{RS} = \text{Vdc} - V_Z \quad \text{voltage drop across the series resistor of a zener regulator} \qquad (14\text{-}9)$$

Practice problems

The problems have the load conditions given and the rms value of the transformer secondary. Use a bridge rectifier circuit with simple capacitive filtering. In all problems, allow a minimum zener current of 10 percent of the full load condition.

Find for each:
Vdc from filter (input to zener regulator), zener voltage, zener current (maximum when the load is disconnected), zener wattage, voltage drop across R_S, value of R_S in ohms and value of R_S in wattage, total circuit current.

1. Load; 12 V/250 mA — Input ac; 24 V rms
2. Load; 9 V/50 mA — Input ac; 18 V rms
3. Load; 6 V/750 mA — Input ac; 12 V rms
4. Load; 15 V/1.5 A — Input ac; 30 V rms
5. Load; 24 V/150 mA — Input ac; 60 V rms

15
Transistors

THE TRANSISTOR IS FORMED BY DOPING A SEMICONDUCTOR CRYSTAL INTO three distinct regions. These three regions are called *emitter, base,* and *collector.* The relative strength of the doping allows the transistor to control the amount of current flow through the transistor, to the load circuit.

NPN and PNP transistors

The three regions are formed as either NPN or PNP, which appears as two diodes, back-to-back, as shown in Fig. 15-1. The polarity of the voltage necessary to forward bias the transistor is determined by the emitter. The polarity must be negative on the N-type emitter and positive on the P-type emitter. The opposite voltage is connected to the collector. The voltage applied to the base will be between the voltage of the collector and emitter.

Biasing of a transistor

Refer to Fig. 15-2 to examine the voltage relationships of a transistor. These relationships are given these names: voltage base-to-emitter (V_{BE}), voltage collector-to-base (V_{CB}) and voltage collector-to-emitter (V_{CE}).

Notice the polarity of the voltages shown in Fig. 15-2. V_{BE} forward biases the emitter-base diode, and V_{CB} reverse biases the base-to-collector diode.

The forward-biased emitter-base allows current to flow easily. The reverse-biased base-collector does not stop current, as would be the case of a rectifier diode. Instead, the amount of reverse voltage can be used to control the amount of current flow. The load is connected in the current path between the collector and emitter circuit, external to the transistor. Therefore, the voltage applied to the base will control the current flow in the load portion of the circuit.

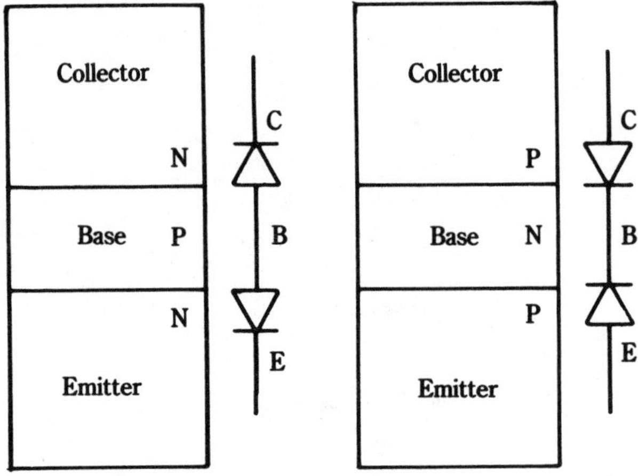

15-1 The three regions of a transistor.

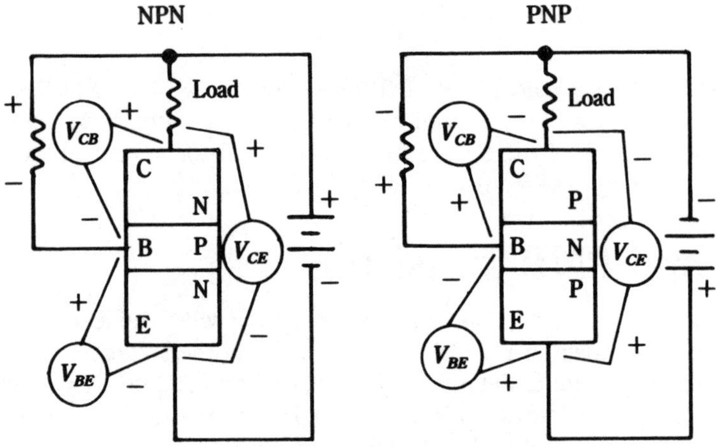

15-2 Voltage relationships of a transistor.

Current relationships in a transistor

The current relationships in a transistor are shown in Fig. 15-3. There are three currents affecting the operation of a transistor: base current (I_B), collector current (I_C) and emitter current (I_E). Base current is very small in comparison to collector current. Emitter current is the sum of base and collector currents.

Emitter current expressed as a formula:

$$I_E = I_B + I_C \tag{15-1}$$

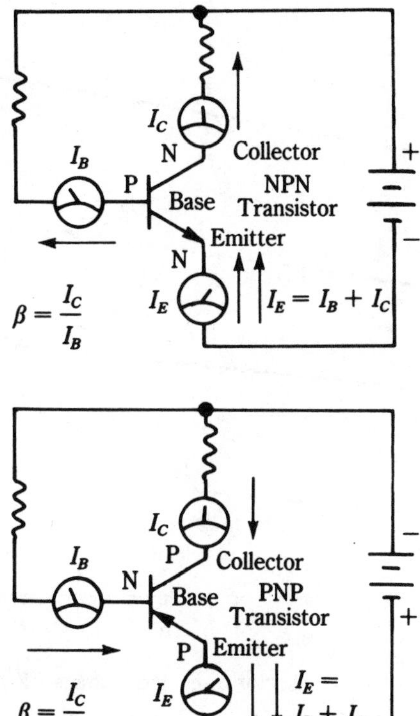

15-3 Current relationships of a transistor.

Beta relationship

Beta (symbol Greek letter β) is the ratio of collector current to base current. Beta is the current gain of the transistor, from base to collector. It is established during the manufacture of the transistor by the relative doping of the base and collector regions. An example of the beta for a small-signal amplifier is approximately 100, and for a large signal amplifier it is 30.

Beta describes the amount of collector current flowing when the base has a certain current flow. The formula for beta is:

$$\beta = \frac{I_C}{I_B} \tag{15-2}$$

Transistor characteristic curves

Figure 15-4 is a sample transistor characteristic curves. This type of graph is used to predict transistor operation. It is used in this chapter to graphically display the transistor load line.

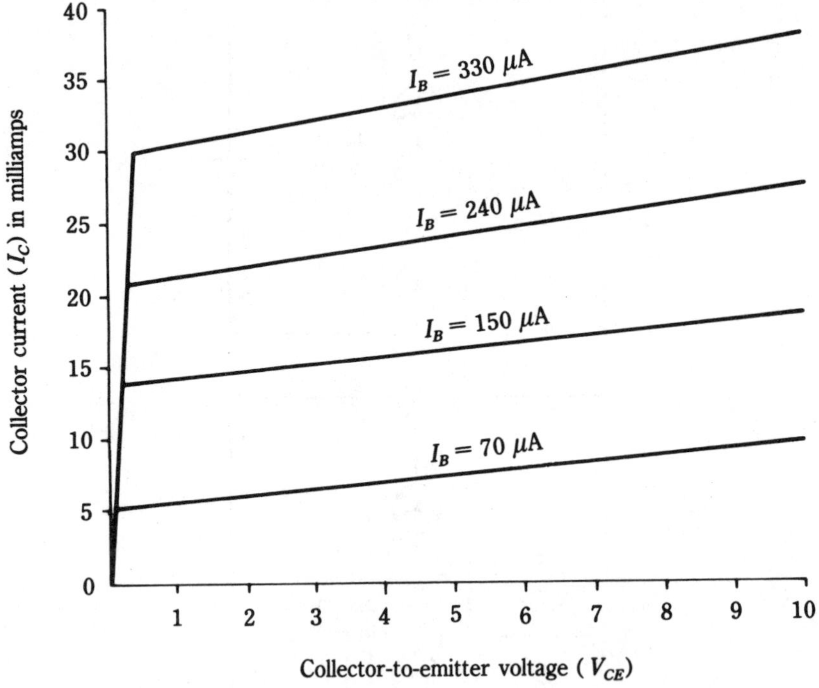

15-4 Sample characteristic curves of a transistor with $\beta = 100$.

The left side of the graph is the value of collector current, measured in milliamps. The bottom of the graph is the value of voltage across the transistor, measured from the collector to emitter (V_{CE}).

The characteristic curve is plotted by first setting a value if I_B. Next, increase the supply voltage, while monitoring V_{CE} at specific values. The collector current will be approximately beta times larger than base current, as stated by rearranging the beta formula:

$$I_C = \beta \times I_B \tag{15-3}$$

Refer to Fig. 15-4. Example:

I_B = 150 μA (0.150 mA)
V_{CE} = 5 A
I_C = 15 mA

Range of transistor operation

A transistor circuit operates within a set of conditions determined by the values of resistors selected to bias the transistor. The transistor displays characteristics similiar to a variable resistor, with the current through the transistor ranging from 0 to maximum and every possible point between. The bias resistors selected determine the operating point.

Definitions of the range of operation

Figure 15-5 represents the range in which the transistor performs. A transistor has the range of operation from saturation to cut-off, with the active region between. In the drawings, V_{CC}, at the top, is the supply voltage, with the triangle at the bottom being the other connection for the supply voltage. *C* represents the location of the collector, *E* is the emitter and *B* is the base.

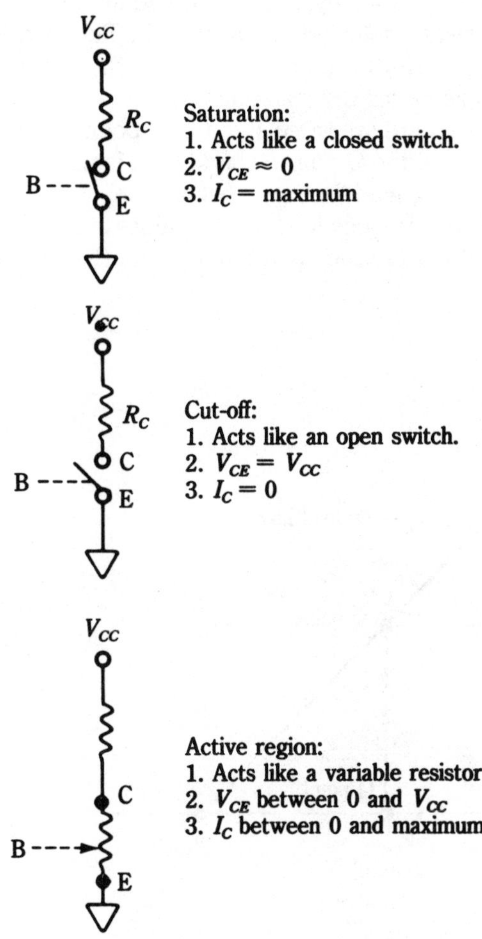

15-5 Definitions of a transistor range of operation.

Saturation is the base current driving the transistor at maximum collector current. Further increase in base current will have no effect. The amount of collector current is limited by the values of the load resistor and supply voltage. The voltage drop across the transistor is very close to zero, as if it were a closed switch.

Cut-off is the voltage applied to the base is not enough to forward bias the base-emitter diode. This results in no base current and no collector current. The entire supply voltage is dropped across the transistor, as if it were an open circuit.

The *active region* is the collector current and voltage across the transistor can vary from zero to maximum. The transistor performs as an amplifier in this region, with the base current having control over the transistor.

dc load line

The dc load line is a graphic analysis of the range of operation of a transistor as determined by the resistors selected for the circuit. The load line is drawn across the characteristic curves. However, for simplicity, in the example shown the curves have been removed, except for the load line.

Refer to Fig. 15-6. There are three significant points on a dc load line: saturation (sat), cut-off (c.o.) and the Q point. The *Q point* is the operating point and can be located anywhere on the load line, even beyond c.o. or sat. Dashed lines are projected from the Q point to the axis to show the values of I_{CP} and V_{CE} at the Q point. These values of I_C and V_{CE} are the operating point of the circuit, also called the dc conditions.

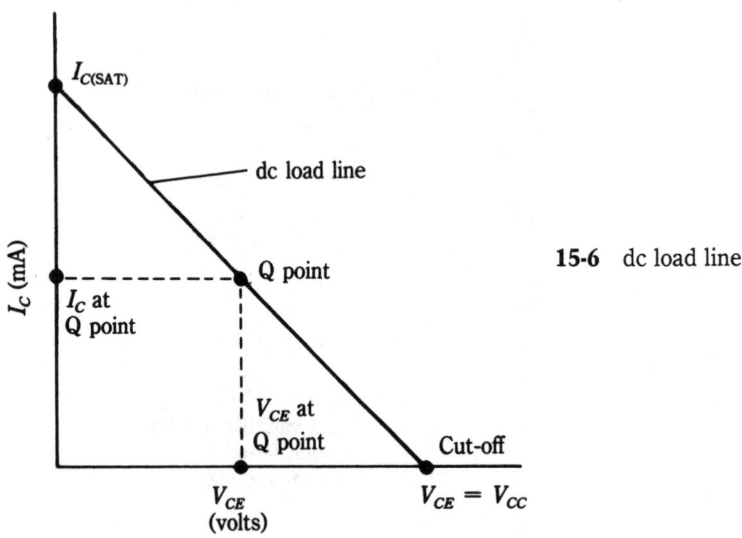

15-6 dc load line.

dc conditions of a transistor amplifier

The dc conditions are the currents and voltages affecting the operation of the transistor. It is also called the biasing and is determined by the placement and value of the biasing resistors in the circuit.

There are many different configurations in which the bias resistors can be placed. Each of the following circuits has a specific procedure to follow when making the calculations. Ohm's Law is the primary formula used, depending on the values known in the circuit.

Base bias

Figure 15-7 is a base bias circuit. A is the circuit with no values given. B is used in the following sample. It shows the given resistor values, circuit currents and voltages to be calculated. C is the load line, and D is a simplified schematic. In D, the voltages at the base and collector are the values measured to ground, which is the same as measuring to the emitter since there is no emitter resistor.

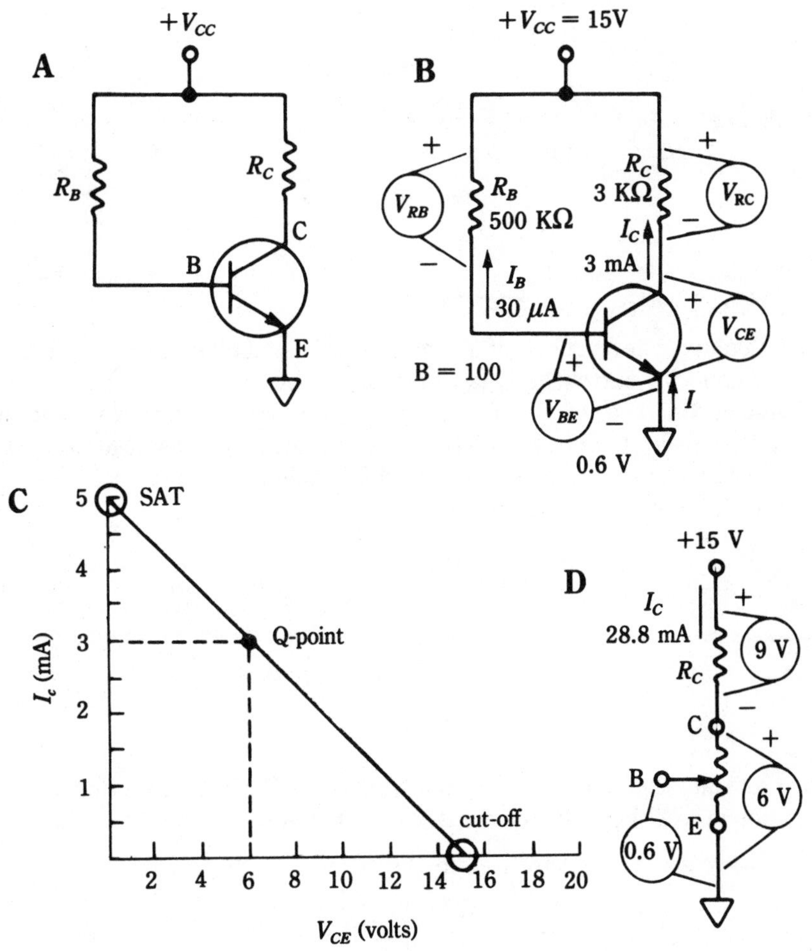

15-7 Base bias, a β-dependent circuit.

Sample calculations of a base bias circuit, Fig. 15-7.

Given: $V_{CC} = 15$ V
$R_B = 500$ kΩ
$R_C = 3$ kΩ
$\beta = 100$

Locate the two ends of the load line (Fig. 15-7C)

a. At saturation the transistor offers no resistance.
$V_{CE} = 0$ and I_C = maximum.

$$I_{C(SAT)} = \frac{V_{CC}}{R_C} \qquad (2\text{-}1A)$$

$$= \frac{15\text{ V}}{3\text{ k}\Omega}$$

$$= 5\text{ mA}$$

b. At cut-off the transistor is an open circuit.
$I_C = 0$ and V_{CE} = maximum.

$$V_{CE}\text{ (cut-off)} = V_{CC}$$
$$= 15\text{ V}$$

Locate the Q point (Fig. 15-7C)

a. Calculate collector current at the Q point. The beta-relationship states that the collector current is beta times larger than base current. The relationship of resistance to current is inverse; therefore, collector resistance is beta times smaller than base resistance. In the formula below $R_{B\beta}$ represents the resistance of the collector circuit, which includes both the collector resistor and the transistor resistance.

$$I_C = \frac{V_{CC}}{R_{B\beta}} \qquad (2\text{-}1A)$$

$$= \frac{15\text{ V}}{500\text{ k}\Omega \times 100}$$

$$= 3\text{ mA}$$

b. Calculate the voltage across the transistor, V_C, by subtracting the voltage drop across the collector resistor from the supply voltage.

Voltage across collector resistor:

$$V_{RC} = I_C \times R_C$$
$$= 3\text{ mA} \times 3\text{ k}\Omega$$
$$= 9\text{ V}$$

Voltage across the transistor:

$$V_{CE} = V_{CC} - V_{RC}$$
$$= 15\text{ V} - 9\text{ V}$$
$$= 6\text{ V}$$

Calculate base current (Fig. 15-7B)

Base current is calculated using the beta relationship of collector current to base current.

$$I_B = \frac{I_C}{\beta} \tag{15-4}$$

$$= \frac{3\text{ mA}}{100}$$

$$= 30\ \mu\text{A}$$

Calculate emitter current

Emitter current, I_E, is found by adding the base and collector currents.

Note Unit multipliers must be the same. This calculation demonstrates that the collector current is approximately equal to the emitter current.

$$\begin{aligned} I_E &= I_B + I_C \\ &= 30\ \mu\text{A} + 3\text{ mA} \\ &= 0.030\text{ mA} + 3\text{ mA} \\ &= 3.03\text{ mA} \end{aligned}$$

Base bias with emitter feedback

Figure 15-8 uses a base bias resistor and includes an emitter resistor. This circuit is beta dependent, with collector current being beta times larger than the base current.

A is the circuit without values given. B shows the circuit with the values and the voltages and currents to be calculated. C is the load line, and D is a simplified circuit showing the voltages. Note: V_B and V_C are measured to ground.

Sample calculations of a base bias with emitter feedback circuit, Fig. 15-8.

$$\text{Given:}\quad \begin{aligned} V_{CC} &= 15\text{ V} \\ R_B &= 2\text{ M}\Omega \\ R_C &= 10\text{ k}\Omega \\ R_E &= 8.2\text{ k}\Omega \\ \beta &= 150 \end{aligned}$$

Locate the two ends of the load line (Fig. 15-8C)

a. At saturation the transistor offers no resistance.
   $V_{CE} = 0$ and I_C = maximum. Both the collector resistor and emitter resistor must be included.

$$I_{C(\text{SAT})} = \frac{V_{CC}}{R_C + R_E} \tag{15-5}$$

$$= \frac{15\text{ V}}{10\text{ k}\Omega + 8.2\text{ k}\Omega}$$

$$= 0.824\text{ mA}$$

b. At cut-off, the transistor is an open circuit.
I_C = 0 and V_{CE} = maximum.

$$V_{CE}\text{ (cut-off)} = V_{CC}$$
$$= 15\text{ V}$$

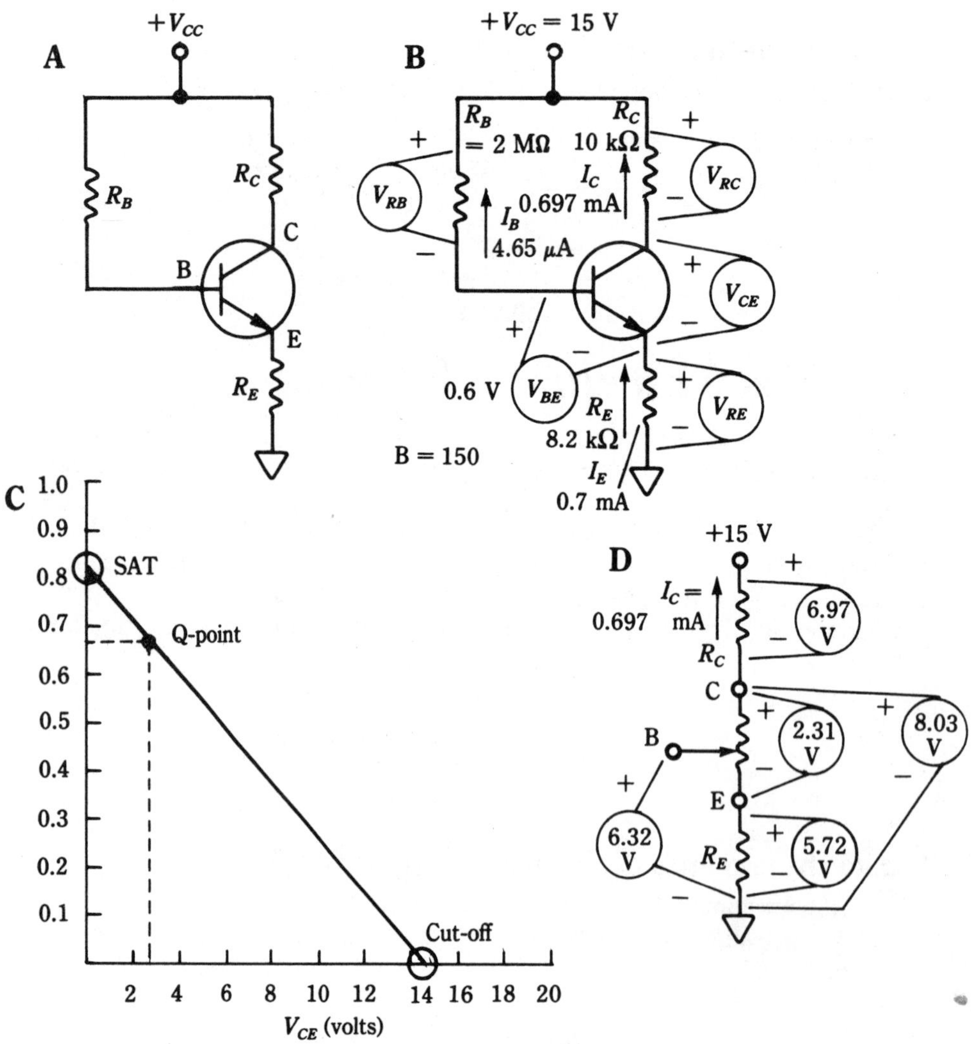

15-8 Base bias with emitter feedback, a β-dependent circuit.

Locate the Q point (Fig. 15-8C)

a. Calculate collector current at the Q point. Calculations are based on the fact that if the collector current is beta times larger than base current, then the base resistance must be beta times larger than the collector resistance. In the formula below, R_B/β represents the collector resistance including the transistor resistance. The emitter resistor must be added because it is in the path of both the base current and collector current.

$$I_C = \frac{V_{CC}}{R_E + (R_B/\beta)} \quad (15\text{-}6)$$

$$= \frac{15\ \text{V}}{8.2\ \text{k}\Omega + (2\text{M}/150)}$$

$$= 0.697\ \text{mA}$$

b. Calculate the voltage across the transistor, V_{CE}, by finding the voltage drops across the emitter and collector resistors and subtracting from the total voltage, V_{CC}.

Voltage across collector resistor:

$$\begin{aligned} V_{RC} &= I_C \times R_C \quad (2\text{-}1) \\ &= 0.697\ \text{mA} \times 10\ \text{k}\Omega \\ &= 6.97\ \text{V} \end{aligned}$$

Voltage across emitter resistor:

$$\begin{aligned} V_{RE} &= I_C \times R_E \\ &= 0.697\ \text{mA} \times 8.2\ \text{k}\Omega \\ &= 5.72\ \text{V} \end{aligned}$$

Voltage across the transistor:

$$\begin{aligned} V_{CE} &= V_{CC} - (V_{RC} + V_{RE}) \\ &= 15\ \text{V} - (6.97\ \text{V} + 5.72\ \text{V}) \\ &= 15\ \text{V} - 12.69\ \text{V} \\ &= 2.31\ \text{V} \end{aligned}$$

Voltages at the collector and base to ground (Fig. 15-8D)

These voltages are important when testing an operational circuit. Collector voltage is found by adding V_{CE} and V_{RE}. Base voltage is found by adding 0.6 to V_{RE}.

Collector voltage to ground:

$$\begin{aligned} V_C &= V_{CE} + V_{RE} \\ &= 2.31\ \text{V} + 5.72\ \text{V} \\ &= 8.03\ \text{V} \end{aligned}$$

Base voltage to ground:

$$\begin{aligned} V_B &= V_{RE} + 0.6 \text{ V} \\ &= 5.72 \text{ V} + 0.6 \text{ V} \\ &= 6.32 \text{ V} \end{aligned}$$

Calculate base current

Use I_C with the beta relationship to find I_B.

$$I_B = \frac{I_C}{\beta} \qquad (15\text{-}4)$$

$$= \frac{0.697 \ \mu\text{A}}{150}$$

$$= 4.65 \ \mu\text{A}$$

Collector feedback

Figure 15-9 uses a base bias resistor connected to the collector, rather than to the supply voltage. This type of connection helps to stabilize the circuit due to heat changing the value of beta. The circuit is beta dependent, with collector current being beta times larger than the base current.

A is the circuit without values given. B shows the circuit with the values and the voltages and currents to be calculated. C is the load line; D is a simplified circuit showing the voltages. Note that V_B and V_C are measured to ground. Sample calculations of a collector feedback circuit, Fig. 15-9.

$$\text{Given: } \begin{aligned} V_{CC} &= 25 \text{ V} \\ R_B &= 330 \text{ k}\Omega \\ R_C &= 2 \text{ k}\Omega \\ B &= 100 \end{aligned}$$

Locate the two ends of the load line (Fig. 15-9C)

a. At saturation the transistor offers no resistance.
$V_{CE} = 0$ and $I_C =$ maximum.
The collector resistor determines saturation current.

$$I_{C(\text{SAT})} = \frac{V_{CC}}{R_C} \qquad (15\text{-}7)$$

$$= \frac{25 \text{ V}}{2 \text{ k}\Omega}$$

$$= 12.5 \text{ mA}$$

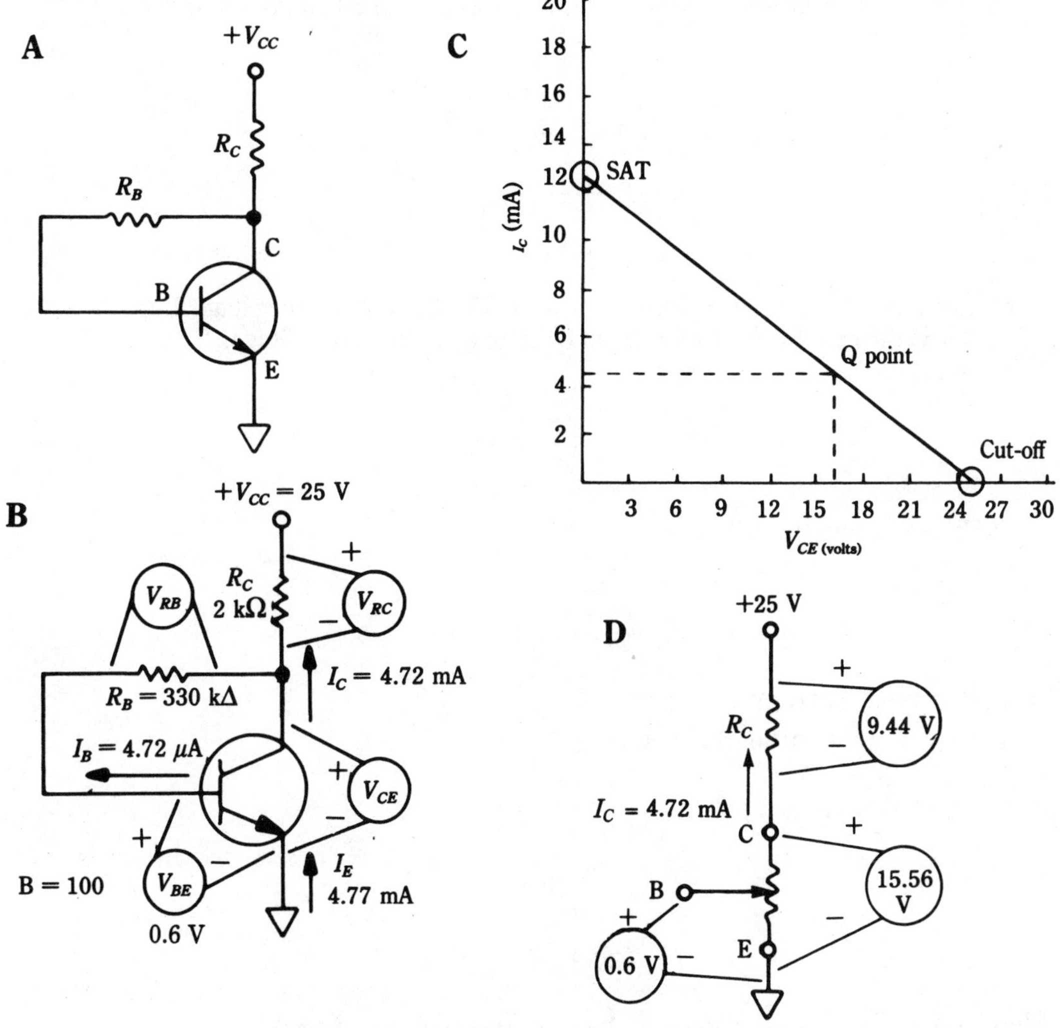

15-9 Collector feedback, a β-dependent circuit.

b. At cut-off the transistor is an open circuit.
$I_C = 0$ and V_{CE} = maximum.

$$V_{CE}\ \text{(cut-off)} = V_{CC}$$
$$V_{CE}\ \text{(cut-off)} = 25\ \text{V}$$

Locate the Q point (Fig. 15-9C)

a. Calculate collector current at the Q point. Calculations are based on the fact that if the collector current is beta times larger than base current, then the base resistance must be beta times larger than the collector resistance. In the formula, R_B/β represents the resistance of the transistor. R_C must be added to the transistor resistance because it is in the path of both base and collector currents.

$$I_C = \frac{V_{CC}}{R_C + (R_B/\beta)} \qquad (15\text{-}6)$$

$$= \frac{25\ \text{V}}{2\ \text{k} + (330\ \text{k}/100)}$$

$$= 4.72\ \mu\text{A}$$

b. Calculate the voltage across the transistor, V_{CE}, by finding the voltage drop across the collector resistor and subtracting from the total voltage, V_{CC}. Voltage across collector resistor:

$$\begin{aligned} V_{RC} &= I_C \times R_C \\ &= 4.72\ \text{mA} \times 2\ \text{k}\Omega \\ &= 9.44\ \text{V} \end{aligned}$$

Voltage across transistor:

$$\begin{aligned} V_{CE} &= V_{CC} - V_{RC} \\ &= 25\ \text{V} - 9.44\ \text{V} \\ &= 15.56\ \text{V} \end{aligned}$$

Calculate base current

Use I_C with the beta relationship to find I_B.

$$I_B = \frac{I_C}{\beta} \qquad (15\text{-}4)$$

$$= \frac{4.72\ \text{mA}}{100}$$

$$= 47.2\ \mu\text{A}$$

Voltage-divider bias—beta independent

The voltage-divider biasing circuit, Fig. 15-10, uses the voltage-divider concept to set the voltage at the base of the transistor, which does not depend on base current. The base current is considered so small in comparison to other currents that it can be ignored. This circuit arrangement results in the circuit being not dependent on beta. Therefore, the collector current does not depend on base current, rather, I_C depends on base voltage.

Sample calculations of voltage divider biasing, Fig. 15-10.

$$\begin{aligned} \text{Given:}\quad V_{CC} &= 10\ \text{V} \\ R_{B1} &= 10\ \text{k}\Omega \\ R_{B2} &= 2.2\ \text{k}\Omega \\ R_C &= 3.6\ \text{k}\Omega \\ R_E &= 1\ \text{k}\Omega \end{aligned}$$

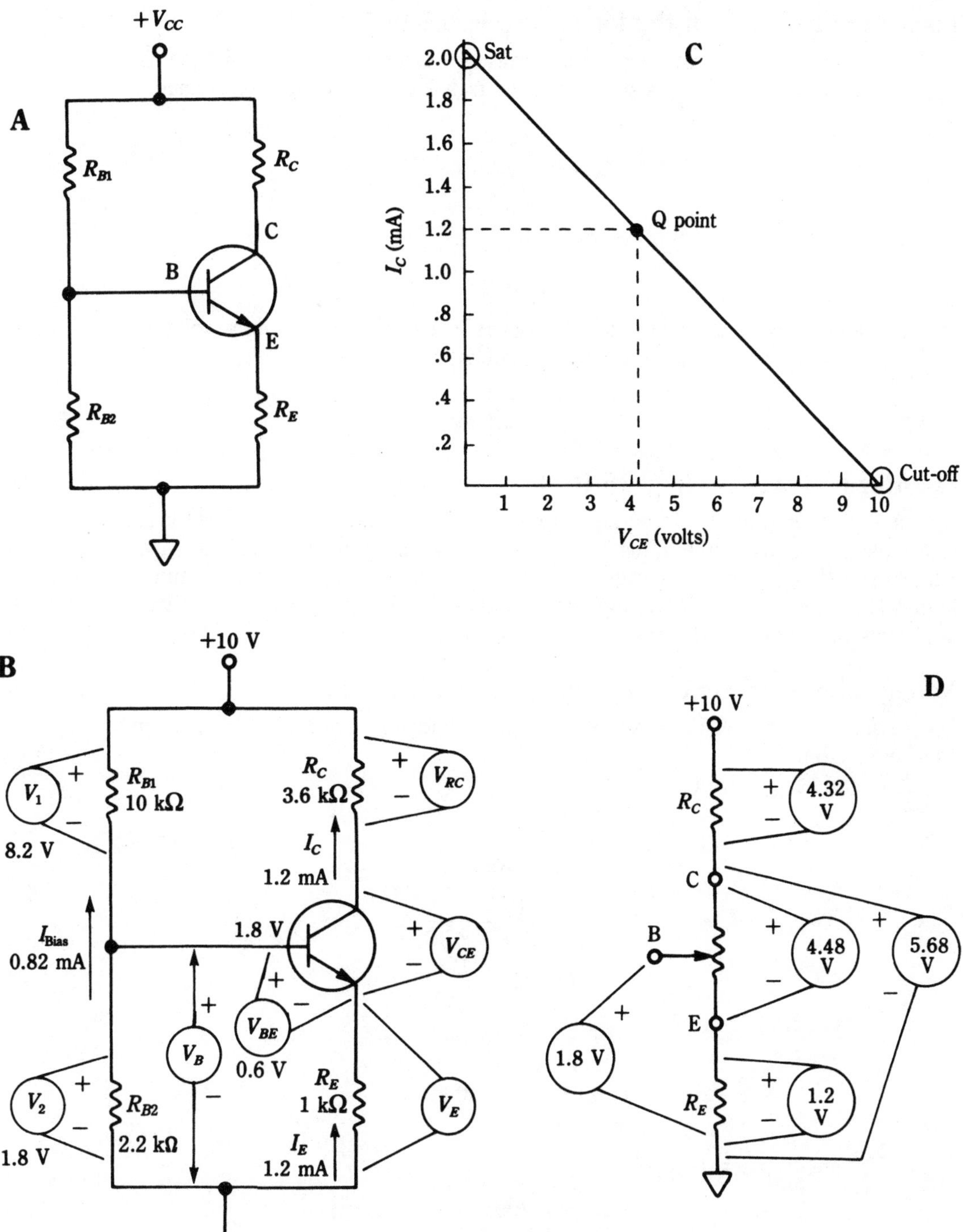

15-10 Voltage-divider bias, a stable circuit, not dependent on β.

Locate the two ends of the load line (Fig. 15-10C)

a. Calculate the collector current at saturation using the supply voltage and collector and emitter resistors, with the transistor having zero resistance.

$$I_{C(SAT)} = \frac{V_{CC}}{R_C + R_E} \tag{15-5}$$

$$= \frac{10 \text{ V}}{3.6 \text{ k}\Omega + 1 \text{ k}\Omega}$$

$$= 2.17 \text{ mA}$$

b. Voltage across the transistor at cut-off is equal to the supply voltage, because of the transistor acting as an open switch.

$$V_{CE} \text{ (cut-off)} = V_{CC}$$
$$= 10 \text{ V}$$

Ignoring base current (Fig. 15-10B)

If a number is ten times, or larger, than another number, the smaller number is so small that it can be ignored. This is the case with setting the voltage at the base. The current flowing in the voltage divider, I_{BIAS}, (see B) is more than ten times larger than base current. The voltage drops across the bias resistor will effectively determine the voltage at the base.

Voltage divider current and voltage drops (Fig. 15-10B)

Ignore the connection of the base to the voltage divider and calculate this portion as a series circuit.

a. Calculate I_{BIAS} using Ohm's Law.

$$I_{\text{BIAS}} = \frac{V_{CC}}{R_{B1} + R_{B2}} \tag{15-8}$$

$$= \frac{10 \text{ V}}{10 \text{ k}\Omega + 2.2 \text{ k}\Omega}$$

$$= 0.82 \text{ mA}$$

b. Voltage drop across R_{B1}.

$$V_1 = I_{\text{BIAS}} \times R_{B1}$$
$$= 0.82 \text{ mA} \times 10 \text{ k}\Omega$$
$$= 8.2 \text{ V}$$

c. Voltage drop across R_{B2}.

$$V_2 = I_{\text{BIAS}} \times R_{B2}.$$
$$= 0.82 \text{ mA} \times 2.2 \text{ k}\Omega$$
$$= 1.8 \text{ V}$$

d. Voltage at the base is found by measuring the voltage from the base to ground. Refering to B of 15-10, base voltage, V_B, is the same as V_2 of the voltage divider.

$$\begin{aligned} V_B &= V_2 \\ &= 1.8\text{ V} \end{aligned}$$

Emitter voltage and current (Fig. 15-10B)

a. If the transistor is assumed to be silicon, the voltage drop across the base-emitter diode, V_{BE}, is 0.6 V. Because the base voltage will be 0.6 V higher than the emitter voltage, calculate the voltage at the emitter by subtracting.

$$\begin{aligned} V_E &= V_B - 0.6\text{ V} \\ &= 1.8\text{ V} - 0.6\text{ V} \\ &= 1.2\text{ V} \end{aligned}$$

b. Emitter current is found using Ohm's Law with the emitter voltage and resistance.

$$I_E = \frac{V_E}{R_E} \qquad (2\text{-}1A)$$

$$= \frac{1.2\text{ V}}{1\text{ k}\Omega}$$

$$= 1.2\text{ mA}$$

Locate the Q point (Fig. 15-10C)

Emitter current is the sum of base current and collector current. As previously shown, collector current is so large in comparison to base current, that when the two are added and the result is rounded, base current is ignored.

a. Collector current at the Q point, for all practical purposes, is the same as emitter current.

$$\begin{aligned} I_C &= I_E \\ &= 1.2\text{ mA} \end{aligned}$$

b. Calculate the voltage drop across the collector resistor using Ohm's Law with the collector current and resistance.

$$\begin{aligned} V_{RC} &= I_C \times R_C \\ &= 1.2\text{ mA} \times 3.6\text{ k}\Omega \\ &= 4.32\text{ V} \end{aligned}$$

c. Calculate the voltage across the transistor, at the Q point by subtracting the voltage drop across the collector and emitter resistors from the supply voltage.

$$\begin{aligned} \text{Note: } V_{RE} &= V_E \\ &= V_{CC} - (V_{RE} + V_{RC}) \\ &= 10\text{ V} - (1.2\text{ V} + 4.32\text{ V}) \\ &= 10\text{ V} - 5.52\text{ V} \\ &= 4.48\text{ V} \end{aligned}$$

d. Calculate the voltage from collector to ground by adding the emitter voltage and transistor voltage.

$$\begin{aligned} V_C &= V_E + V_{CE} \\ &= 1.2\text{ V} + 4.48\text{ V} \\ &= 5.68\text{ V} \end{aligned}$$

ac conditions of a common-emitter amplifier

Biasing of a transistor is setting its dc voltages and currents, which determine the Q point. The ac conditions are changing voltages and currents caused by introducing an input signal to the base and taking an output signal from the collector.

A transistor amplifies the input signal by a small change in base current, which results in a large change in collector current. The large change in collector current results in a large change in output voltage.

Capacitive coupling an ac signal

Figure 15-11 is the same voltage-divider circuit whose dc conditions were calculated in Fig. 15-10. In Fig. 15-11, an input signal is capacitor coupled to the base. The output is capacitor coupled from the collector.

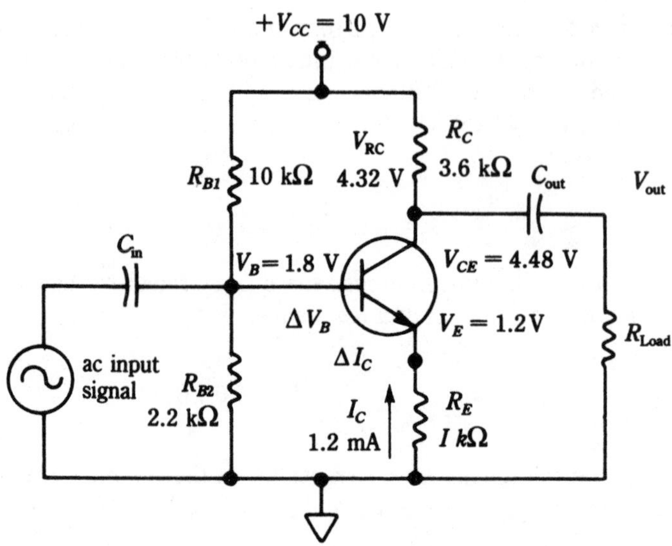

15-11 Voltage-divider biasing used as a common-emitter amplifier.

The capacitors labeled C_{IN} and C_{OUT} are coupling capacitors. A capacitor does not allow dc current to pass, but it offers very low resistance to a changing signal. The coupling capacitors do not allow the connections from the input or output to affect the dc bias voltages.

Passing an ac signal through a transistor

The ac signals shown in Fig. 15-11 are expanded in Fig. 15-12. The triangle (Δ) preceding a voltage or current is the Greek letter delta, which represents change.

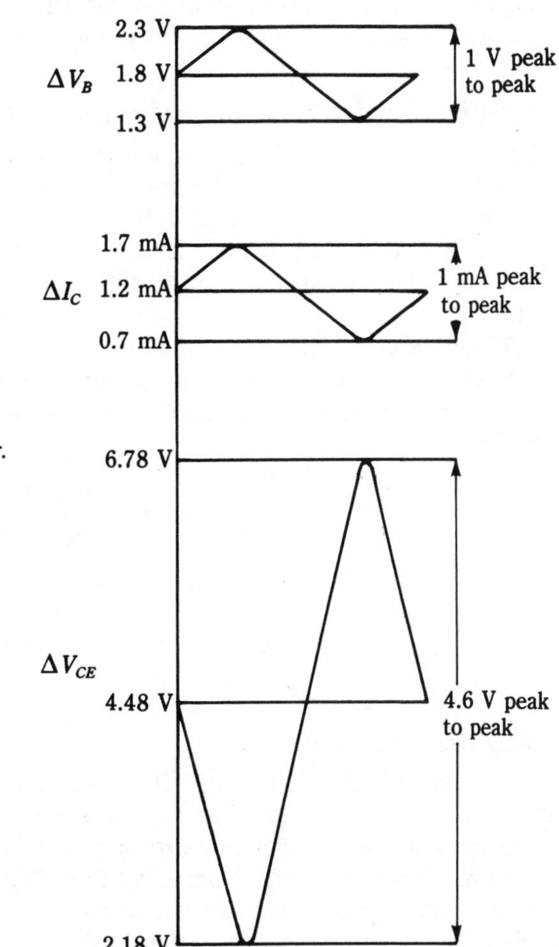

15-12 ac signal in a transistor.

The ac input signal has a voltage of 0.5 V peak or 1 V peak to peak. This voltage is applied to the base, which has a dc bias voltage of 1.8 V. The ac signal will swing the base voltage up to 2.3 V and down to 1.3 V by adding and subtracting 0.5 V peak, producing 1 V peak to peak.

Collector current is determined these steps:

a. find V_E by subtracting 0.6 V from V_B.
b. find I_E using Ohm's Law with V_E and R_E.
c. $I_C = I_E$

I_C has a bias level of 1.2 mA. It will swing 0.5 mA peak, 1 mA peak to peak following the changing base voltage. As I_C changes, the voltage drops across the collector and emitter resistors also change, as determined by Ohm's Law. The voltage across the transistor, V_{CE} will change in the opposite direction because it is found by subtracting the voltage drops from the supply voltage. V_{CE} has a bias voltage of 4.48 V, and the changing I_C will produce a 4.6 V peak-to-peak change in V_{CE}.

Amplification with 180-degree phase shift

Refer to Figs. 15-11 and 15-12. The output signal is taken at the collector. The voltage at the collector is the sum of the voltage drop across the emitter resistor and the voltage across the transistor. The transistor voltage will decrease with an increase in collector current and increase with a decrease in collector current. The transistor voltage reacts opposite to the collector current, or 180-degree phase shift.

Amplification of the input signal occurs by comparing the 1 V peak-to-peak input to the 4.6 V peak to peak of the output. Gain is calculated by dividing the output by the input. The letter symbol for voltage gains is *Av*. Gain is a ratio, with no units.

$$Av = \frac{\text{output}}{\text{input}}$$

$$= \frac{4.6\text{ V}}{1\text{ V}}$$

$$= 4.6$$

Using the load line

The load line calculated in Fig. 15-10 is used in Fig. 15-13 to the input and output sine waves. The input, shown on the left as ΔI_C, is projected to the load line. The maximum and minimum points are projected down to give the output voltage waveform, shown as ΔV_{CE}. Notice, with the Q point near the center of the load line, the transistor does not saturate or cut-off and the output signal is the same shape as the input, without distortion.

Figure 15-14 demonstrates the effects of placing the Q point either too close to cut-off (A) and too close to saturation (B). Misplacing the Q point results in clipping the output waveform, producing distortion.

In circuits where distortion is not a consideration, the Q point can be placed at or even beyond the end points.

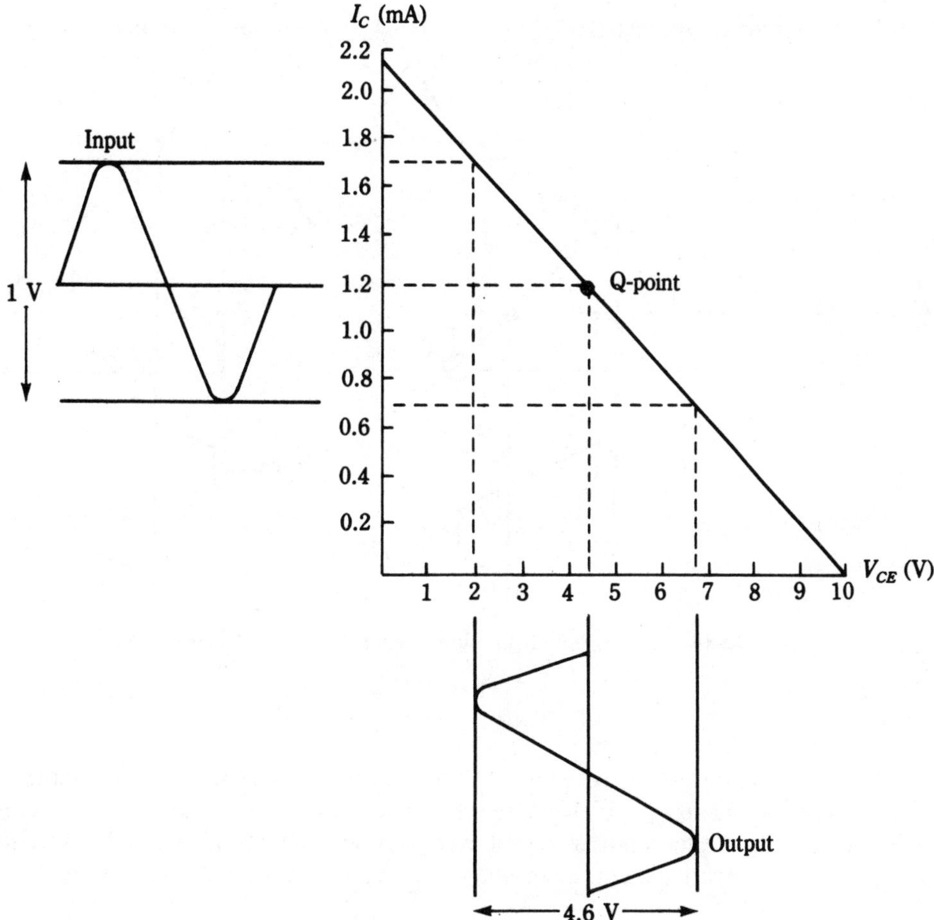

15-13 Load line with input and output waveforms.

Examples:

1. A transistor can be used as a switch, either on or off. "OFF" has no base voltage and "ON" has enough base voltage to drive the transistor into saturation. The Q point is placed at saturation.
2. Placing the Q point at cut-off (or below) requires the input sine wave to be large enough to lift the base voltage to start collector current flow.

Amplifier configurations

The transistor can be connected as an amplifier in three different configurations; common emitter, common collector, common base. To this point, the only configuration discussed has been the common emitter amplifier.

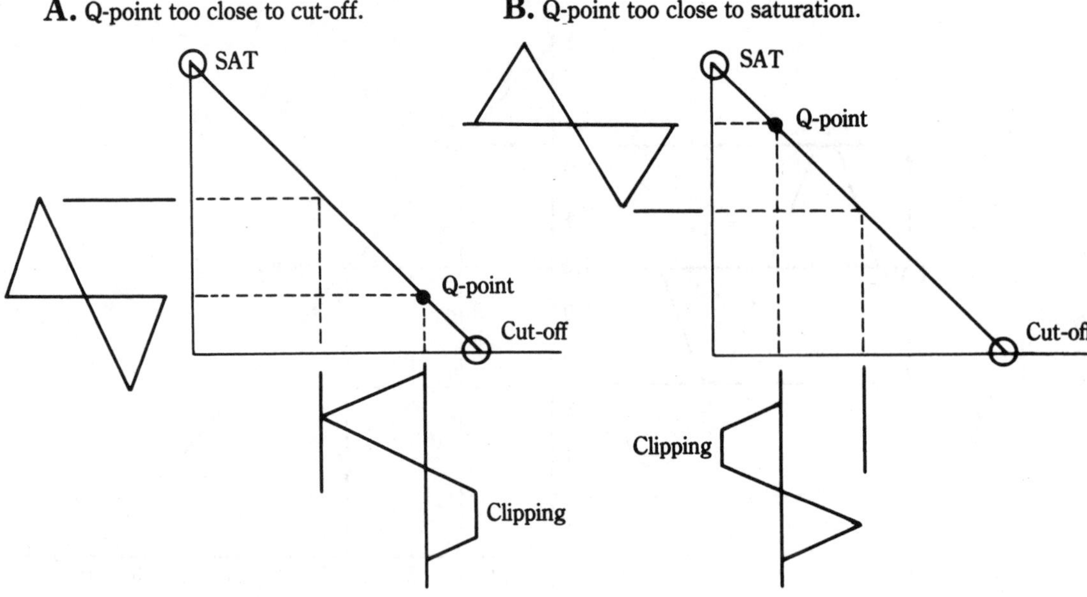

15-14 Effects on output signal when the Q point is not in the center.

The physical difference between the three types is where the input and output is connected. Refer to Fig. 15-15 to see the different connections. Notice, the element not used as either input or output is considered common. The amplifiers also have different operational characteristics, making each useful in different applications.

Common-emitter characteristics

1. Input on base—output on collector.
2. Highest power gain, with both voltage and current gains.
3. Only amplifier with 180-degree phase inversion.
4. Medium input and output impedance.
5. Applications: wide range, especially as an amplifier.

Common-collector characteristics

1. Input on base—output on emitter.
2. Voltage gain is less than 1.
3. Input and output are in phase.
4. Highest input impedance with lowest output impedance.
5. Applications: used as impedance matching between other amplifier stages.

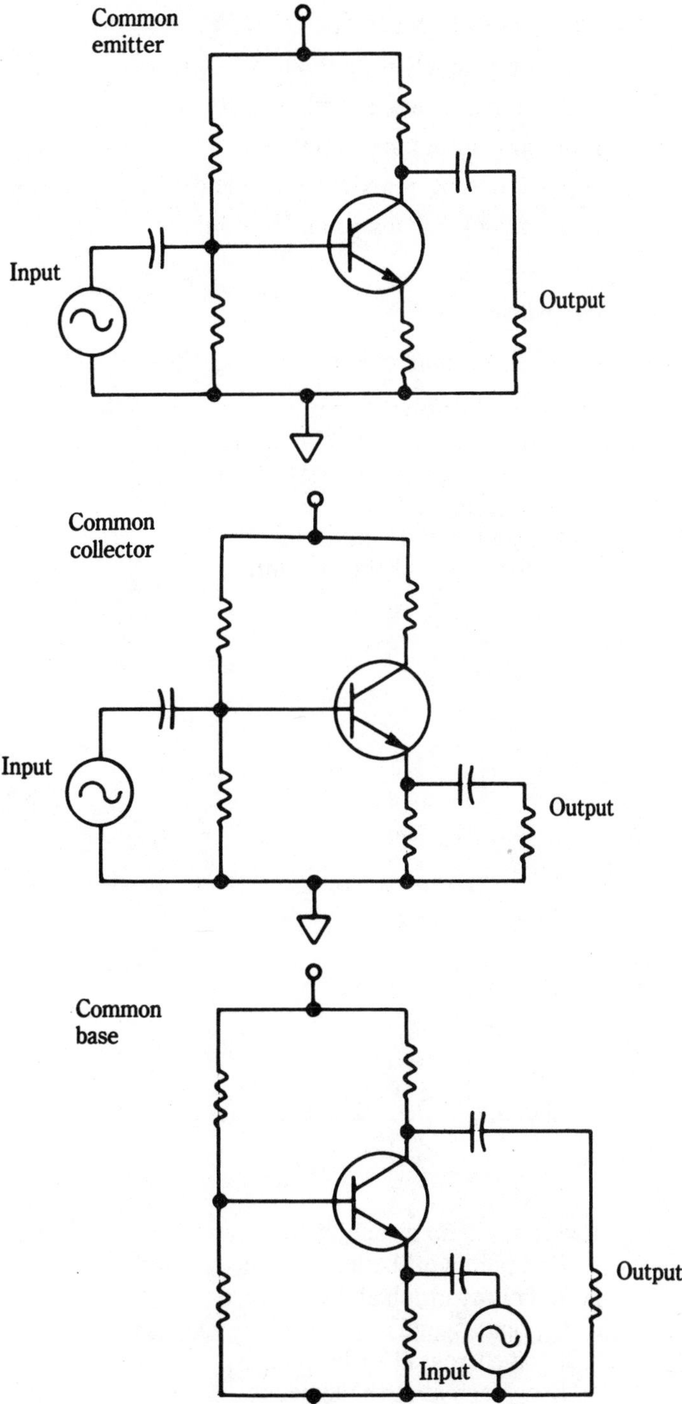

15-15 Amplifier configurations.

Common-base characteristics

1. Input on emitter—output on collector.
2. Current gain is less than 1.
3. Input and output are in phase.
4. Highest output impedance with lowest input impedance.
5. Applications: mainly as an RF (radio frequency) amplifier.

Practice problems

Show all formulas. In all circuits, assume V_{BE} = 0.6 V.

1. Use the schematic shown to answer the following questions:
   a. End points of the load line; $I_{C(SAT)}$ and cut-off.
   b. Label the graph and draw load line.
   c. Q point; I_C and V_{CE}.
   d. Base current at saturation.
   e. Base current at the Q point.

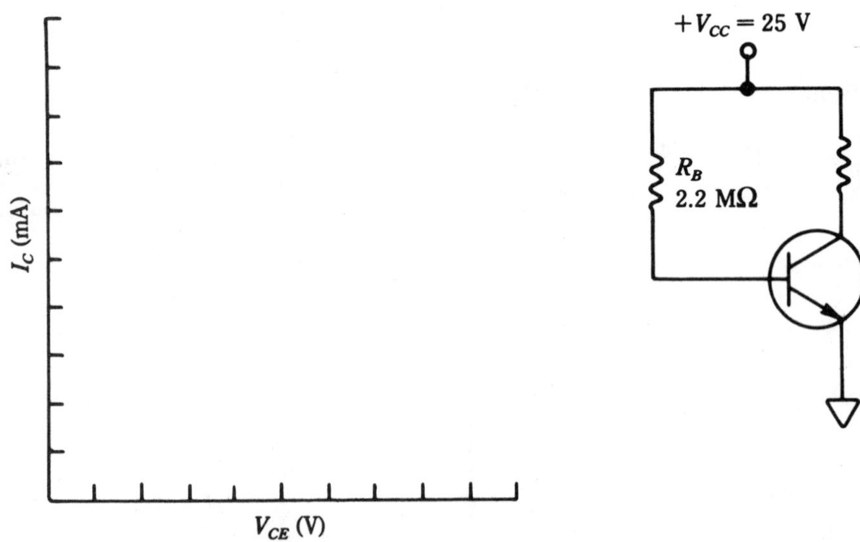

2. Use the schematic shown to answer the following questions:
   a. End points of the load line; $I_{C(SAT)}$ and cut-off.
   b. Label the graph and draw load line.
   c. Q point; I_C and V_{CE}.
   d. Base current at saturation.
   e. Base current at the Q point.

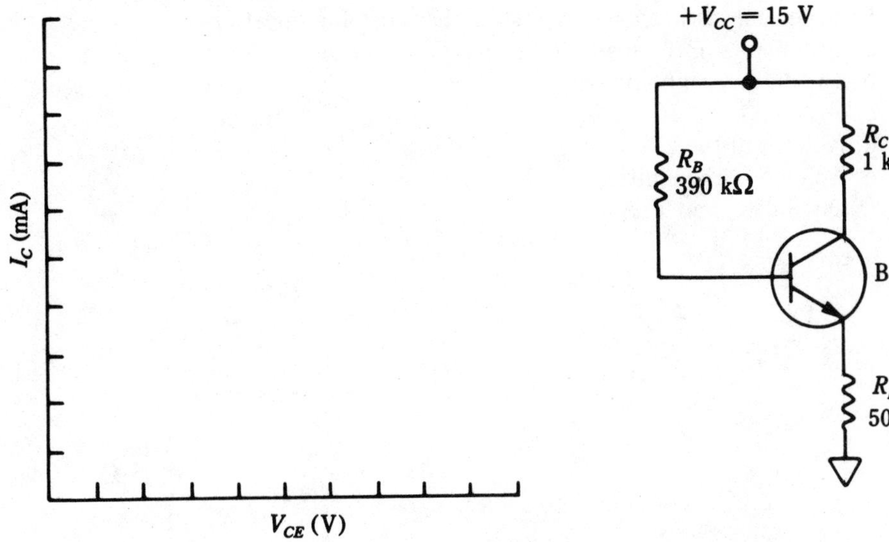

3. Use the schematic shown to answer the following questions:
   a. End points of the load line; $I_{C(SAT)}$ and cut-off.
   b. Label the graph and draw load line.
   c. Q point; I_C and V_{CE}.
   d. Base current at saturation.
   e. Base current at the Q point.

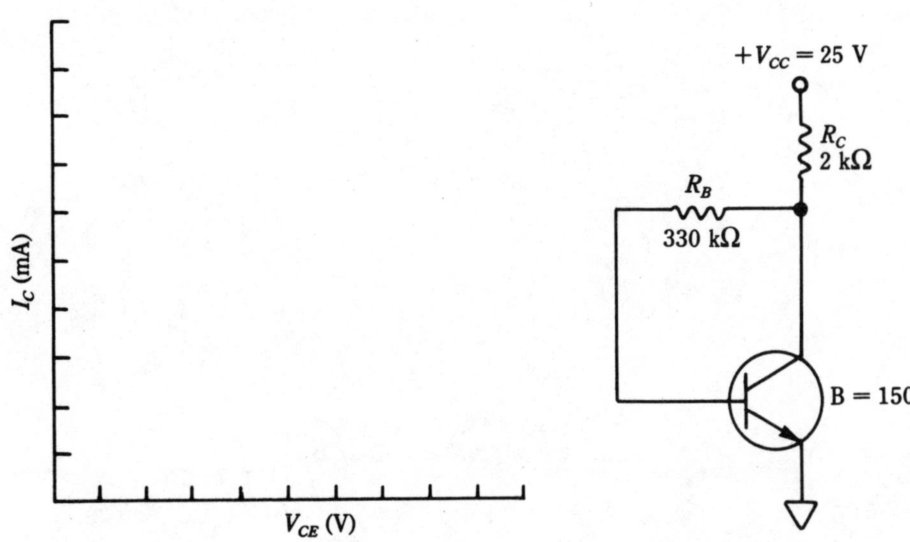

4. Use the schematic shown to answer the following questions:
   a. End points of the load line; $I_{C(SAT)}$ and cut-off.
   b. Label the graph and draw load line.
   c. I_{BIAS}.
   d. Voltage at the base.
   e. Voltage at the emitter.
   f. Q point; I_C and V_{CE}.

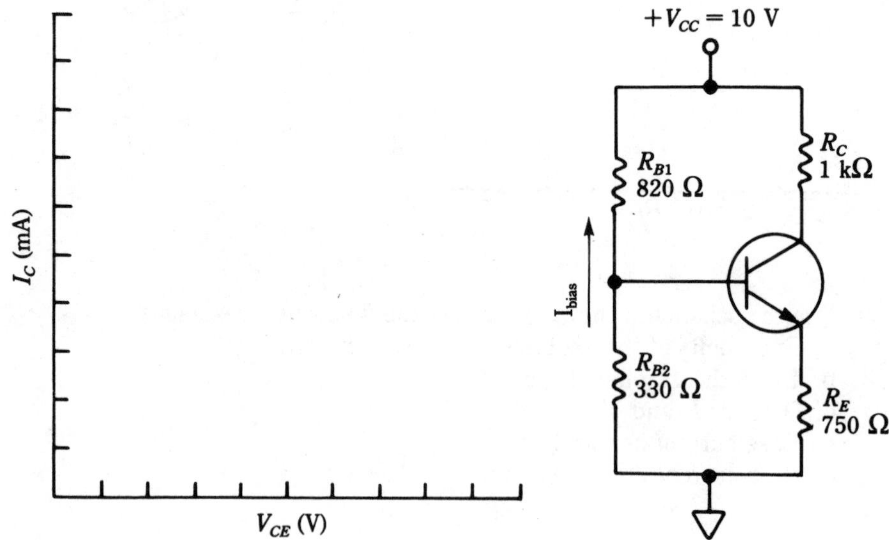

16

Binary, octal, and hexadecimal numbering systems

HUMANS DEAL WITH NUMBERS IN THE DECIMAL SYSTEM. COMPUTERS, however, cannot use the decimal system, so instead, they use binary, octal, or hexadecimal numbering systems.

Digits used

The digits used in a numbering system are determined by the base of the system. The base is the number of digits used including the digit 0. When counting in a numbering system, count up to the value of the base, then go up to the next place value and repeat the sequence. Hint in reading numbers in a different system: read the digits separately, rather than using the familiar names used with the decimal system.

Decimal system, base 10

There are 10 digits:

0, 1, 2, 3, 4, 5, 6, 7, 8, 9

Beyond 10:

9, 10, 11, ... 19, 20, 21, ... 99, 100, 101

Binary system, base 2

There are 2 digits:

0, 1

Beyond 2:

1, 10, 11, 100, 101, 110, 111, 1000

Octal system, base 8

There are 8 digits:

0, 1, 2, 3, 4, 5, 6, 7

Beyond 8:

7, 10, 11, ... 17, 20, 21, ... 77, 100, 101

Hexadecimal system, base 16

There are 16 digits:

0, 1, 2, 3, 4, 5, 6, 7, 8, 9, A, B, C, D, E, F

Beyond 16:

F, 10, 11, ... 1F, 20, 21, ... FF, 100, 101

Place values

The position a digit holds in a number determines the value of the digit. The location has a place value, which is the numbering system base, raised to a power.

A number raised to a power, such as 5^3, is the number multiplied by itself the number of times indicated by the power: $5^3 = 5 \times 5 \times 5 = 125$.

With all numbering systems, starting on the right, and increasing to the left, the place value power starts with zero and counts up. This method of raising the base to a power will give the decimal-system equivalent of the place value. The reason other systems are converted to the decimal system is that other numbering systems have very little meaning to human beings.

Each digit in a number is multiplied by its place value and added together. In the examples below, the subscript written with the number is the base of the numbering system. These examples are shown to make it clear which system is used.

Decimal-system place values

10^5	10^4	10^3	10^2	10^1	10^0
100,000	10,000	1,000	100	10	1

Example: $720{,}315_{10}$ =

$$
\begin{aligned}
7 \times 100{,}000 &= 700{,}000 \\
2 \times 10{,}000 &= 20{,}000 \\
0 \times 1{,}000 &= 0{,}000 \\
3 \times 100 &= 300 \\
1 \times 10 &= 10 \\
5 \times 1 &= 5 \\
\hline
& \quad 720{,}315_{10}
\end{aligned}
$$

Binary-system place values

2^5	2^4	2^3	2^2	2^1	2^0
32	16	8	4	2	1

Example: 110011_2 =

$$\begin{aligned}1 \times 32 &= 32\\ 1 \times 16 &= 16\\ 0 \times 8 &= 0\\ 0 \times 4 &= 0\\ 1 \times 2 &= 2\\ 1 \times 1 &= \underline{1}\\ & \quad 51_{10}\end{aligned}$$

Octal-system place values

8^5	8^4	8^3	8^2	8^1	8^0
32,768	4,096	512	64	8	1

Example: 547023_8 =

$$\begin{aligned}5 \times 32{,}768 &= 163{,}840\\ 4 \times 4{,}906 &= 16{,}384\\ 7 \times 512 &= 3{,}584\\ 0 \times 64 &= 0\\ 2 \times 8 &= 16\\ 3 \times 1 &= \underline{3}\\ & \quad 183{,}827_{10}\end{aligned}$$

Hexadecimal-system place values

16^5	16^4	16^3	16^2	16^1	16^0
1,048,576	65,536	4,096	256	16	1

Example: $3A0B19_{16}$ =

$$\begin{aligned}3 \times 1{,}048{,}576 &= 3{,}145{,}728\\ A \times 65{,}536 &= 655{,}360\\ 0 \times 4{,}096 &= 0\\ B \times 256 &= 2{,}816\\ 1 \times 16 &= 16\\ 9 \times 1 &= \underline{9}\\ & \quad 3{,}803{,}929_{10}\end{aligned}$$

Converting from one system to another

The place value method is used here to convert from decimal to binary and binary to decimal. Both the octal and hexadecimal systems will first use binary to make the conversions.

Decimal to binary

To convert from decimal to binary follow these steps:

1. Make a binary place-value chart.
2. Determine the largest place value that can be subtracted from the decimal number.
3. Subtract the place value and record a "1" on the place-value chart. The result from the subtraction will be used with the next place value.
4. Continue subtracting the largest place values, placing a "1" for each place value used and a "0" for each place value not used.

Binary place values:

2^8	2^7	2^6	2^5	2^4	2^3	2^2	2^1	2^0
256	128	64	32	16	8	4	2	1

Example:
Convert 350_{10} to binary:

$$
\begin{aligned}
350 - 256 &= 94\ (2^8)\\
94 - 64 &= 30\ (2^6)\\
30 - 16 &= 14\ (2^4)\\
14 - 8 &= 6\ (2^3)\\
6 - 4 &= 2\ (2^2)\\
2 - 2 &= 0\ (2^1)
\end{aligned}
$$

Binary place-value chart. Place a 1 or 0 with each place value:

2^8	2^7	2^6	2^5	2^4	2^3	2^2	2^1	2^0
1	0	1	0	1	1	1	1	0

Result: $350_{10} = 101011110_2$
Example:
Convert 100_{10} to binary:

$$
\begin{aligned}
100 - 64 &= 36\ (2^6)\\
36 - 32 &= 4\ (2^5)\\
4 - 4 &= 0\ (2^2)
\end{aligned}
$$

Place a 1 or 0 with each place value:

2^8	2^7	2^6	2^5	2^4	2^3	2^2	2^1	2^0
		1	1	0	0	1	0	0

Result: $100_{10} = 1100100_2$

Binary to decimal

To convert from binary to decimal, follow this procedure:

1. Write the binary number in the proper location on the place-value chart.
2. Add all of the place values for each 1 in the binary number.
3. The sum is the decimal equivalent.

Example:
Convert 10111001_2 to decimal:
Write the binary number under the place-value chart:

2^8	2^7	2^6	2^5	2^4	2^3	2^2	2^1	2^0
0	1	0	1	1	1	0	0	1

Add the place value:

$$\begin{aligned} 2^7 &= 128 \\ 2^5 &= 32 \\ 2^4 &= 16 \\ 2^3 &= 8 \\ 2^0 &= \underline{1} \\ & \quad 185_{10} \end{aligned}$$

Result: $10111001_2 = 185_{10}$

Example:
Convert 10110011_2 to decimal:
Write the binary number under the place-value chart:

2^8	2^7	2^6	2^5	2^4	2^3	2^2	2^1	2^0
1	0	1	1	0	0	1	1	1

Add the place value:

$$\begin{aligned} 2^8 &= 256 \\ 2^6 &= 64 \\ 2^5 &= 32 \\ 2^2 &= 4 \\ 2^1 &= 2 \\ 2^0 &= \underline{1} \\ & \quad 359_{10} \end{aligned}$$

Result: $10110011_2 = 359_{10}$

Converting octal

It is possible to convert the octal numbers to decimal numbers using the place-value chart; however, an easier method is to first convert to binary.

Converting octal to binary

To convert octal to binary, follow these steps:

1. Treat each octal digit as an individual.
2. Change each octal digit to a three-bit binary number, using the binary place-value chart. Note: the place-value chart needs only three places.
3. Combine the three-bit binary numbers into one final binary number.

Use this place values for each octal digit:

2^2	2^1	2^0
4	2	1

Example:
Convert 7024_8 to binary:

Octal number:	7	0	2	4
3-bit binary:	101	000	010	100

Results: $7\text{-}24_8$ = 101 000 010 100_2

Note A space is left between groups of three bits to allow for easier reading.

Example:
Convert 5310_8 to binary:

Octal number:	5	3	1	0
3-bit binary:	101	011	001	000

Result: 5310_8 = 101 011 001 00_2

Converting binary to octal

To convert a binary number to octal, follow these steps:

1. Write the binary number in groups of three binary bits. Add 0s on the left if needed.
2. Change each 3-bit binary to its decimal equivalent. Note that these are actually octal because only the digits 0 – 7 will be used.
3. Combine the octal equivalents for the final result.

Example:
Convert 10100011_2 to octal:

3-bit binary:	010	100	011
Octal number:	2	4	3

Result: $10100011_2 = 243_8$

Converting hexadecimal

Hexadecimal numbers are related to binary in a manner similiar to octal, except hexadecimal is in 4-bit binary groups.

Converting hexadecimal to binary

To convert a hexadecimal number to binary, follow these steps:

1. Treat each hexadecimal digit as an individual.
2. Change each hexadecimal digit to a four-bit binary number, using the binary place-value chart. Note that the place-value chart needs only four places.
3. Combine the four-bit binary numbers into one final binary number.

Use these place values for each octal digit:

2^3	2^2	2^1	2^0
8	4	2	1

Example:
Convert $4C02_{16}$ to binary:

Hexadecimal:	4	C	0	2
4-bit binary:	0100	1100	0000	0010

Result: $4C02_{16} = 0100\ 1100\ 0000\ 0010_2$

Converting binary to hexadecimal

To convert a binary number to hexadecimal, follow these steps:

1. Write the binary number in grouped of four binary bits. Add 0s on the left if needed.
2. Change each 4-bit binary to its hexadecimal equivalent.
3. Combine the hexadecimal equivalents for the final result.

Example:
Convert 11001001111101_2 to hexadecimal:

4-bit binary:	0011	0010	0111	1101
Hexadecimal	3	2	7	D

Result: $11001001111101_2 = 327D_{16}$

Combine octal, hexadecimal, binary, and decimal

The previous sections show that decimal, octal, and hexadecimal numbers can all be converted to binary using the binary place-value chart. Also, using the binary place-value chart, binary can be converted to each numbering system. Therefore, to convert from one system to another, first convert to binary, then to the desired system.

Practice problems

Convert the number given to each of the other systems.

	decimal	binary	octal	hexadecimal
(1)	56	________	________	________
(2)	________	10111000	________	________
(3)	________	________	132	________
(4)	________	________	________	7D
(5)	75	________	________	________
(6)	________	1010001	________	________
(7)	________	________	102	________
(8)	________	________	________	31
(9)	37	________	________	________
(10)	________	1100101	________	________
(11)	________	________	161	________
(12)	________	________	________	19
(13)	50	________	________	________
(14)	________	10000000	________	________
(15)	________	________	135	________
(16)	________	________	________	4D
(17)	84	________	________	________
(18)	________	101101	________	________
(19)	________	________	100	________
(20)	________	________	________	100

Appendix
Answers to practice problems

Chapter 1

Page 3	Page 4	Page 5	Pages 7–8
1. 5610	1. −2	1. 6	1. 8.76×10^5
2. 5610	2. 10	2. 20	2. 1.03×10^9
3. 5600	3. 6	3. −15	3. 3.2×10^4
4. 29900	4. −4	4. −18	4. 2.5×10^1
5. 360,000	5. −4	5. −6	5. 5.8×10^0
6. 569	6. 16	6. 2	6. 3.0×10^{-2}
7. 573	7. 6	7. 0	7. 5.6×10^{-4}
8. 333	8. −7	8. 0.2	8. 4.05×10^{-3}
9. 12.1	9. −7	9. −0.18	9. 1.0×10^{-7}
10. 23.2	10. −2	10. 4	10. 2.0×10^{-1}
11. 54.5	11. 0	11. −8	11. 1.3×10^9
12. 57.0	12. −1	12. −5	12. 4.5×10^4
13. 8.2	13. −11	13. −5	13. 4.5×10^4
14. 8.43	14. 20	14. −5	14. 3.9×10^{-2}
15. 9.00	15. 0.5	15. 5	15. 5.6×10^{-3}
16. 0.00333	16. 5.3	16. −4	16. 5.2×10^1
17. 0.900	17. 7.96	17. $-1/6$	17. 3.2×10^{-3}
18. 0.000932	18. −10.76	18. 6	18. 4.6×10^{-9}
19. 0.0100	19. −9.11	19. 3	19. 7.05×10^{-5}
20. 0.200	20. $-1\,3/8$	20. 27	20. 4.0×10^{-14}
21. 0.667			
22. 0.00556			
23. 745			
24. 0.0909			
25. 0.0125			

Page 8

1. 480,000
2. 785
3. 8.9
4. 346
5. 457,000
6. 0.1
7. 0.00201
8. 0.00000003
9. 1
10. 0.1
11. 35
12. 4.0
13. 0.709
14. 5,600,000
15. 65
16. 98
17. 0.000109
18. 0.00000078
19. 10
20. 10

Page 10

1. 2,400 kΩ (2.4 MΩ)
2. 353 kHz (0.353 MHz)
3. 2,500,000 mW (2.5 kW)
4. 25,000 mV (0.025 kV)
5. 10,000 μH (10 mH)
6. 1500 Ω (0.0015 MΩ)
7. 25,000 kHz (0.025 GHz)
8. 30 MHz (30,000,000 Hz)
9. 56,000 kV (56,000,000 V)
10. 75,000 W (0.075 MW)
11. 25,000 μA (0.025 A)
12. 0.0015 μA (1.5 mA)
13. 1,000,000,000 pF (0.001 F)
14. 0.00000001 μF (0.000001 nF)
15. 25 μV (0.000025 V)
16. 7.5 W (7,500,000 μW)
17. 0.5 kV (500,000 mV)
18. 2.4 V (0.0024 kV)
19. 0.001 kA (1000 mA)
20. 10,000 mV (0.01 kV)

Page 12

1. 100 (10^{0})
2. 60 kilo (10^{3})
3. 15 kilo (10^{3})
4. 8 milli (10^{-3})
5. 30 nano (10^{-9})
6. 5 milli (10^{-3})
7. 2 kilo (10^{3})
8. 4 micro (10^{-6})
9. 5 kilo (10^{3})
10. 62.8 milli (10^{3})
11. 628 kilo (10^{3})
12. 0.159 (10^{0})
13. 15.9 mega (10^{6})
14. 1 giga (10^{9})
15. 5 milli (10^{-3})

Page 13

1. 44 kilo (10^{3})
2. 914 (basic units, 10^{0})
3. 76.2 kilo (10^{3})
4. 1.0 mega (10^{6})
5. 725 kilo (10^{3})
6. 1350 (basic units, 10^{0})
7. 12.5 (basic units, 10^{0})
8. 25 kilo (10^{0})
9. 86 milli (10^{-3})
10. 1.05 milli (10^{3})
11. 6 milli (10^{-3})
12. 9.99 milli (10^{-3})
13. 10 micro (10^{-6})
14. 251.5 micro (10^{-6})
15. 25.1 nano (10^{-9})
16. 0.011 micro (10^{-6})
17. 40 pico (10^{-12})
18. 0.0115 micro (10^{-6}) or 11.5 nano (10^{-9})
19. 1. (basic units, 10^{0})
20. 221.1 (basic units, 10^{0})

Chapter 2

Page 18

1. E = 20 V
2. E = 50 V
3. E = 20 V
4. E = 300 V
5. E = 12 V
6. I = 0.1 A
7. I = 1 A
8. I = 10 mA
9. I = 10 mA
10. I = 10 mA
11. R = 10 Ω
12. R = 10 Ω
13. R = 2 kΩ
14. R = 2 MΩ
15. R = 5 kΩ
16. I = 0.1 A
17. E = 30 V
18. I = 1 A
19. R = 1.2 MΩ
20. R = 1 Ω

Page 21 (answers are underlined)

Voltage	Current	Resistance	Power
1. E = 30 V	I = 15 A	R = 2 Ω	P = 450 W
2. E = 2.5 V	I = 2 A	R = 1.25 Ω	P = 5 W
3. E = 100 V	I = 0.01 A	R = 10 kΩ	P = 1 W
4. E = 250 V	I = 50 A	R = 5 Ω	P = 12,500 W
5. E = 15 V	I = 6.67 mA	R = 2250 Ω	P = 100 mW
6. E = 3.16 V	I = 3.16 mA	R = 1 kΩ	P = 10 mW
7. E = 250 V	I = 25 mA	R = 10 kΩ	P = 6.25 W
8. E = 12.5 V	I = 2 mA	R = 6.25 kΩ	P = 25 mW
9. E = 50 V	I = 5 mA	R = 10 kΩ	P = 250 mW
10. E = 5 V	I = 2 mA	R = 2500 Ω	P = 10 mW
11. E = 0 V	I = 0 A	R = 100 Ω	P = 0 W
12. E = 10 V	I = 1 A	R = 10 Ω	P = 10 W
13. E = 0.1 V	I = 10 μA	R = 10 kΩ	P = 1 μW
14. E = 10 V	I = 100 mA	R = 100 Ω	P = 1 W
15. E = 25 V	I = 1 A	R = 25 Ω	P = 25 W
16. E = 20 V	I = 100 mA	R = 200 Ω	P = 2 W
17. E = 1 V	I = 1 A	R = 1 Ω	P = 1 W
18. E = 10 V	I = 10 mA	R = 1000 Ω	P = 0.1 W
19. E = 10 kV	I = 1000 mA	R = 10,000 Ω	P = 10 kW
20. E = 10 V	I = 100 mA	R = 100 Ω	P = 1 W

Pages 27–30

1. $I_T = 0.2$ A, $R_T = 250\ \Omega$, $E_{R1} = 20$ V, $E_{R2} = 30$ V, $P_{R1} = 4$ W, $P_{R2} = 6$ W
2. $I_T = 50$ mA, $R_T = 300\ \Omega$, $E_{R1} = E_{R2} = 5$ V, $P_{R1} = P_{R2} = P_{R3} = 0.25$ W
3. $R_T = 280\ \Omega$, $I_T = 50$ mA
4. $I_T = 3$ A, $V = 150$ V
5. $R_T = 80\ \Omega$, $I_T = 1.5$ A, $V = 120$ V
6. $R_T = 2.5$ kΩ, $R_1 = 1.5$ kΩ
7. $R_T = 800\ \Omega$, $R_3 = 100\ \Omega$, $V = 80$ V
8. $I_T = 25$ mA, $R_2 = 320\ \Omega$, $E_{R1} = 4$ V
9. $V = 32$ V, $I_T = 400$ mA
10. $V = 120$ V, $I_T = 2$ A, $R_2 = 20\ \Omega$

Note Because of rounding and the fact that some calculations are used to make other calculations, your answers might be slightly different.

Pages 36–38

1. $R_T = 13.6\ \Omega$, $I_T = 1.83$ A, $I_A = 1$ A, $I_B = 0.5$ A, $I_C = 0.33$ A
2. $R_T = 512.8\ \Omega$, $I_T = 29$ mA
3. $R_T = 40\ \Omega$
4. $R_T = 125\ \Omega$
5. $R_3 = 1500\ \Omega$
6. $R_4 = 33.3\ \Omega$
7. $R_3 = 1000\ \Omega$, $I_T = 100$ mA, $R_T = 200\ \Omega$

8. R_T = 60 Ω, I_A = 0.333 *A*, I_B = 0.2 *A*, I_C = 0.1333 A
9. R_T = 200 Ω, I_A = 50 mA, I_B = 16.7 mA, I_C = I_D = 33.3 mA
10. R_T = 9.1 Ω, I_A = 1 A, I_B = 0.364 A, I_C = 0.637 A, I_D = 0.228 A, I_E = 0.409 A, I_F = 0.228 A, I_G = 0.182 A

Note Because of rounding and the fact that some calculations are used for other calculations, your answers might be slightly different.

Pages 45–47

1. R_T = 2.2 k, I_T = 20 mA, E_{R1} = 20 V, E_{R2} = 24 V, E_{R3} = 24 V, I_{R2} = 12 mA, I_{R3} = 8 mA
2. R_T = 4 k, I_T = 5 mA, E_{R1} = 5 V, E_{R2} = 1.5 V, E_{R3} = 1.5 V, E_{R4} = 13.5 V, I_{R2} = 2.21 mA, I_{R3} = 2.68 mA
3. R_T = 50 Ω, I_T = 100 mA, E_{R1} = 1 V, E_{R2} = E_{R3} = 0.5 V, E_{R4} = 1 V, I_{R5} = E_{R6} = 0.5 V, E_{R7} = E_{R8} = 1 V, I_{R2} = I_{R5} = 50 mA
4. R_T = 314.5 Ω, I_T = 31.8 mA, I_{R1} = 31.8 mA, I_{R2} = 21.2 mA, I_{R3} = 10.6 mA, I_{R4} = 31.8 mA, I_{R5} = 20 mA, I_{R6} = 8 mA, I_{R7} = 4 mA, E_{R1} = 3.18 V, E_{R2} = E_{R3} = 5.72 V, E_{R4} = 0.7 V, E_{R5} = E_{R6} = E_{R7} = 0.4 V
5. R_T = 133.3 Ω, I_T = 75 mA, I_{R2} = 50 mA, I_{R3} = 25 mA, I_{R1} = 7.5 V, E_{R2} = 2.5 V
6. R_T = 16.9 Ω
7. R_T = 100 Ω
8. R_T = 1 k
9. R_T = 275 Ω, I_T = 200 mA, E_{R1} = E_{R2} = 20 V, E_{R3} = 15 V, E_{R4} = 2.5 V, E_{R5} = 5 V, E_{R6} = 7.5 V, E_{R7} = E_{R8} = E_{R9} = 2.5 V, I_{R1} = I_{R2} = 200 mA, I_{R3} = I_{R4} = I_{R5} = 100 mA, I_{R6} = I_{R7} = I_{R8} = I_{R9} = 50 mA
10. R_T = 2520 Ω, I_T = 39.7 mA

Note Because of rounding and the fact that some calculations are used for other calculations, your answers might be slightly different.

Pages 56–58

1. R_T = infinity (open circuit), I_T = 0, E_{R1} = E_{R2} = 0
2. R_T = 25 kΩ (notice the short), I_T = 1 mA, E_{R1} = E_{R2} = , E_{R3} = 25 V
3. R_T = 1000 Ω, I_T = 30 mA, V_a = 15 V, V_b = 6 V
4. R_T = 667, I_T = 18 mA, I_{R1} = 6 mA, I_{R2} = 12 mA
5. R_T = 25 Ω, I_T = 1 A, I_{R1} = 0.5 A, I_{R2} = 0.25A, I_{R3} = 0.25 A
6. R_T = 62.5, I_T = 1.92 A, P_T = 230.4 W
7. R_T = 55.3 Ω, I_T = 2.17 A, P_T = 260 W, R_1 = 144 Ω, R_2 = 192 Ω, R_3 = 576 Ω, R_4 = 240 Ω
8. R_T = 76.7 Ω, I_T = 0.196 A, I_{R1} = 0.196 A, I_{R2} = 0.130 A, I_{R3} = 0.065A, E_{R1} = 11.7 V, E_{R2} = 3.24 V, E_{R3} = 3.24 V % of I_T in R_2 = 66 %
9. R_T = 40 Ω, I_T = 1 A, I_{R1} = 1A, I_{R2} = I_{R3} = I_{R4} = I_{R5} = 0.5A, I_{R6} = 1A, E_{R1} = 10 V, E_{R2} = 7.5 V, E_{R3} = 2.5 V, E_{R4} = E_{R5} = 5 V, E_{R6} = 20 V
10. I in meter = 0 . . . This is a balanced bridge

Chapter 3

Page 66

Practice Meter 1

a. range = 50 DCV
b. set of numbers on scale:
   0, 10, 20, 30, 40, 50
c. value of each scale line = 1
d. value the meter is reading:
   needle A = 7
   needle B = 21
   needle C = 35

Page 67

Practice Meter 2

a. range = 2.5 DCmA
b. set of numbers on scale:
   0, 0.5, 1.0, 1.5, 2.0, 2.5
c. value of each scale line = 0.05
d. value the meter is reading:
   needle A = 0.45
   needle B = 1.35
   needle C = 2.15

Pages 67–68

Practice Meter 3

a. range = 1000 ACV
b. set of numbers on scale:
   0, 200, 400, 600, 800, 1000
c. value of each scale line = 20
d. value the meter is reading:
   needle A = 200
   needle B = 460
   needle C = 880

Page 68

Practice Meter 4

a. range = 0.5 DCV
b. set of numbers on scale:
   0, 0.1, 0.2, 0.3, 0.4, 0.50
c. value of each scale line = 0.01
d. value the meter is reading:
   needle A = 0.1
   needle B = 0.22
   needle C = 0.39

Page 69

Practice Meter 5

a. range = 10 ACV
b. set of numbers on scale:
   0, 2, 4, 6, 8, 10
c. value of each scale line = 0.2
d. value the meter is reading:
   needle A = 1.0
   needle B = 3.6
   needle C = 8.2

Page 69

Practice Meter 6

a. range = 25 DCmA
b. set of numbers on scale:
   0, 5, 10, 15, 20, 25
c. value of each scale line = 0.5
d. value the meter is reading:
   needle A = 1.0
   needle B = 4.0
   needle C = 10.0

Page 71

Practice Meter 7

a. range = 500 ACV
b. set of numbers on scale:
0, 100, 200, 300, 400, 500
c. value of each scale line = 10
d. value the meter is reading:
needle A = 130
needle B = 4.26
needle C = 450

Page 71

Practice Meter 8

a. range = 250 DCV
b. set of numbers on scale:
0, 50, 100, 150, 200, 250
c. value of each scale line = 5
d. value the meter is reading:
needle A = 70
needle B = 115
needle C = 155

Page 72

Practice Meter 9—OHMs scale

needle	each line	reading	range	measurement
a	–	infinity	x1 k	infinity
b	2	34	x1 k	34 k
c	0.5	6.5	x1 k	6.5 k

Page 72

Practice Meter 10—OHMs scale

needle	each line	reading	range	measurement
a	20	160	x10 k	1600k
b	1	17	x10 k	170 k
c	0.5	3.0	x10 k	30 k

Chapter 5

Page 92

1. a) rms = 21.2 b) p to p = 60 c) avg = 19.1
2. a) p to p = 200 b) avg = 63.6 c) rms = 70.7
3. a) rms = 28.3 b) peak = 40 c) avg = 25.4
4. a) peak = 130 b) avg = 82.7 c) rms = 91.9
5. a) peak = 99 b) p to p = 198 c) avg = 63
6. a) avg = 10.8 b) p to p = 34 c) peak = 17
7. a) rms = 17.8 b) peak = 25.2 c) p to p = 50.3
8. a) peak = 189 b) p to p = 377 c) rms = 133
9. a) peak = 120 b) p to p = 240 c) avg = 76.5
10. a) p to p = 140 b) avg = 44.5 c) rms = 49.5
11. a) p to p = 40 b) peak = 20 c) rms = 14.1 d) avg = 12.7
12. a) p to p = 70 b) peak = 35 c) rms = 24.7 d) avg = 22.3
13. a) p to p = 400 b) peak = 200 c) rms = 141 d) avg = 127
14. a) p to p = 120 b) peak = 60 c) rms = 42.4 d) avg = 38.2
15. a) p to p = 80 b) peak = 40 c) rms = 28.3 d) avg = 25.4

Pages 94–95

1. 0.1 s	6. 1 μs	11. 20 Hz	16. 60,000 cm
2. 0.01 s	7. 0.1 μs	12. 200 Hz	17. 6,000 cm
3. 1 ms	8. 0.01 μs	13. 2 kHz	18. 600 cm
4. 0.1 ms	9. 1 ns	14. 20 kHz	19. 6 cm
5. 0.01 ms	10. 0.1 ns	15. 2 Hz	20. 0.06 cm

Page 99 – 103

1. a. 400 mV
   b. 200 mV
   c. 141 mV
   d. 127 mV
   e. 6 ms
   f. 167 Hz

2. a. 7.6 V
   b. 3.8 V
   c. 2.69 V
   d. 2.42 V
   e. 200 μs
   f. 5 kHz

3. a. 22 V
   b. 11 V
   c. 7.78 V
   d. 7 V
   e. 120 ms
   f. 8.33 Hz

4. a. 1 V
   b. 500 mV
   c. 354 mV
   d. 318 mV
   e. 360 ms
   f. 2.78 Hz

5. a. 11.6 V
   b. 5.8 V
   c. 4.1 V
   d. 3.69 V
   e. 2.4 ms
   f. 417 Hz

6. a. 32 V
   b. 20 ms
   c. 50 Hz
   d. 48 V
   e. 24 ms
   f. 41.7 Hz

7. a. 1.2 V
   b. 480 μs
   c. 2080 Hz
   d. 200 mV
   e. 400 μs
   f. 2500 Hz

8. a. 30 mV
   b. 15 ms
   c. 66.7 Hz
   d. 18 mV
   e. 50 ms
   f. 20 Hz

9. a. 1.2 V
   b. 200 ms
   c. 5 Hz
   d. 2 V
   e. 100 ms
   f. 10 Hz

10. a. 240 mV
    b. 60 μs
    c. 17 kHz
    d. 30 V
    e. 60 μs
    f. 17 kHz

Note Because these drawings are difficult to read accurately, the answers have been rounded considerably.

Chapter 6

Page 107 (answers are underlined)

	Turns ratio (pri:sec)	Primary volts (rms)	Secondary volts (rms)	Primary current (A)	Secondary current (A)
1	10:1	120	12	0.1	1
2	6:1	120	20	0.1	0.6
3	1:5	30	150	1	0.2
4	1:4	24	96	4	1
5	3:1	81	27	0.2	0.6
6	1:7	17	119	0.21	0.03
7	2:5	10	25	0.3	0.12
8	3:4	120	160	1.7	1.3
9	4.5:2	120	53.3	0.5	1.1
10	7:6	28	24	0.18	0.21

Chapter 7

Pages 114–115

1. 1200 μH
2. 90 mH
3. 470 mH
4. 3.65 H
5. 66.7 mH
6. 200 mH
7. 220 mH
8. 180 mH
9. 220 mH
10. 180 mH

Page 121

1. $C_T = 5\ \mu F$
2. $C_T = 1015$ pF
3. $C_T = 0.231\ \mu F$
4. $C_T = 141\ \mu F$
5. $C_T = 500\ \mu F$
6. $C_T = 0.667\ \mu F$, $V_{C1} = 6.67$ V, $V_{C2} = 3.33$ V
7. $C_T = 3.33$ pF, $V_{C1} = 6.67$ V, $V_{C2} = 3.33$ V
8. $C_T = 0.0091\ \mu F$, $V_{C1} = 4.55$ V, $V_{C2} = 0.45$ V
9. $C_T = 3.33\ \mu F$, $V_{C1} = 5$ V, $V_{C2} = 5$ V, $V_{C3} = 5$ V
10. $C_T = 0.067\ \mu F$, $V_{C1} = 13.3$ V, $V_{C2} = 6.67$ V

Chapter 8

Pages 136–137

Note The times shown here for charge and discharge are all for one time constant. To show full charge and discharge, multiply by 5.

1. Charge time = 0.02s, full charge current = 0.15 A
   discharge time = 0.02s, discharge voltage = 15 V
2. Charge time = 2 ms, charge current = 15 mA
   discharge time = 0.2s, discharge voltage = 0.15 V
3. Charge time = 0.2s, charge current = 1.5 A
   discharge time = 0.2 ms, discharge voltage = 15,000 V
4. Charge time = 5 μs, charge current = 0.75 mA
   discharge time = 5 μs, discharge voltage = 15 V
5. Charge time = 30 ms, charge current = 0.12 A
   discharge time = 6.6 μs, discharge voltage = 54,000 V
6. Charge time = 5 ms, charge voltage = 12 volts
   discharge time = 5 ms, discharge current = 0.12 A
7. Charge time = 10 ms, charge voltage = 12 V
   discharge time = 25 ms, discharge current = 48 mA
8. Charge time = 375 ms, charge voltage = 15 V
   discharge time = 150 ms, discharge current = 15 mA

9. Charge time = 10 ms, charge voltage = 15 V
   discharge time = 20 μs, discharge current = 0.75 A
10. Charge time = 1.25s, charge voltage = 1.5 V
    discharge time = 500 μs, discharge current = 1.5 A

Chapter 10

Page 151

1. 628 Ω	2. 15.9 mH	3. 3.2 mH	4. 1990 Hz	5. 5970 Hz
6. 62.8 kΩ	7. 6.4 mH	8. 9950 Hz	9. 942 Ω	10. 0 Ω

Page 160

1. L_T = 6H, X_{L1} = 377 Ω, X_{L2} = 754 Ω, X_{L3} = 1131 Ω, X_{LT} = 2262 Ω

2. L_T = 140 mH, X_{L1} = 7.54 Ω, X_{L2} = 15.1 Ω, X_{L3} = 30 Ω, X_{LT} = 52.6 Ω

3. Z = 141 Ω, I = 0.71 A, V_R = 70.7 V, V_L = 70.7 V, Θ = 45°

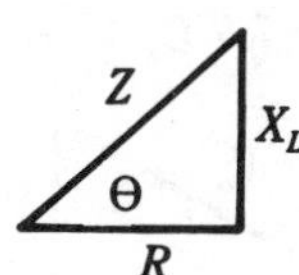

4. Z = 55.9 Ω, I = 1.79 A, V_R = 44.7 V, V_L = 89.4 V, Θ = 63.4°

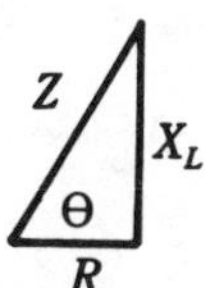

5. Z = 100.5 Ω, I = 0.99 A, V_R = 9.9 V, V_L = 99.5 V, Θ = 84.3°

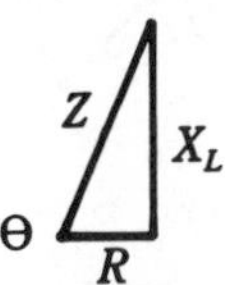

6. Z = 75.2 Ω, I = 1.33 A, V_R = 99.8 V, V_L = 6.65 V, Θ = 3.8°

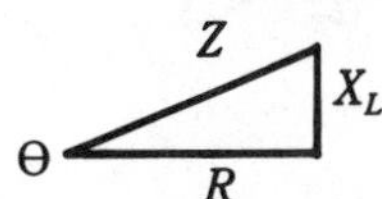

7. $Z = 79\ \Omega$, $I = 1.26$ A, $V_R = 94.9$ V, $V_L = 31.6$ V, $\Theta = 18.4°$

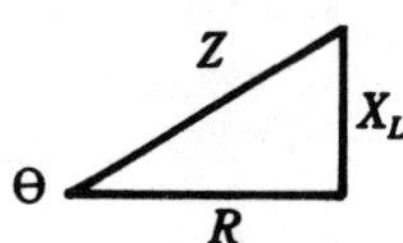

8. $Z = 50\ \Omega$, $I = 2$ A, $V_R = 0$ V, $V_L = 100$ V, $\Theta = 90°$

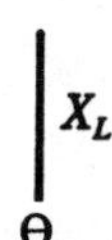

9. $R = 80\ \Omega$, $X_L = 160\ \Omega$, $Z = 179\ \Omega$, $V_a = 44.7$ V, $\Theta = 63.4°$

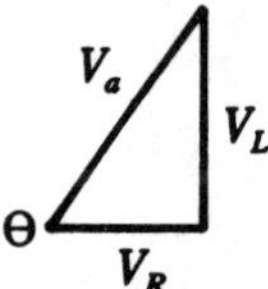

10. $R = 75\ \Omega$, $X_L = 45\ \Omega$, $Z = 87.7\ \Omega$, $V_a = 29.2$ V, $\Theta = 31°$

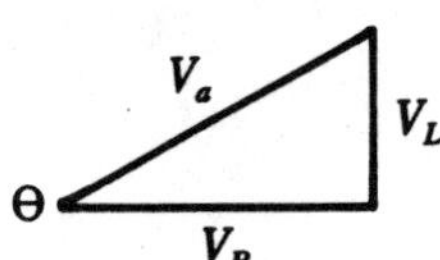

Pages 165 – 166

1. $L_T = 0.55$ H, $X_{L1} = 314\ \Omega$, $X_{L2} = 628\ \Omega$, $X_{L3} = 942\ \Omega$, $X_{LT} = 171\ \Omega$

2. $L_T = 11.4$ mH, $X_{L1} = 6.28\ \Omega$, $X_{L2} = 12.6\ \Omega$, $X_{L3} = 25\ \Omega$, $X_{LT} = 3.6\ \Omega$

3. $I_R = 1$ A, $I_L = 1$ A, $I_T = 1.4$ A, $\Theta = -45°$, $Z = 70.7\ \Omega$

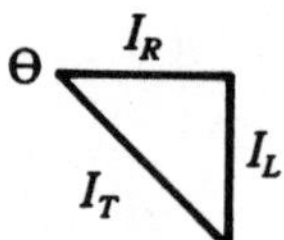

4. $I_R = 4$ A, $I_L = 2$ A, $I_T = 4.5$ A, $\Theta = -26.6°$, $Z = 22.4\ \Omega$

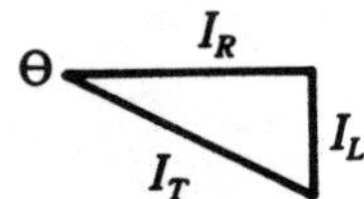

5. $I_R = 10$ A, $I_L = 1$ A, $I_T = 10.05$ A, $\Theta = -5.7°$, $Z = 10\ \Omega$

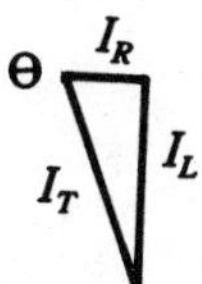

6. $I_R = 1.3$ A, $I_L = 20$ A, $I_T = 20.04$ A, $\Theta = -86.3°$, $Z = 5\ \Omega$

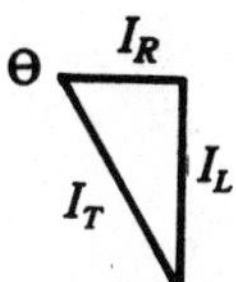

7. $I_R = 1.3$ A, $I_L = 4$ A, $I_T = 4.2$ A, $\Theta = -72°$, $Z = 23.8\ \Omega$

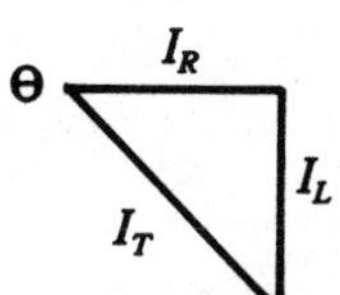

8. $I_R = 20$ A, $I_L = 20$ A, $I_T = 28.3$ A, $\Theta = -45.°$, $Z = 3.5\ \Omega$

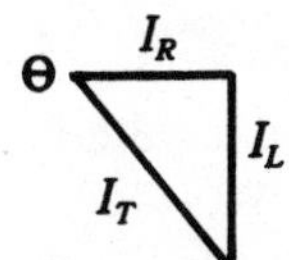

9. $R = 100\ \Omega$, $X_L = 33.3\ \Omega$, $I_T = 3.16$ A, $\Theta = -71.6$, $Z = 31.6\ \Omega$

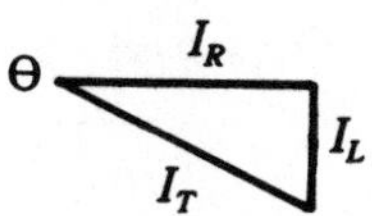

10. $R = 4$ kΩ $= 4$ kΩ, $X_L = 10$ kΩ, $I_T = 27$ mA, $\Theta = -21.8°$, $Z = 3.7$ kΩ

Page 169

1. $Z = 707\ \Omega$, $I = 0.141$ A, $V_R = 70.7$ V, $V_L = 70.7$ V, $\Theta = 45°$, $P_R = 10$ W, $P_X = 10$ VARS, $P_A = 14.1$ VA, $PF = 0.707$

2. $Z = 269\ \Omega$, $I = 0.372$ A, $V_R = 37.2$ V, $V_L = 92.9$ V, $\Theta = 68.2°$, $P_R = 13.8$ W, $P_X = 34.6$ VARS, $P_A = 37.2$ VA, $PF = 0.371$

3. $Z = 1250\ \Omega$, $I = 0.08$ A, $V_R = 80$ V, $V_L = 60$ V, $\Theta = 36.9°$, $P_R = 6.4$ W, $P_X = 4.8$ VARS, $P_A = 8$ VA, $PF = 0.8$

4. $Z = 1118\ \Omega$, $I = 0.089$ A, $V_R = 89.4$ V, $V_L = 44.7$ V, $\Theta = 26.6°$, $P_R = 8$ W, $P_X = 4$ VARS, $P_A = 8.94$ VA, $PF = 0.894$

5. $Z = 22.4\ \Omega$, $I = 4.47$ A, $V_R = 44.7$ V, $V_L = 89.4$ V, $\Theta = 63.4°$, $P_R = 200$ W, $P_X = 400$ VARS, $P_A = 446$ VA, $PF = 0.448$

6. $I_R = 0.2$ A, $I_L = 0.2$ A, $I_T = 0.283$ A, $Z = 353\ \Omega$, $\Theta = -45°$, $P_R = 20$ W, $P_X = 20$ VARS, $P_A = 28.3$ VA, $PF = 0.707$

7. $I_R = 1$ A, $I_L = 0.4$ A, $I_T = 1.08$ A, $Z = 92.8\ \Omega$, $\Theta = -21.8°$, $P_R = 100$ W, $P_X = 40$ VARS, $P_A = 108$ VA, $PF = 0.928$

8. $I_R = 0.1$ A, $I_L = 0.133$ A, $I_T = 0.166$ A, $Z = 600\ \Omega$, $\Theta = -53.1°$, $P_R = 10$ W, $P_X = 13.3$ VARS, $P_A = 16.6$ VA, $PF = 0.6$

9. $I_R = 0.1$ A, $I_L = 0.2$ A, $I_T = 0.224$ A, $Z = 447\ \Omega$, $\Theta = -63.4°$, $P_R = 10$ W, $P_X \neq$ *20 VARS*, $P_A = 22.4$ VA, $PF = 0.447$

10. $I_R = 10$ A, $I_L = 5$ A, $I_T = 11.2$ A, $Z = 8.94\ \Omega$, $\Theta = -26.6°$, $P_R = 1000$ W, $P_X = 500$ VARS, $P_T = 1120$ VA, $PF = 0.894$

Pages 175–179

1. a. 10 div.
   b. 36°
   c. 1.4 div.
   d. 50.4°
2. a. 10 div.
   b. 36°
   c. 1.0 div.
   d. 36°
3. a. 9 div.
   b. 40°
   c. 1.0 div.
   d. 40°
4. a. 9 div.
   b. 40°
   c. 0.5 div.
   d. 20°
5. a. 8 div.
   b. 45°
   c. 1.2 div.
   d. 54°
6. a. 8 div.
   b. 45°
   c. 0.4 div.
   d. 18°
7. a. 7 div.
   b. 51.4°
   c. 1.0 div.
   d. 51.4°
8. a. 7 div.
   b. 51.4°
   c. 1.5 div.
   d. 77°
9. a. 6 div.
   b. 60°
   c. 0.4 div.
   d. 24°
10. a. 6 div.
    b. 60 °
    c. 0.8 div.
    d. 48°

Chapter 11

Page 184

1. 159 kΩ
2. 0.1 μF
3. 10 kHz
4. 0.01 μF
5. 1 kHz
6. 0.106 Ω
7. infinity
8. 26.5 Ω
9. 31.8 Ω
10. 1.59 kΩ

Pages 191 – 192

1. C_T = 0.009 μF, X_{C1} = 2650 Ω, X_{C2} = 265,000 Ω, C_{C3} = 26,500 Ω, X_{CT} = 294,000 Ω

2. C_T = 26.4 μF, X_{C1} = 56.4 Ω, X_{C2} = 17.7 Ω, X_{C3} = 26.5 Ω, X_{CT} = 100 Ω

3. Z = 141 Ω, I = 0.71 A, V_R = 70.7 V, V_C = 70.7 V, Θ = −45°

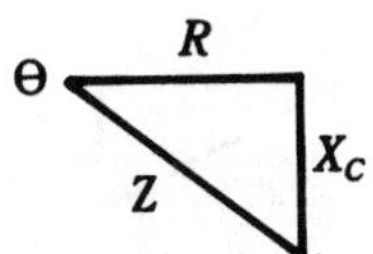

4. Z = 55.9 Ω, I = 1.79 A, V_R = 44.7 V, V_C = 89.4 V, Θ = −63.4°

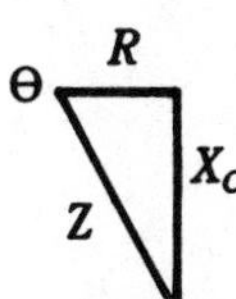

5. Z = 100.5 Ω, I = 0.99 A, V_R = 9.9 V, V_C = 99.5 V, Θ = −84.3°

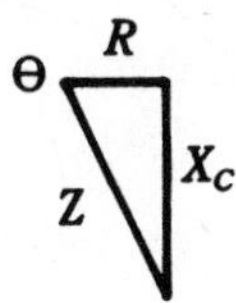

6. Z = 75.2 Ω, I = 1.33 A, V_R = 99.8 V, V_C = 6.65 Θ = −3.8°

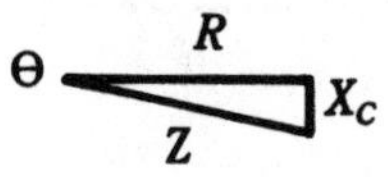

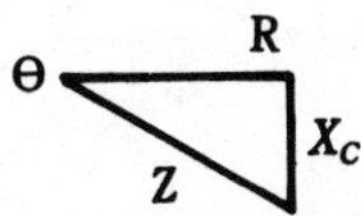

7. Z = 79 Ω, I = 1.26 A, V_R = 94.9 V, V_C = 31.6 V, Θ = −18.4°

8. $Z = 50\ \Omega$, $I = 2$ A, $V_R = 0$ V, $V_C = 100$ V, $\Theta = -90°$

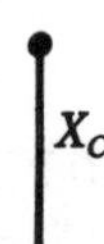

9. $R = 80\ \Omega$, $X_C = 160\ \Omega$, $Z = 179\ \Omega$, $V_a = 44.7$ V, $\Theta = -63.4°$

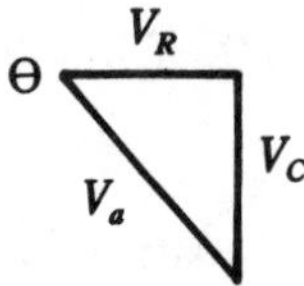

10. $R = 75\ \Omega$, $X_C = 45\ \Omega$, $Z = 87.5\ \Omega$, $V_a = 29.2$ V, $\Theta = -31°$

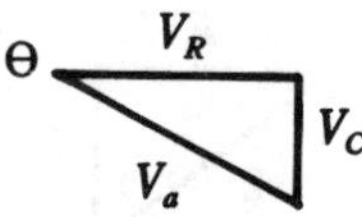

Pages 197 – 198

1. $C_T = 6\ \mu$F, $X_{C1} = 1590\ \Omega$, $X_{C2} = 796\ \Omega$, $X_{C3} = 531\ \Omega$, $X_{CT} = 265\ \Omega$

2. $C_T = 225\ \mu$F, $X_{C1} = 31.8\ \Omega$, $X_{C2} = 15.9\ \Omega$, $X_{C3} = 21.2\ \Omega$, $X_{CT} = 7.07\ \Omega$

3. $I_R = 1$ A, $I_C = 1$ A, $I_T = 1.4$ A, $\Theta = 45°$, $Z = 70.7\ \Omega$

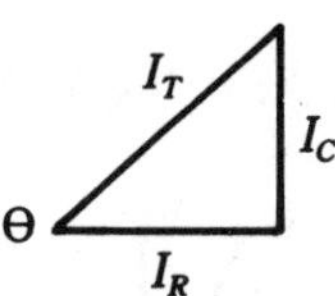

4. $I_R = 4$ A, $I_C = 2$ A, $I_T = 4.5$ A, $\Theta = 26.6°$, $Z = 22\ \Omega$

I_T I_C Θ I_R

5. $I_R = 10$ A, $I_C = 1$ A, $I_T = 10.05$ A, $\Theta = 5.7°$, $Z = 10\ \Omega$

I_T I_C Θ I_R

6. I_R = 1.3 A, I_L = 20 A, I_T = 20.04, Θ = 86.3°, Z = 5 Ω

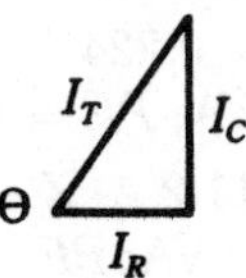

7. I_R = 1.3 A, I_C = 4 A, I_T = 4.2 A, Θ = 72°, Z = 23.8 Ω

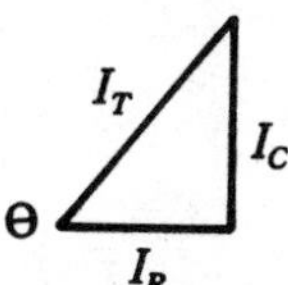

8. I_R = 20 A, I_C = 20 A, I_T = 28.3 A, Θ = 45°, Z = 3.5 Ω

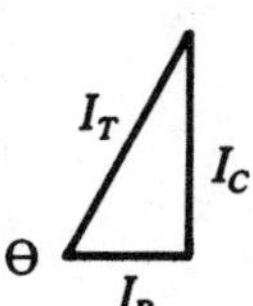

9. R = 100 Ω, X_C = 33.3 Ω, I_T = 3.16 A, Θ = 71.6°, Z = 3.16 Ω

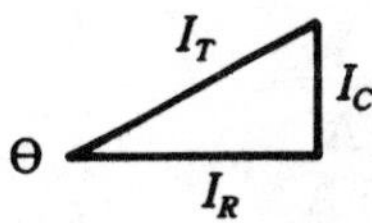

10. R = 4 kΩ, X_C = 10 kΩ, I_T = 27 mA, Θ = 68.2°, Z = 3.7 kΩ

I_T / I_C / Θ / I_R

Pages 199–200

1. Z = 707 Ω, I = 0.141 A, V_R = 70.7 V, V_C = 70.7 V, Θ = −45°, P_R = 10 W, P_X = 10 VARS, P_A = 14.1 VA, PF = 0.707
2. Z = 269 Ω, I = 0.371 A, V_R = 37.1 V, V_C = 92.9 V, Θ = −68.2°, P_R = 13.8 W, P_X = 34.5 VARS, P_A = 37.1 VA, PF = 0.371
3. Z = 1250 Ω, I = 0.08 A, V_R = 80 V, V_C = 60 V, Θ = −36.9°, P_R = 6.4 W, P_x = 4.8 VARS, P_A = 8 VA, PF = 0.8
4. Z = 1118 Ω, I = 0.089 A, V_R = 89.4 V, V_C = 44.7 V, Θ = −26.6°, P_R = 8 W, P_X = 4 VARS, P_A = 8.94 VA, PF = 0.894

5. $Z = 22.4\ \Omega$, $I = 4.47$ A, $V_R = 44.7$ V, $V_C = 89.4$ V, $\Theta = -63.4°$, $P_R = 200$ W, $P_x = 400$ VARS, $P_A = 447$ VA, $PF = 0.447$
6. $I_R = 0.2$ A, $I_C = 0.2$ A, $I_T = 0.283$ A, $Z = 354\ \Omega$, $\Theta = 45°$, $P_R = 20$ W, $P_X = 20$ VARS, $P_A = 28.3$ VA, $PF = 0.707$
7. $I_R = 1$ A, $I_C = 0.4$ A, $I_T = 1.08$ A, $Z = 92.6\ \Omega$, $\Theta = 21.8°$, $P_R = 100$ W, $P_X = 40$ VARS, $P_A = 108$ VA, $PF = 0.928$
8. $I_R = 0.1$ A, $I_C = 0.133$ A, $I_T = 0.167$ A, $Z = 600\ \Omega$, $\Theta = 53.1°$, $P_R = 10$ W, $P_X = 13.3$ VARS, $P_A = 16.7$ VA, $PF = 0.6$
9. $I_R = 0.1$ A, $I_C = 0.2$ A, $I_T = 0.224$ A, $Z = 447\ \Omega$, $\Theta = 63.4°$, $P_R = 10$ W, $P_X = 20$ VARS, $P_A = 22.4$ VA, $PF = 0.448$
10. $I_R = 10$ A, $I_C = 5$ A, $I_T = 11.2$ A, $Z = 8.94\ \Omega$, $\Theta = 26.6°$, $P_R = 1000$ W, $P_X = 500$ VARS, $V_T = 1120$ VA, $PF = 0.894$

Chapter 12

Page 219

1. $X_{net} = 75\ \Omega$
   $I_T = 0.111$ A
   $Z = 90\ \Omega$
   net L
   $\Theta = +56.3°$

2. $X_{net} = 150\ \Omega$
   $I_T = 0.055$ A
   $Z = 180\ \Omega$
   net c
   $\Theta = -56.3°$

3. $X_{net} = 5$ kΩ
   $I_T = 1.28$ mA
   $Z = 7.81$ kΩ
   net c
   $\Theta = -39.8°$

4. $X_{net} = 10\ \Omega$
   $I_T = 0.446$ A
   $Z = 22.4\ \Omega$
   net C
   $\Theta = -26.6°$

5. $X_{net} = 200\ \Omega$
   $I_T = 0.035$ A
   $Z = 282\ \Omega$
   net L
   $\Theta = +45°$

6. $I_C = 1$ A
   $I_L = 0.5$ A
   $I_{net} = 0.5$ A
   $X_{net} = 20\ \Omega$
   $I_R = 1$ A
   $I_T = 1.12$ A
   $\Theta = +26.6°$
   net C

7. $I_C = 0.05$ A
   $I_L = 0.1$ A
   $I_{net} = 0.05$ A
   $X_{net} = 200\ \Omega$
   $I_R = 0.1$ A
   $I_T = 0.112$ A
   $\Theta = -26.5°$
   net L

8. $I_C = 0.02$ A
   $I_L = 0.005$ A
   $I_{net} = 0.015$ A
   $X_{net} = 667\ \Omega$
   $I_R = 0.01$ A
   $I_T = 0.018$ A
   $\Theta = +56.3°$
   net C

9. $I_{C1} = 0.5$ A
   $I_{C2} = 0.5$ A
   $I_{L1} = 0.25$ A
   $I_{L2} = 0.2$ A
   $I_{net} = 0.55$ A
   $X_{net} = 18.2\ \Omega$
   $I_R = 1$ A
   $I_T = 1.14$ A
   $\Theta = +28.8°$
   net C

10. $I_{C1} = 0.2$ A
    $I_{C2} = 0.25$ A
    $I_{L1} = 0.5$ A
    $I_{L2} = 1$ A
    $I_{net} = 1.05$ A
    $X_{net} = 9.5\ \Omega$
    $I_R = 0.5$ A
    $I_T = 1.16$ A
    $\Theta = -64.5°$
    net L

Pages 229–230

1. $Z = 178\underline{/47.7}$ $I = 2.47\underline{/-47.7}$
2. $I = 10\underline{/53.1}$ $Z = 10\underline{/-53.1}$ $V_R = 60\underline{/53.1}$ $V_C = 80\underline{/-36.9}$

3. $Z = 318\angle -56.6$ $I = 0.472\angle 56.6$
4. $Z = 123\angle -35.8$ $I = 1.79\angle 35.8$ $V_R = 179\angle 35.8$ $V_L = 236\angle 125.8$ $V_C = 356\angle -54.2$
5. $Z = 660\angle -83.5$ $I = 0.33\angle 83.5$
6. $I_R = 6\angle 0$ $I_L = 8\angle -90$ $I_T = 10\angle -53.1$ $Z = 12\angle 53.1$
7. $I_R = 6\angle 0$ $I_C = 8\angle 90$ $I_T = 10\angle -53.1$ $Z = 12\angle 53.1$
8. $Z = 1560\angle 38.7$
9. $Z = 44\angle 22.9$
10. $Z = 77.2\angle -29.2$
11. $Z = 53.2\angle 90$
12. $Z = 144\angle 1.52$
13. $Z = 40.2\angle 41.5$
14. $Z = 39.2\angle 25$
15. $Z = 38.2\angle 12.5$

Chapter 13

Page 243

1. $f_r = 12.6$ Hz
2. $f_r = 1.624$ MHz
3. a. $Q = 100$
   b. $X_C = X_L = 1500\ \Omega$
   c. $I = 1$ mA
   d. $V_C = 1.5$ V
4. $L = 254\ \mu$H at 1 MHz
   $L = 15.9\ \mu$H at 4 MHz
5. $C_{max} = 32.7$ pF
   $C_{max} = 32.7$ pF
   $C_{min} = 21.7$ pF

Chapter 14

Page 269

1. $V_{dc} = 28.8$ V, $I_T = 275$ mA, $V_Z = 12$ V, $I_Z = 275$ mA, $P_Z = 3.3$ W, $V_{R2} = 16.8$ V, $R_S = 61.1\ \Omega$, $P_{RS} = 4.62$ W
2. $V_{dc} = 21.6$ V, $I_T = 55$ mA, $V_Z = 9$ V, $I_Z = 55$ mA, $P_Z = 49.5$ mW, $V_{RS} = 12.6$ V, $R_S = 229\ \Omega$, $P_{RS} = 69.3$ mW
3. $V_{dc} = 14.4$ V, $I_T = 825$ mA, $V_Z = 6$ V, $I_Z = 825$ mA, $P_Z = 4.95$ W, $V_{RS} = 8.4$ V, $R_S = 10.2\ \Omega$, $P_{RS} = 6.93$ W
4. $V_{dc} = 36$ V, $I_T = 1.65$ A, $V_Z = 15$ V, $I_Z = 1.65$ A, $P_Z = 24.75$ W, $V_{RS} = 21$ V, $R_S = 12.7\ \Omega$, $P_{RS} = 34.65$ W
5. $V_{dc} = 72$ V, $I_T = 165$ mA, $V_Z = 24$ V, $I_Z = 165$ mA, $P_Z = 3.96$ W, $V_{RS} = 48$ V, $R_S = 291\ \Omega$, $P_{RS} = 7.92$ W

Chapter 15

Pages 294–296

1. (a) $I_{C(SAT)} = 3.68$ mA
   V_{CE} *(cut-off)* = 25 V
   (c) $I_{C(Q)} = 1.99$ mA
   $V_{CE(Q)} = 11.5$ V
   (d) $I_{B(SAT)} = 21\ \mu A$
   (e) $I_{B(Q)} = 11.4\ \mu A$

2. (a) $I_{C(SAT)} = 10$ mA
   V_{CE} *(cut-off)* = 15 V
   (c) $I_{C(Q)} = 4.84$ mA
   $V_{CE(Q)} = 7.74$ V
   (d) $I_{B(SAT)} = 66.7\ \mu A$
   (e) $I_{B(Q)} = 32.3\ \mu A$

3. (a) $I_{C(SAT)} = 12.5$ mA
   V_{CE} *(cut-off)* = 25 V
   (c) $I_{C(Q)} = 5.95$ mA
   $V_{CE(Q)} = 13.1$ V
   (d) $I_{B(SAT)} = 83.3\ \mu A$
   (e) $I_{B(Q)} = 39.7\ \mu A$

4. (a) $I_{C(SAT)} = 5.71$ mA
   V_{CE} *(cut-off)* = 10 V
   (c) $I_{BIAS} = 8.7$ mA
   (d) $V_B = 2/87$ V
   (e) $V_E = 2.27$ V
   (f) $I_{C(Q)} = 3.03$ mA
   $V_{CE(Q)} = 4.7$ V

Chapter 16

Page 304

Note The given number is in ( ).

	decimal	binary	octal	hexadecimal
(1)	(56)	111000	70	38
(2)	184	(10111000)	270	B8
(3)	90	1011010	(132)	5A
(4)	125	1111101	175	(7D)
(5)	(75)	1001011	113	4B
(6)	81	(1010001)	121	51
(7)	66	1000010	(102)	42
(8)	49	110001	61	(31)
(9)	(37)	100101	45	25
(10)	101	(1100101)	145	65
(11)	113	1110001	(161)	71
(12)	25	11001	31	(19)
(13)	(50)	110010	62	32
(14)	128	(10000000)	200	80
(15)	93	1011101	(135)	5D
(16)	77	1001101	115	(4D)
(17)	(84)	1010100	124	54

(18)	45	(101101)	55	2D
(19)	64	1000000	(100)	40
(20)	256	100000000	400	(100)

Index

A

ac circuits
  complex circuits, *j*-operator calculations, 227-230
  parallel configurations, *j*-operator calculations, 226-227
  series configuration, *j*-operator calculations, 223-226
ac conditions, common emitter amplifier, transistors, 288-291
active region, transistors, 275-276
addition
  engineering-notation numbers, 13
  signed numbers, 3-4
amplifiers
  180-degree phase shift, 290-291
  common-base, 294
  common-collector, 292
  common-emitter, 292
  configurations, transistors, 291-294
amplitude, 90
apparent power, 167, 168, 199, 219
average value, sine waves, 90, 91

B

back emf, 111
bandwidth, resonance/resonant circuits, 240-242
base bias, transistors, 277
  emitter feedback, 279-282
base current, transistors, 279, 282, 284, 286
base region, transistors, 271-272
batteries, dual-battery dc circuits, 50-51
bell curve, resonance/resonant circuits, 240-242
beta ratio, transistors current relationship, 273
bias
  PN junction diodes, 246-248
  transistors, 271-272
  transistors, base, 277
  transistors, base, emitter feedback, 279-282
  transistors, voltage-divider bias, beta independent, 284-288
binary number system and conversions, 297-304
bridge circuits, Wheatstone bridge, 52-53
bridge rectifiers, 252-253
  filtering, 261

C

capacitance/capacitors, 116-121
  capacitive reactance (*see* capacitive reactance)
  current flow, 116
  dielectrics, 116
  farad measurement of capacitance, 117
  formulas, 122
  high-current production, RC circuits, 134-135
  integration, RC time constant circuits, 139-140
  leakage, 117
  net reactance, 213-218
  parallel configurations, 118
  RC time constant, 132-134
resonance/resonance circuit (*see* resonance/resonance circuits)
  series configurations, 118-119
  voltage charge/store, 116
  voltage dividers, 119
capacitive coupling, ac signals, transistors, 288-289
capacitive reactance, 183-211
  apparent power, 199
  formulas, 211
  impedance, 189
  parallel configurations, 185
  parallel configurations and resistance (R), 192-198
  parallel configurations and resistance (R), complete calculations, 195
  phase angle, oscilloscope measurement/calculation,

capacitive reactance (*cont.*)
200-210
phase relationships, 186-189, 192-195
phasor diagrams, 186-189, 192-195
power factors, 199
power measurements, 198-200
reactive power, 199
real or true power, 198
series configuration and resistance (R), complete calculations, 189
series configuration and resistance (R), impedance, 189
series configurations, 185
series configurations and resistance (R), 185-192
vector analysis, sine waves, 186-189, 192-195
characteristic curves, transistors, 273-274
charge, 111
universal time constant curves, 124-126
coefficient of coupling, transformers, 107
collector feedback, transistors, 282-284
collector region, transistors, 271-272
common-base amplifier, 294
common-collector amplifier, 292
common-emitter amplifier, 292
complex dc circuits, series-and-parallel, 38-47
current flow, 38-44
power dissipation, 41-44
resistance, 38-44
voltage, 38-44
complex numbers
calculation rules, 223, 232
polar form, *j*-operator, 222-223
rectangular form, *j*-operator, 221-222
converting
binary/decimal/hex/octal numbers, 297-304
coupling
capacitive, ac signals, transistors, 288-289
coefficient of coupling, transformers, 107
current
base current, transistors, 279, 282, 284, 286
capacitance/capacitors flow, 116
complex dc circuits, 38-44
direct relationship with power, resistance, voltage, 22-23
dual-battery dc circuits, 51
emitter current, transistors, 279, 287
high-current production, RC circuits, 134-135
indirect relationship with power, resistance, voltage, 22-23
inductance, 111-112
Ohm's law, 15-18
parallel-circuit (dc) flow, 30-31
power formulas, 18-21
series-circuit (dc) flow, 23-30
transformers, current ratio, 106-107
transistors, current relationships, 272-274
volt-ohm-milliammeters measurements, 59-73
current ratio, transformers, 106-107
cut-off, transistors, 275
cycle, sine waves, 88

D

dc circuits, 15-58
balancing a Wheatstone bridge, 53
complex circuits, series-and-parallel, 38-47
dual-battery circuits, 50-51
Ohm's law, 15-18
parallel circuits, 30-38
power formulas, 18-21
power transfer, maximum, 53-54
series circuits, 23-30
voltage-divider circuits, 47-50
Wheatstone bridges, 52-53
dc conditions, transistor amplifiers, 276-279
dc equivalent value (*see* rms value)
decimal number system and conversions, 297-304
depletion zone, PN junction diode, 246-248
dielectrics, capacitance/capacitors, 116
differentiation, RC time constant circuits, 139-140
diodes, PN junction, 245-248
discharge, universal time constant curves, 126-128
division
engineering-notation numbers, 10-12
signed numbers, 4-5
doping, semiconductors, 245
dual-battery dc circuits, 50-51

E

effective value (*see* rms value)
efficiency, transformers, 108-109
electromotive force (emf), back emf, 111
emitter current, transistors, 279, 287
emitter region, transistors, 271-272
emitter voltage, transistors, 287
engineering notation, 1, 8-13
addition and subtraction, 13
multiplication and division, 10-12
writing numbers, 8-9

F

farad, unit of capacitance, 117
feedback
collector feedback, transistors, 282-284
emitter feedback, transistors, 279-282
figure of merit (*see* Q measurement)
filters, power supplies, rectifier circuits, 256-261
forward bias, PN junction diode, 246-248
frequency
inductive reactance, 149-151
oscilloscope measurement, 98
resonance/resonant circuits (*see* resonance/resonant circuits)
sine waves, 88-89, 93
full-wave rectifiers, 249-251
filtering, 260-261

G

generators, ac, sine waves generation, 85-87

H

half-wave rectifiers, 248-249
filtering, 258
henry, unit of inductance, 112
hexadecimal number system and

conversions, 297-304

I

impedance
  capacitive reactance, 189
  inductive reactance, 156-157
inductance/inductors, 111-115
  back emf, 111
  formulas, 122
  henry measurement of inductance, 112
  inductive reactance (*see* inductive reactance), 149-181
  L/R time constant, 128-131
  mutual inductance, 112-113
  net reactance, 213-218
  parallel configurations, 114-115
  phasing dots indicate polarity, 113
  resonance/resonant circuits (*see* resonance/resonant circuits)
  RL circuits, high-voltage production, 131-132
  series configurations, 112-113
inductive reactance
  apparent power, 167, 168
  formulas, 180-181
  frequency relationships, 149-151
  impedance, 156-157
  parallel configuration and resistance (R), 160-166
  parallel configurations, 151-152
  phase angle, oscilloscope measurement/calculation, 169-179
  phase relationships, 154-156, 160-166
  phasor diagrams, 154-156, 160-166
  power factors, 167, 168
  power measurements, 166-169
  reactive power, 166-167
  resistive power, 167, 168
  series configuration and resistance (R), 152-160
  series configuration and resistance (R), complete calculations, 157
  series configuration and resistance (R), impedance, 156-157
  series configurations, 151-152
  sine wave analysis, series configuration and R, 153-154
  true or real power, 166
  vector analysis, sine waves, 154-156, 160-166
integration, RC time constant circuits, 139-140

J

j-operator, 213, 220-231
  complex ac circuits, 227-230
  complex numbers, calculation rules, 223, 231
  complex numbers, polar form, 222-223
  complex numbers, rectangular form, 221-222
  parallel ac circuits, 226-227
  series ac circuits, 223

L

lag voltage, 154, 186, 192
leakage, capacitance/capacitors, 117
load line, dc, transistors, 276, 278, 279, 282, 286, 290-291
long time constant circuit, RC, 144-147
losses, transformers, 108-109

M

medium time constant circuit, RC, 143-144
multiplication
  engineering-notation numbers, 10-12
  signed numbers, 4-5
mutual inductance, 112-113

N

negative numbers
  addition and subtraction, 3-4
  multiplication and division, 4-5
net reactance, 213-218
NPN-type transistors, 271-272
numbers/numbering systems, 297-304
  complex numbers, *j*-operator, 221-231
  conversions between systems, 300-304
  digits used, 207-298
  engineering notation, 8-13
  negative numbers, 3-5
  numbering systems, 297-304
  place values, 298-922
  positive numbers, 3-5
  rounding up and down, 2-3
  scientific notation, 5-8
  significant figures, 1-2

O

octal number system and conversions, 297-304
Ohm's law, 15-18
  power formulas used with Ohm's law, 19-21
ohmmeters, transformers testing, 109-110
oscillators, sine waves generation, 85
oscilloscopes, 75-83
  frequency measurements (sine waves), 98
  phase angle measurement/calculation, 169-179, 200-210
  sample measurements, 76-83
  screen divisions or blocks, 75
  sine waves measurement, 95-103
  voltage measurements, ac and dc, 75-76, 95-98

P

parallel circuits, dc, 30-38
  current flow, 30-31
  power dissipation, 30
  resistance, 31-35
  voltage, 30
parallel resonance/resonant circuits, 238-239
peak value, sine waves, 89, 90
peak-to-peak value, sine waves, 89, 91
period, sine waves, 88, 93
phase angle, oscilloscope measurement/calculation, 169-179, 200-210
phase shift, transistors, amplification with 180-degree phase shift, 290
phasing dot, inductance/inductors polarity, 113
phasor diagrams, 154-156, 160-166, 186-189, 192-195
PN junction diodes, 245-248
  biases, forward and reverse, 246-248
  depletion zone, 246-248
  rectifiers, bridge, 252-253
  rectifiers, full-wave, 249-251
  rectifiers, half-wave, 248-249
  zener regulators, 261-269

PNP-type transistors, 271-272
positive numbers
  addition and subtraction, 3-4
  multiplication and division, 4-5
power
  apparent, 167, 168, 199, 219
  capacitive reactance, 198-200
  complex dc circuits, 41-44
  direct relationship with current, resistance, voltage, 22-23
  indirect relationship with current, resistance, voltage, 22-23
  inductive reactance, reactive power, 166
  maximum transfer, dc circuits, 53-54
  parallel-circuit (dc) dissipation, 30
  power factor, 167, 168, 199, 219
  power formulas, 18-21
  reactive, 166, 199, 219
  real or true, 198, 219
  resistive, 167, 168
  RLC circuits, 219-220
  series-circuit (dc) dissipation, 23-30
  transformers power relationships, 108
power factors, 167, 168, 199, 219
power formulas, dc circuits, 18-21
  Ohm's law used with power formulas, 19-21
power supplies, 245-269
  block diagram, 263
  PN junction diodes, 245-248
  rectifiers, bridge, 252-253
  rectifiers, bridge, filtering, 261
  rectifiers, filtering, 256-261
  rectifiers, full-wave, 249-251
  rectifiers, full-wave, filtering, 260-261
  rectifiers, half-wave, 248-249
  rectifiers, half-wave, filtering, 258
  rectifiers, output calculations, 252-255
  zener regulators, 261-267
  zener regulators, calculating capacity, 264-266
  zener regulators, value selections, 266-267
powers of ten, scientific notation, 6

Q

Q measurement,
  resonance/resonant circuits, 239-240
Q point, transistors, 276, 278, 281, 283, 287
quality (*see* Q measurement)

R

RC time constant circuits, 132-134, 139-147
  differentiation, 139-140
  integration, 139-140
  long time constant circuit, 144-147
  medium time constant circuit, 143-144
  short time constant circuits, 140-142
reactance
  capacitive reactance (*see* capacitive reactance)
  inductive reactance (*see* inductive reactance), 149-181
  net reactance, 213-218
  sine waves, 85
reactive power, 166, 199, 219
real power, 166, 198, 219
rectifiers
  bridge, 252-253
  bridge, filtering, 261
  filtering, power supplies, 256-261
  full-wave, 249-251
  full-wave, filtering, 260-261
  half-wave, 248-249
  half-wave, filtering, 258
  output calculations, power supplies, 252-255
regulators, zener voltage regulators, 261-269
resistance (*see also* reactance)
  capacitive reactance (*see* capacitive reactance)
  complex dc circuits, 38-44
  differentiation, RC time constant circuits, 139-140
  direct relationship with current, power, voltage, 22-23
  dual-battery dc circuits, 50-51
  indirect relationship with current, power, voltage, 22-23
  inductive reactance (*see* inductive reactance), 149-181
  L/R time constant, 128-131
  maximum power transfer, dc circuits, 54
  Ohm's law, 15-18
  ohmmeter testing, transformers, 109-110
  parallel circuits (dc), 31-35
  RC time constant, 132-134
  RL circuits, high-voltage production, 131-132
  series circuits (dc), 24-25
  sine waves (*see also* reactance), 85-104
  volt-ohm-milliammeters measurements, 59-73
  voltage-divider circuits, dc, 47-50
  Wheatstone bridge, 52-53
resistive power, 167, 168
resonance/resonant circuits, 215, 233-243
  bandwidth and bell curve, 240
  parallel resonance, 238-239
  Q measurement, 239-240
  resonant frequency, 233-236
  resonant frequency, C calculated, 235
  resonant frequency, effect above and below, 236
  resonant frequency, L calculated, 235
  resonant frequency, LC combinations at 1000 kHz, 234
  series resonance/resonant circuits, 236-238
resonant frequency, 233-236
  C component calculations, 235
  effects above and below resonant frequency, 236
  L component calculations, 235
  LC combinations at 1000 kHz, 234
reverse bias, PN junction diode, 246-248
RL circuits, high-voltage production, 131-132
RLC circuits, 214-220
  apparent power, 219
  net reactance, 214-220
  parallel configurations, 217-218
  parallel configurations, vector relationships, 215-216
  power factors, 219
  power measurements, 219-220
  reactive power, 219
  resonant circuits, 215
  series configurations, 216-217
  series configurations, vector

relationships, 215-216
true or real power, 219
vector relationships, series and parallel configurations, 215-216
rms value, sine waves, 90, 91
rounding numbers, 2-3

S

saturation, transistors, 275
scientific notation, 1, 5-8
powers of ten, table, 6
semiconductors (*see also* transistors)
doping, 245
PN junction diodes, 245-248
series circuits, dc, 23-30
current flow, 23
power dissipation, 23, 26
resistance, 24-25
voltage drop, 23, 25
voltage-divider circuits, 47-50
series resonance/resonant circuits, 236-238
short time constant circuits, RC, 140-142
signed numbers
addition and subtraction, 3-4
multiplication and division, 4-5
significant figures, 1-2
sine waves, 85-104
amplitude, 90
average value, 90, 91
calculating values, 87
converting values: frequency/period/wavelength, 93
converting values: rms/peak/peak-to-peak/average, 90-95
cycle, 88
formulas, summary, 104
frequency, 88-89, 93
frequency, measurement on oscilloscope, 98
generator (ac) to produce sine waves, 85-87
inductive reactance, series configuration and R, 153-154
oscillator to produce sine waves, 85
oscilloscope measurement, 95-103
peak value, 89, 90
peak-to-peak value, 89, 91
period, 88, 93
phase angle
measurement/calculation on oscilloscope, 169-179, 200-210
phase relationships, 154-156, 160-166, 186-189, 192-195
phasor diagrams, 154-156, 160-166, 186-189, 192-195
plotting value on graphs, 87
rms values, 90, 91
vector analysis, 154-156, 160-166, 186-189, 192-195
voltage measurement on oscilloscope, 95-98
wavelength, 88, 93-94
step-up/step-down transformers, 106
subtraction
engineering-notation numbers, 13
signed numbers, 3-4

T

time constants, 123-137
charge curve plotting, universal curve, 124-126
discharge curve plotting, universal curve, 126-128
L/R time constant, 128-131
long time constant circuit, RC, 144-147
medium time constant circuit, RC, 143-144
RC time constant, 132-134
RL circuits, high-voltage production, 131-132
short time constant circuit, RC, 140-142
universal time constant curve, 123-131
transformers, 105-110
coefficient of coupling, 107
current ratio, 106-107
current, primary, power relationship, 108
efficiency rating, 108-109
losses, 108-109
power relationships, 108
step-up/step-down, 106
testing with ohmmeter, 109-110
turns ratio, 105-106
voltage ratio, 106
voltage, primary, power relationship, 108
transistors, 271-296
ac conditions of common emitter amplifier, 288-291
ac signal passing, 289-290
active region, 275-276
amplifier configurations, 291-294
base bias, 277
base bias, emitter feedback, 279-282
base current, 279, 282, 284, 286
base region, 271-272
beta relationship, current, 273
biasing, 271-272
capacitive coupling, ac signals, 288-289
characteristic curves, 273-274
collector feedback, 282-284
collector region, 271-272
common-base amplifier, 294
common-collector amplifier, 292
common-emitter amplifier, 292
current relationships, 272-274
cut-off, 275
dc conditions of transistor amplifiers, 276-279
dc load line, 276, 278, 279, 282, 286, 290-291
emitter current, 279
emitter region, 271-272
emitter voltage and current, 287
NPN-type, 271-272
phase shift, amplification and 180-degree phase shift, 290
PNP-type, 271-272
Q point, 276, 278, 281, 283, 287
range of operation, 274-276
saturation, 275
voltage drop, 286
voltage, collector and base to ground, 281
voltage-divider bias, beta independent, 284-288
true power, 166, 198, 219
turns ratio, transformers, 105-106

U

universal time constant curves, 123-131

V

vector analysis, 154-156, 160-166, 186-189, 192-195
volt-ohm-milliammeters, 59-73
ohms scale readings, 61-62
range, 59
sample readings, 62-73

scale arc or line markings, 61
scale line value, 61
scale or meter face (DCV, ACV, DCmA), 59-61
scale with decimal adjusted, 60-61

voltage
capacitance/capacitors charge/store, 116
complex dc circuits, 38-44
direct relationship with current, power, resistance, 22-23
drop, transistors, 286
dual-battery dc circuits, 50-51
emitter voltage, transistors, 287
high-voltage production, RL circuits, 131-132
indirect relationship with current, power, resistance, 22-23
inductance (induced voltage), 111-112
inductive reactance (*see* inductive reactance), 149-181
lag voltage, 154, 186, 192
Ohm's law, 15-18
oscilloscopes measurements, ac and dc voltage, 75-76, 95-98
parallel-circuit (dc), 30
power formulas, 18-21
series-circuit (dc) drop, 23-30
sine waves, 85-104
transformers voltage ratio, 106
transistors, collector and base to ground, 281
volt-ohm-milliammeters measurements, 59-73
voltage-divider circuits, biasing, 284-288
voltage-divider circuits, capacitive, 119-120
voltage-divider circuits, dc, 47-50
waveforms, oscilloscopes measurement, 75-83
Wheatstone bridge, 52-53
zener regulators, 261-267

voltage ratio, transformers, 106

voltage-divider circuits, 47-50
biasing, transistors, 284-288
capacitive, 119-120, 119

W

waveforms
amplitude, 90
average values, 90
cycle, 88
frequency, 88-89
oscilloscopes measurement/monitoring, 75-83
peak values, 89
peak-to-peak values, 89
period, 88
rms values, 90
sine waves (*see* sine waves), 85-104
wavelength, 88
waveshaping circuits, RC, 139-147

wavelength, sine waves, 88, 93-94

waveshaping circuits, RC, 139-147

Wheatstone bridge, 52-53

Z

zener regulators, 261-267
capacity calculations, power supplies, 264-266
power supplies block diagram and use, 263
value selections, power supplies, 266-267

Other Bestsellers of Related Interest

HOW TO READ ELECTRONIC CIRCUIT DIAGRAMS—2nd Edition—Robert M. Brown, Paul Lawrence, and James A. Whitson

In this updated edition of a classic handbook, the authors take an unhurried approach to the task. Basic electronic components and their schematic symbols are introduced early. More specialized components, such as transducers and indicating devices, follow—enabling you to learn how to use block diagrams and mechanical construction diagrams. Before you know it, you'll be able to identify schematics for amplifiers, oscillators, power supplies, radios, and televisions. 224 pages, 213 illustrations. **Book No. 2880, $14.95 paperback, $20.95 hardcover**

ELECTRONIC DATABOOK—4th Edition —Rudolf F. Graf

If it's electronic, it's here—current, detailed, and comprehensive! Use this book to broaden your electronics information base. Revised and expanded to include all up-to-date information, this fourth edition makes any electronic job easier and less time-consuming. You'll find information that will aid in the design of local area networks, computer interfacing structure, and more! 528 pages, 131 illustrations. **Book No. 2958, $24.95 paperback, $34.95 hardcover**

MASTER HANDBOOK OF 1001 PRACTICAL ELECTRONIC CIRCUITS—Solid-State Edition —Edited by Kendall Webster Sessions

Tested and proven circuits that you can put to immediate use in a full range of practical applications! You'll find circuits ranging from battery chargers to burglar alarms, from test equipment to voltage multipliers, from power supplies to audio amplifiers, from repeater circuits to transceivers, transmitters, and logic circuits. Whatever your interest or electronics specialty, the circuits you need are here, ready to be put to immediate use. 420 pages, 632 illustrations. **Book No. 2980, $19.95 paperback, $28.95 hardcover**

THE COMPLETE ELECTRONICS CAREER GUIDE—Joe Risse

Here is all the information you need to get started in a satisfying career in electronics! This book presents electronics from a career perspective. It shows what jobs are out there, what it takes to get one, and whether a career in electronics is right for you. Just a few of the careers covered include broadcasting and communications, robotics, electronic assembly, sales engineering, aerospace and aviation electronics instrumentation, computer servicing, and others. 208 pages, 64 illustrations. **Book No. 3110, $12.95 paperback, $19.95 hardcover**

THE MASTER HANDBOOK OF IC CIRCUITS —2nd Edition—Delton T. Horn

Crammed into this hefty handbook are 979 different circuits, using more than 200 popular ICs! This revised bestseller offers the latest IC designs and project ideas. All can be constructed in your own home workshop at an amazingly low cost. You'll find circuit design descriptions, applications, pinout diagrams, schematics, and step-by-step instructions. 592 pages, 960 illustrations. **Book No. 3185, $24.95 paperback, $34.95 hardcover**

ELECTRONICS EQUATIONS HANDBOOK —Stephen J. Erst

Here is immediate access to equations for nearly every imaginable application! In this book, Stephen Erst provides an extensive compilation of formulas from his 40 years' experience in electronics. He covers 21 major categories and more than 600 subtopics in offering the over 800 equations. This broad-based volume includes equations in everything from basic voltage to microwave system designs. 280 pages, 219 illustrations. **Book No. 3241, $16.95 paperback, $24.95 hardcover**

INDUSTRIAL ELECTRONICS FOR TECHNICIANS—Sam Wilson

This guide provides an effective overview of the topics covered in the industrial electronics CET test, as well as being a valuable reference on industrial electronics in general. It covers the theory and applications of industrial hardware from the technician's perspective, giving you the explanations you need to fully understand all of the areas needed to qualify for CET accreditation. This book is also an ideal workbench companion. 350 pages, 120 illustrations. **Book No. 3321, $16.95 paperback, $24.95 hardcover**

THE MASTER IC COOKBOOK—2nd Edition —Clayton L. Hallmark and Delton T. Horn

"A wide range of popular experimenter/hobbyist linear ICs is given in this encyclopedic book."

—Popular Electronics, on the first edition

Find complete information on memories, audio amplifiers, RF amplifiers, and related devices, in addition to other sections on TTL, CMOS, special-purpose CMOS, and other linear devices. Circuits come complete with pinouts, specifications, and a concise description of the IC and its applications. 576 pages, 390 illustrations. **Book No. 3550, $22.95 paperback, $34.95 hardcover**